REVISED EDITION

Logging Practices

REVISED EDITION
Logging Practices

Principles of Timber Harvesting Systems

Steve Conway

Miller Freeman Publications, Inc.

Contents

Section Four

Section Five

Section Six

Preface

It has now been 20 years since I began learning logging principles from the jaggered end of a choker. A career in the wood products industry has allowed me to view the logging business from the hands-on perspective of a woods worker, as a supervisor, and as an administrator. This personal experience, plus the tutoring of many experienced professionals in the business, has taught me some valuable lessons.

Of all those lessons, one stands out boldly. It was best articulated by Dave Elkin, a good friend and professional logger from southwest Arkansas. Dave has been heard to say over and over again—to foremen, superintendents, and anyone else who would listen—"You have to stick to basics." To run a logging operation successfully, whether it is large or small, you have to apply the basics forcefully and consistently: These basics are the principles of logging practices and systems.

The times we are living in provide a further strong incentive for getting back to basics. There are many externals impinging on the industry. The demand for environmental quality, particularly water quality, has caused logging operators to do work that does not contribute to the productivity and profitability of their operations. The energy crisis of the 1970s has had a severe impact on the cost of logging. Escalating costs of new equipment have caused many a logging manager to spend more on maintenance or to rebuild equipment rather than purchase new replacement equipment.

In general, the inflationary spiral of the past 10 years has attached a sense of urgency to the need for productivity improvement. These improvements are not coming from new technology, as has been the case in other industries. Rather, the improvements will come through efforts to do our job better—more efficiently and more effectively. This means doing all the right things and doing things right: back to basics.

If there is one thing certain in our world today, it is change. The logging industry is not different in this respect. Therefore, there are certainly some omissions in this work despite a dedicated effort to avoid them. Any cursory examination of industry journals or a tour through R&D facilities will reveal new concepts and new ideas. Change is unavoidable—except in a book, which stops developing when the last page is completed. This is a revised edition, to be sure, but it was still revised within a time frame and it still has a final page.

The pursuit of knowledge continues with each new association and each new set of responsibilities. The list of people who truly contributed to the success of this book is long indeed. There are naturally some who helped in this revised edition more than others. Several friends and business associates were especially helpful. Bryce Webster, Warren Hootman, and Ken Sheaffer reviewed the operations chapters. Tim McClain and Bob Jenkins, both professional safety experts, reviewed and criticized Chapter 18.

Special thanks go to Dave Pease, executive editor of *Forest Industries* magazine. Dave was extremely helpful and patient in researching and editing this revision. In addition, some of the photographs used in the text were taken from the *Forest Industries* photo file.

Steve Conway
North Bend, Oregon
January 1982

Preface

to the First Edition

Thirteen years ago the author of this book began learning logging principles from the jaggered end of a choker while working on a highlead logging crew. Later, as a student, he was exposed to another kind of education in principles—out of textbooks. But no matter how hard the writers of these books worked, or how deeply they researched their subject, their books lacked the tiny practical details learned in the brush.

Perhaps the problem is one of perspective, or slant. There are books about logging by foresters, by college professors, and by engineers. This is a book about logging by a logger. Hopefully it approaches that happy medium which is found somewhere between the traditional textbook approach and the crisp, expletive-studded explanations of a tough, old hooktender.

The objective of this book is to present the principles of timber harvesting systems so that the reader, student or otherwise, will understand not only what happens, but how and why it happens. As a principles book there naturally will be found some gaps in topics not directly related to logging—for example, timber acquisition and forestry. Engineering too is a subject not covered in detail, for that is properly the subject for an entire book in itself—and best written by engineers who are expert in the field. This is a book, however, that an aspiring engineer or forester might read in order to better understand the problems of logging production and the methods applied in solving them.

Just as a book is a matter of perspective, it is also a matter of the times. Earlier logging texts had a place for hay rack loading, horse logging, and hand cutting. Not so in this book. Although some rather primitive methods are still used, even in the United States and Canada, the emphasis and focus of this book is the here-and-now. The reasons should be obvious. Natives in the jungles of Indonesia or Africa, the Kuda-Kuda loggers, will not read the book and would not, could not, change if they did. You, the reader, however, are looking for a definition of modern logging practices and a basic understanding of the principles involved. This text has been designed to meet your needs.

The book takes a systematic approach to the timber harvesting business. The various components of the harvesting system are presented in an ordered fashion. And each component is broken down into its major elements. Finally, the elements are reconstructed with a fairly detailed description of what happens in each component.

The question of sufficient or insufficient detail is probably debatable. For a prospective logger, learning is doing, as it should be; nevertheless a great deal can be learned through a text such as this. For a prospective manager, engineer, or timber operations accountant, learning takes place through observing, listening, and questioning—or, in this case, reading about what others have learned. Most readers will no doubt find that there is sufficient detail to give them a deeper understanding of logging principles.

One subject that receives a great deal of emphasis is planning. An entire chapter is devoted to the subject in a general way. However, each part of the logging system, from skidding to log transportation, reacts differently to operating variables. The impact may be greater or lesser. A variable which is of critical importance to one system component may have no effect on another component. In some cases different variables come into play. For these reasons, and more, each of the major operational subjects covered—timber cutting, primary transportation, and secondary transportation—deals with its own aspect of planning, with specific reference to the operations involved.

Safety is one important area in the text which has been found to be inadequately treated in most other books about logging. If it is treated at all, the emphasis comes in the form of do's and don'ts. Such an approach is more appropriate between foremen and workmen than in a logging text. Although some of the do's and don'ts are mentioned throughout the text, they are generally avoided in the safety chapter. Safety is of critical importance in the logging industry, and it is the job of managers and supervisors not only to teach their charges how to work safely, but also to promote the appropriate sense of urgency which is inherent in safe operations. The safety chapter, therefore, presents a rationale for the management of safety and describes the role which managers and supervisors must play. The emphasis is on why the reader should be vitally concerned about this key activity and how an effective safety program can be developed.

A principles book always seems to presuppose a certain elemental level of understanding and, in some cases, even an advanced level. The reason for this phenomenon is not entirely clear to the author, but having been influenced by it in his student days, he has attempted to avoid it in this text. This book therefore, makes no assumptions of specific levels of knowledge. Each chapter builds on the knowledge acquired in the previous chapter. Some references are duplicated in the interest of emphasis and clarity. Examples are used whenever possible to further illuminate a written explanation. Finally, the bibliography contains a long list of references. The purpose is not to impress the reader with the research effort. The author remembers the difficulties he experienced in finding secondary sources in some areas. The reader will benefit from a complete bibliography in two ways. First, he will be able to consult other sources besides this book. Second, he will be in the position to do further research into the subject of logging practices should he so desire.

A note of caution. The world and all that is in it moves continuously— growing, changing, improving. This book, like any other, stopped developing when the last page was completed. Dedicated effort was made to keep the book up-to-date during the writing, but of course there is a practical limit to such effort. Only a short time ago, the author was exposed to a slide presentation introducing new technological advances made in Canada. It was too late to include this information in this text, but early enough to remind the author that there are some gaps.

Any omissions in this book are as much a loss to the writer as to the reader. When the objective of the book was stated a little earlier, it was not covered fully. Before the book was even an idea in the author's mind he began to accumulate knowledge and experience just because he wanted to know more about the business he was in. That pursuit of knowledge continued after the book became a reality and still continues now that the book is published. The accumulation of knowledge began with the first hooktender the author worked for 13 years ago and continued with each new association that developed. Choker setters and timber fallers, contract loggers and corporate executives—these are the men from whom I have learned and who are the real authors of this book. I have simply written what I learned on the job and what those men shared with me.

There are naturally some, however, who "helped" more than others. Several business associates and friends read the chapters and offered constructive criticism—a difficult and touchy job at best. Special thanks go to Bryce Webster, Warren Hootman, and Ken Sheaffer, who were especially helpful with the operations chapters. Joe Beckman and Lou Hoelscher, both experts on industrial safety, were helpful in giving advice on the subject and criticizing the contents of Chapter 18.

Chapter 15, Aerial Logging, was written by Dr. Jens Jorgensen, Associate Professor at the University of Washington. Jerry Gilles and Pat Mulligan, both personal friends and men experienced in aerial logging, were valued contributors to the contents of this chapter.

The final chapter, Cost and Production Control, was written with the able assistance of Gene Christensen, whose help was greatly appreciated.

Many companies and manufacturers, too numerous to list here, contributed photographs for use in the book. Where the photographs are used, however, the companies are credited.

Special thanks also go to the publishers, Miller Freeman Publications. The three chapters on timber cutting are excerpted from the author's previous book, *Timber Cutting Practices*, also published by Miller Freeman. In addition, the publisher contributed some of the photographs used in the text, taken from the files of *Forest Industries* magazine.

Steve Conway
Washington, North Carolina
March 1976

12   Logging practices

Introduction

It has not been many years since the sole objective of the logger was to let daylight in the swamp. The swamp was the forest and letting in daylight simply meant to clear the trees. Logging was defined as those activities which reduced an area of forest to stumps and saw the resulting logs delivered to a mill pond at the least possible cost. Words like *products, harvest,* and *crops* were reserved by the loggers for the nation's farmers: men who systematically prepared the soil, planted seeds, tended young plants, harvested mature crops, and, finally, began the entire cycle again.

The timber industry has since changed. To be sure, a great deal of daylight has been let into the swamp. But today that daylight shines on a new crop of trees. The men and corporations that own and control timber tend to view themselves as stewards of the land. Their business is not cutting trees but growing them. The farmers are the foresters, the land managers, the geneticists—those in the timber industry who plan, plant, and grow a timber crop to maturity. The logger is the harvester.

This book is about logging methods—the systems, equipment, and practices applied in North America to harvest timber crops. The book will introduce the reader, whether student, management trainee, or interested observer, to the principles of timber-harvesting systems. No book on logging methods can be anything but introductory. Total understanding and expertise comes only through experience and the practice of the methods described in these pages.

The meaning of harvesting

Harvesting timber products—sawlogs, veneer logs, pulpwood, poles, and piling—can be logically classified in either of two ways. First, it can be regarded as an end in itself, the desired result of a timber management plan. This is the viewpoint of the professional forester or private landowner who is growing timber as a cash crop. Second, and equally acceptable, timber production can be regarded as a part of the total manufacturing process. In the first case the timber is sold to a second party for conversion. In the second case the timber is either owned or purchased for conversion.

For many companies, including the large, integrated wood products firm, both views are appropriate. These companies are growing and harvesting their own timber on a sustained yield basis with the objective of supplying a market with raw materials. In any case, the harvest operation is an essential part of the timber-growing business. In this text, harvesting is regarded as the initial step in the manufacturing process—a view, incidentally, not shared by all logging managers.

Unfortunately, the logger, producer, or independent contractor often views his work as a process independent of either manufacturing or forestry. The attitude, even of many conscientious loggers, is that once the logs are delivered to the mill they become the mill's problem. However, this is not an acceptable attitude and can only result in suboptimizing the value of both the timber and the land. Moreover, this myopic view of objectives is also in direct conflict with other social objectives, as the industry discovered not too long ago.

The management of a forest is a business, and the rational objective of a business is to maximize total benefits and values. The definition of benefits and values, however, sometimes poses a problem. A forest can be managed for timber harvest, recreation, watershed protection, or wildlife uses. Though these objectives are not necessarily in conflict, special interest groups often view them as such.

For example, today several factions refuse to accept recreation as being compatible with timber harvesting. In fact, more often than not, these groups refuse to recognize timber harvesting as being compatible with any other use.

Environmental issues

This apparent incompatibility was brought forcefully to the industry's attention during the 1960s, when the industry came face to face with a new social issue—the *environmental crisis*. The preservationist was the motivating force behind this crisis. But the public has become the moving force. The crisis has matured into a continuing fact of life that has had a tremendous impact on the industry and its ability to supply the projected increase in consumer demand for wood products.

Although the preservationists are relatively few, they have been able to arouse the public merely by focusing attention on industry's mistakes. They look at the burns, poorly stocked lands, ugly clearcuts, erosion, and misplaced forest roads and point an accusing finger. It does not matter who is actually to blame—the logger, a governmental agency, or a careless tourist—the entire wood products industry gets the blame. This is a contention that is easy for much of the public and many legislators in urban and non-timber-producing areas to accept.

The environmentalist and preservationist have a point; the industry has not been doing the best job possible. Yet as a result of the hard-hitting attacks begun during the sixties and continuing into the present, industry leaders have

Aerial view of a large clearcut that was tractor logged. Note the skid road pattern. This practice is a thing of the past.

begun to solve some of the problems. At times the solutions are painful, requiring industry to accept the full responsibility for some glaring errors of judgment. However, there are numerous examples of how the overall effect of these attacks has been good for both the public and the industry.

When the Forest Service began a study in 1970 to identify and promote skyline logging applications on national forest lands in the Northwest, it took steps that would have been considered unnecessary only a few years before. Teams of men composed of engineers, foresters, landscape architects, soil scientists, and biologists were assigned to carry out the program of planning and introducing skyline systems. The objectives of the program were several and included controlling erosion, improving water quality, and preserving aesthetics. Another objective was minimizing logging road construction. This objective served the industry well by reducing road-building costs and preserving forest lands for forest production.

Also in 1970, at the Oregon Logging Conference, one industry representative, J. Dean Prater of Crown Zellerbach Corporation, focused on the problem of stream pollution. He suggested smaller logging areas, clean logging, buffer strips between logging areas and streams, and more carefully chosen road locations. There were 29 specific points mentioned in his speech and each was aimed, in some way, at controlling stream pollution—through water temperature control, reduced siltation, and elimination of stream blockage (Prater

It is both illegal and socially irresponsible to log across creeks.

16   Logging practices

1970). These practices are now commonplace in the industry, some adopted voluntarily and others written into tighter federal and state regulations.

Environmental quality has become an important factor in planning the timber harvest. Yet cleaning up the environment is a two-edged sword. On the one hand, the issue reaches into every phase of the industry and sometimes requires costly solutions. On the other hand, the issue has served as a catalyst for other important issues that have been lying dormant for many years. Restrictions on herbicide and insecticide applications are an example of how such broad issues affect forest management.

Timber harvesting has long been an organized, systematic activity. But during the past several years, there have been more changes, technologically speaking, than ever before. These changes are being forced on the industry by concomitant changes in social structure, economics, politics, and myriad other areas. And it is surprising how many of those areas are either directly or indirectly influenced by environmental issues.

Timber supply

Timber supply is another problem area that quite literally grasps at the industry's heart. While it is true that the timber supply is being threatened by adverse tax laws, conservation legislation, and public opinion, it is also true that loggers, lumbermen, and timberland owners have to accept at least a part of the burden for the supply problem.

When the industry was in its infancy it more or less wasted the resource. The approach to cutting timber was similar to that of mining—exploitation. Those were the days when "cutting around 40 acres" meant the operator left a neat square containing approximately 40 acres and cut everything around it for as long as he could or until he was caught. At the time the "cut out and get out" philosophy was a matter of expedience and economic necessity.

One contributing cause was timber taxes. They were so high as to discourage a timber owner from holding his investment. The condition is summed up in the following quotation from *Timber and Men* (Hidy, Hill, and Nevins 1963, p. 134):

> Having obtained lands, the legitimate investor was under irresistible pressures to harvest ripe timber as soon as possible. The states, with few exceptions, had tax laws which made the cost of long-term retention of growing trees prohibitive. Capital costs were heavy, interest had to be earned on the investment, and virgin timber could always be purchased at low prices. The states had practically no fire-prevention laws to protect lumbermen from sudden calamity. Nobody could predict the future course of the market, but fear of such a drop as took place in the depression years of 1893-1897 was always before the lumberman, impelling him to harvest his trees.

The conditions described were accompanied by an insatiable demand for more wood. Most cities were growing, other cities had been destroyed and needed to be rebuilt. The great city fires—Chicago, Boston, Baltimore, San Francisco—put a strain on wood supplies. From one point of view, the strain was welcome since it provided a needed push for the industry. When San

Francisco burned in 1906, the demand for lumber in that area jumped from 200 million board feet to 300 million board feet. The number of sawmills in the Northwest increased from 557 in 1905 to 1,036 in the spring of 1907 (Lucia 1965, pp. 58-60).

Demand alone was not the only problem; another problem was the grade of lumber demanded. Lumbermen, especially in the Northwest, have been criticized for the great volume of wood they left on the ground. The loggers took only the best timber, often leaving in the woods all of the trees above the first limbs. This was not a matter of waste and it certainly does not happen anymore. At the time there simply was no market for the lumber that could have been manufactured from the upper portions of the tree. If there was no market there was no sense in harvesting.

Fire also had a terrible effect on timber supply. No region was immune. The South, the Lake States, the Northwest, Canada—every section of the continent has its charred records to remind it of the great waste of the timberlands.

The disastrous effects of fire are not a thing of the past. They continue to plague industry and reduce our raw material base. In 1894 it was Hinckley, Minnesota; in 1933 it was Tillamook, Oregon. In 1980 a series of fires destroyed timber on 8 million acres in Canada (*Forest Industries* 1980, p. 15).

The industry's history of managing its resource has been pretty bleak. There were reasons for this, often good ones; but the fact remains the industry did not always use its powers wisely. Of course, the consuming public was no less guilty. Until the early 1900s, and for some time thereafter, both the general public and the government looked upon the forests as a means to an end. Hundreds of thousands of acres were traded to get the iron rails across the continent. Hundreds of thousands more were burned to make room for farms and settlements.

So these are some of the problems that have contributed to the current raw material crisis—high taxes, growing demand, wastefulness, negligent management, and fire. Still the problem grows.

In the past the public, government, and industry have been guilty of thinking of the forests as an unlimited resource. Today the story is different. Social conditions change, opinions change—the pendulum is swinging the other way.

The importance of national forest timber as a major supply source has been underlined by recent controversies between the U.S. Forest Service and industry over management policies that sacrifice dying old-growth timber to a rigid formula equalizing harvest and growth volumes. Productive federal timberland has been withdrawn into wilderness status to a degree many times that of the goals declared when the Wilderness Act of 1964 was passed. RARE II, the second Roadless Area Review and Evaluation conducted by the Forest Service, produced a lengthy and unresolved controversy between industry, government, and preservationists about how much designated wilderness is enough and how much weight should be given economic impacts.

To put the national forest resource into perspective, this ownership accounts for 18 percent of the commercial timberland in the United States, containing 51 percent of the standing softwood sawtimber inventory. Yet, national

The condition shown here no longer exists—logging debris in a stream.

forests provide only 23 percent of the annual softwood sawtimber harvest.

A withdrawal does not have to be a wilderness area of thousands of acres to be significant. Buffer strips along streams, for example, appear to be of minor impact when considered individually but have a tremendous cumulative effect on volume lost to production.

The industry in the Northwest has long depended on an old-growth economy. The sawmills, plywood mills, even the pulp industry, relied on the unlimited supply of old-growth Douglas fir. Today there is a scant 15 years of old-growth left on privately owned lands. If the national forests—our public timber holdings—were contributing at a properly managed rate, that 15 years could possibly be extended to 30 years—still a relatively short time. But public lands are being withdrawn from commercial use, and the withdrawals are causing no small amount of concern with industry and government.

The southern United States is in no better condition. During 1977 the South produced more than 32 million cords of roundwood pulpwood, representing an 18.5 percent increase over a 10-year period (*Forest Industries* 1979, p. 58). The demand for both pulpwood and roundwood has been increasing rapidly. The plywood industry in the South has grown dramatically since 1964, putting pressure on pulpwood industry supplies of raw material.

In the southern and southeastern United States the old-growth has long since disappeared. The trees being harvested now are second and third generation. In those parts of the country, the industry is beginning to learn how to live off the growth rate. And the industry's success is entirely based on the ability to adapt to changing conditions, to develop new technology, and to work around restrictions.

Labor supply

Although timber supply is extremely important, it is only one part of the equation. Labor is equally important although not quite as perplexing as it once was. The problem of labor supply has always been with the industry to some degree along with the peaks and valleys following the supply-and-demand characteristics of the economy. During the past 20 years, the industry has at times voiced fear and trepidation over the quantity of labor supply. During the same two decades, there have also been expressions of satisfaction over labor supply.

It appears that industry leaders are correct who suggest that as long as there is demand for forest products the industry will be able to find the workers to produce them. However, this may not always be as easy to accomplish as it once was. In contrast to those who suggest that labor supply will always be available, there are other industry leaders who insist the supply will be available *only* if the incentives and the quality of work life are acceptable. Perhaps the latter is a more accurate definition of the situation. High wages, competitive benefits, and job security are all necessary. But something more must be done than simply applying a carrot-and-stick concept of management.

Logging is dirty work, often dangerous, and always physically hard. There are many occupations and many professions, even within the industry, that have a lot more to offer. Simply offering top wages is not enough to attract and keep quality help. The workers must be challenged by their work and feel like they are a part of the company. There are too many other alternatives where the work place is more pleasant.

Competition for labor has had its impact on the labor supply. This is especially true in areas of the country where wages are relatively lower and working conditions are poor. Until the mid-sixties the logging industry was not highly mechanized. There was a great deal of hard labor, skidding was done with horses or mules, and the work was seasonal. When big industry—automobiles, steel, and textiles—began to boom, the woods workers left for the mills and factories. There wages were higher, working conditions more favorable, and the worker did not have to suffer the stigma of being a "bull-worker."

There were and are many causes for the labor problem—both quality and quantity. The causes included low wages, seasonal work, adverse working conditions, and the lack of opportunity. Industry's first attempt at solving the problem was mechanization. Logging bulls were replaced by cable systems; the horses, by logging tractors and skidders; and cutters, by mechanized harvesters. The transition did not take place all at once but occurred over many years. And industry found, in some cases, that mechanization did not solve the problem at all. It traded labor costs for maintenance costs.

Mechanization does not solve all the problems, but it has helped. For one thing, the view of the logger is changing. Because of the machines, the industry can offer better, higher paying jobs with more status. It is easier to say to your children, "I operate a big logging tractor," than it is to admit you spend your days loading pulpwood onto a truck by hand. The industry has achieved a degree of differentiation that will allow it to compete more effectively in the labor market.

While mechanization probably solved more problems than it created, the industry soon discovered that sophisticated machines require a quality work force. Further, as the investment per employee increases, a very different worker attitude is required to maintain the productivity growth that is required. While many factors have impacts on productivity, labor is the one factor that management can control. This is the newest challenge of industry—how to make the labor resource more productive. The answers are in the sociology text and plain common sense. It has been said that a wage will guarantee only that the worker will come to work—most of the time. What he does after arriving depends on the environment management presents for him to work in. This is the lesson industry must now learn if it is to succeed in the economic environment of the eighties.

Synergy

As time goes on, all the factors discussed—environment, raw material, and labor—will have a synergistic effect. The environmental issues that to a large extent are causing the withdrawal movement are accomplishing more than a simple shrinking of the raw material base. A decrease in timber supply, for whatever reason, when accompanied by an increased demand for wood products results in an increase in stumpage prices. This means higher prices for the raw material to the mill. Stumpage prices are further pressed by the need to maintain a clean environment. The pressures on supply and price can be partially offset by improved productivity and reduced costs of production. A complete understanding of logging systems, the relationship between components of the system, and the relationship between the system and the external environment is necessary before both productivity and costs can be improved.

While this book does dwell upon the different parts of the environment, its main objective is to make clear the functions of the logging system and its components. The emphasis is on operational functions and practical applications. To accomplish its objectives the book is divided into sections which act as building blocks for each successive section. Likewise, each chapter depends on previous chapters for complete understanding.

Section One

2.

Forest Resources

Several hundred years ago, before any commercial cutting had been done, the continent of North America was covered with what appeared to be an unending supply of timber. Even after harvesting began the loggers were sure the forests would never be cleared. They thought, for instance, that the white pine forests of Maine extended clear to the North Pole, and for years that state was considered the center of the timber industry. The loggers and the industry eventually moved west into New York, Pennsylvania, and the Lake States. Everywhere they went the supply seemed limitless. If, as the years passed, all the good and easily accessible timber disappeared, what was the problem?—the industry could always move farther west into a new frontier and a new limitless stand of timber.

Timber was as cheap as it was plentiful. When loggers moved into the Lake States in the 1830s and 1840s, they bought prime timberlands from the government for $1.25 per acre. In the late 1870s, when the lumbermen declared the Lake States cut out, they moved into the South and West. In the South, pine tracts were purchased for between $1.25 and $3.00 per acre. The hardwood forests were going for a whopping $10.00 per acre. In 1900, the Weyerhaeuser Timber Company bought 900,000 acres of now-valuable western timberlands from the Northern Pacific Railroad for only $6.00 per acre (Hidy, Hill, and Nevins 1963, pp. 208, 212).

Supply and demand

Even in those days of "cheap and plenty" the forces of change were at work. The agricultural community was growing; the timber harvest was followed by a second harvest—food for a growing nation. In 1920 the population of the United States was slightly over 100 million. The compound annual growth rate between 1910 and 1930 was 1.4 percent. By 1930 the population had grown to about 123 million. Households increased, income increased, and the demand for timber products increased. The preliminary census total for 1980 was 226 million, and the U.S. Bureau of the Census estimates a population of 260.4 million in 2000 and 300.3 million in 2030, in the "medium" of three forecasts (U.S. Department of Commerce, Bureau of the Census 1977).

In 1900, total wood products consumption in the United States was 8.8 billion cubic feet; 10 years later, in 1910, consumption was 9.5 billion cubic feet. Wood consumption varied in the United States between 1910 and the roaring fifties. From a high of 9.5 billion cubic feet in 1910, it dropped to a low of 5.9 billion cubic feet in 1935. But by the early 1950s, the industry was once again in high gear and the loggers were working the forests in earnest. In 1952, the total consumption of roundwood, including domestic production and net imports, was 12.268 billion cubic feet. Ten years later, in 1962, consumption had dropped slightly, to 11.800 billion cubic feet, but the upward trend was set.

Consumption rebounded in 1963 to 13.8 billion cubic feet. In 1977, the figure was 13.7 billion cubic feet. Projections by the U.S. Forest Service center on roundwood demand estimates of 22.7 billion cubic feet in 2000, rising to 28.3 billion cubic feet in 2030. Much of the projected increase in demand is represented by pulp products, with pulpwood accounting for 45 percent of the total in 2030 versus 33 percent in 1976 (*An Assessment of the Forest and Range Land Situation in the United States,* U.S. Department of Agriculture, Forest Service 1980, p. 336).

The Forest Service conducted comprehensive studies of timber supply and demand in the United States in 1952 and 1965. Passage of the Forest and Rangeland Renewable Resources Planning Act of 1974, commonly called the RPA, required the Forest Service, through the Administration, to submit regular assessments and natural resource management program updates to Congress for review.

The 1965 study, *Timber Trends in the United States* (U.S. Department of Agriculture, Forest Service 1965), presented the industry with some startling facts. It made projections of timber supply and demand conditions through the year 2000, at which time the shortfall for sawtimber was estimated at 16 percent, given a continuation of then-current levels of forest management. Improved management techniques on commercial forests, it was suggested, could narrow the gap.

The 1980 RPA Assessment indicated a similar gap between domestic forest supply and demand for 2000, increasing to 18.4 percent by 2030. Considering only softwood roundwood, however, the difference between the demand on forests of the United States and the volumes they are capable of supplying is estimated at 23.3 percent in 2000 and 27.6 percent in 2030 (Table 2.1).

The Assessment, as did *Timber Trends,* suggests that improved forest management techniques can improve the supply picture (U.S. Department of Agriculture, Forest Service 1980, p. 411). These include:

- Accelerated regeneration
- Increased use of genetically improved planting stock
- Changing species compostion and site condition of some lands

Table 2.1 *Roundwood Demand versus Supply,
U.S. Forests (in billions of cubic feet)*

Softwood	Demand	Supply	Difference
1976	9.2	9.2	0.0%
2000	13.8	11.1	24.3%
2030	15.7	12.3	27.6%
Hardwood	Demand	Supply	Difference
1976	2.9	2.9	0.0%
2000	6.0	6.0	0.0%
2030	9.4	8.9	5.6%

Source: U.S. Department of Agriculture, Forest
Service 1980, pp. 397, 404.

- Improving the scheduling of harvest cuts and intermediate removals
- Reducing losses from mortality, fire, insects, and diseases
- Harmonizing the production of timber with other benefits

The Assessment also lists opportunities for extending timber supplies through improved utilization and research, including increased use of residues, additions to timber harvest, more efficient processing techniques, and improvements in end-use applications.

Projections, however, are just guesses of what's likely to happen in the future. Production estimates in *Timber Trends* were substantially exceeded during the early years of the projection period, and by 1969 lumber production in the United States had exceeded the 1980 estimates in *Timber Trends* by roughly 300 million board feet.

The 1980 RPA Assessment is not without its critics. Forest industry spokesmen have complained that the government's outlook fails to recognize the production potential of federally owned timberlands under higher levels of management. From another quarter, Dr. John H. Beuter, Professor of Forest Management at Oregon State University, noted that "supply responses are allowed only in the private sector; the public sector supply is fixed by current allowable cut policies" (Forest Products Research Society 1979, p. 44).

Residential construction is the major market for solid wood products, and the health of the lumber and plywood industries is directly affected by the strength of the housing market. A housing boom in the early 1970s gave false promise that goals of the 1968 Housing Act, which called for 26 million new homes by 1978, would be met. But economic factors unrelated to housing demand stepped into the path of the housing industry's performance. Federal fiscal policies aimed at reducing the inflation rate resulted in severe housing slumps in 1973-1974 and 1980-1981.

Even so, housing production during the 1970s averaged 2.1 million conventional and mobile home units annually, according to Robert J. Sheehan,

Associate Chief Economist and Director of Economic Research for the National Association of Home Builders, who indicated the association had forecast an average of 1.9 million units a year over the 1979–1988 period (Forest Products Research Society 1979, p. 23).

Medium projections (of three ranges: high, medium, and low) by the Forest Service place average annual production of new conventional housing units, excluding mobile homes, at 2.25 million for the decade of the 1980s, declining slightly thereafter (U.S. Department of Agriculture, Forest Service 1980, p. 319). The composition of the starts is nearly as important as the absolute numbers since it dictates demand for lumber and plywood. For instance, a multifamily dwelling uses only 3/5 as much plywood as a single-family unit. Among conventional units, multifamily starts averaged over the 1970–1977 period represented 32.3 percent of the total. In a reversal of earlier projections, the outlook for the 1980s and beyond calls for fewer multifamily units in proportion to the total (U.S. Department of Agriculture, Forest Service 1980, p. 319).

As total wood demand increases, so does demand for products that make up that total. Examples are indicated in Table 2.2.

Note the sharp increase in demand estimated for paper products. Pulpwood consumption in U.S. mills was 77.6 million cords in 1977, of which 31.8 million cords was represented by mill residues. Per capita consumption of paper and board rose from 145 pounds in 1920 to 611 pounds in 1977, and is projected to reach 948 pounds in 2000 and 1,296 pounds in 2030 (U.S. Department of Agriculture, Forest Service 1980, p. 329).

All of these estimates simply indicate what future demand for timber will be. Whether the timber will be provided is still another matter.

Forest land base

A primary consideration when discussing timber supply and future demand on that supply is the forest land base. Unlike other resources, such as iron and coal, our forests are renewable, which means we can grow more timber after an initial harvest. The *land base* refers to the acres of forest that are actually

Table 2.2 *Consumption of Wood Products in the United States (medium levels projected, calculated on base-level price trends)*

	1976	2000	2030
Lumber (billion bd ft)	42.7	59.9	67.3
Plywood (billion sq ft, 3/8-in. basis)	20.7	30.1	34.1
Board[1] (billion sq ft, 3/8-in. basis)	13.5	27.3	37.3
Paper, paperboard, building board (million tons)	66.2[2]	123.4	194.4

Source: U.S. Department of Agriculture, Forest Service 1980, pp. 325–327.

1. Particleboard, hardboard, and insulation board.
2. 1977 consumption.

available. And this involves future trends not only in forest growth but also in deletion from the land base.

The total number of acres of forest land in Canada and the United States is 1,584 million. Canada has approximately 844 million acres and the United States has 740 million. These figures, however, are no measure of what is available for timber production. In Canada, more than 90 percent of the total forest land area is allocated to wood production. In the United States, the figure is closer to 65 percent. Tables 2.3 and 2.4 give the breakdown for both countries.

As can be seen from Table 2.3, the lion's share of Canada's forest lands are owned by the Crown. Only 42,997,000 acres are privately owned. This means, of course, that the Crown controls the timber resources and, to a certain extent, the industry. In British Columbia, for instance, the provincial government owns 98 percent of the forest land.

Pulp and paper mills in Canada lease approximately 120 million acres of Crown forests for pulpwood production. These leases are called *Crown limits* and represent areas on which temporary use has been granted for timber production. Unlike the federal lands in the United States, where relatively small parcels of timber are sold at one time to a single buyer, Crown limits are generally quite large and allow the user to cut standing timber in areas ranging in size from parcels as small as 100 acres to in excess of 10,000 square miles. The largest limits are found in eastern Canada and are licensed or leased in perpetuity to the user.

There is no current wood shortage in Canada. However, on a regional basis, availability ranges from full utilization of timber resources to vast surpluses. In some areas, industry is cutting primarily virgin forests, while in others it is cutting on the fourth rotation. Some problem is expected with individual

Table 2.3 Forest Land in Canada

Type of land	Acres
Allocated to wood production (Crown)	732,681,000
Allocated to wood production (private)	42,997,000
Reserved forest land[1]	23,723,000
Non-productive and other unused forest land	44,974,000
Total forest land	844,375,000

Source: Canadian Pulp and Paper Association 1981, p. 10. Data from Environment Canada, Canadian Forestry Service.

Note: Forest land refers to land capable of producing stands of trees 4 in. or 10 cm DBH (diameter breast height) and larger on 10% or more of the area. Shelter beds of forest of 5 acres or less are excluded. Figures are latest available: 1976.

1. Land in parks, game refuges, water conservation areas, and nature preserves where, by legislation, wood production is not the primary use.

Table 2.4 *Forest Land in the United States*

Type of land	Acres
Reserved forest land (productive)	19,531,000
Deferred forest land (productive)	4,626,000
Non-commercial forest land	228,264,000
Commercial forest land	487,726,000
Total forest land	740,147,000

Source: U.S. Department of Agriculture, Forest Service 1977, pp. 1-2.

Note: Commerical forest land is land capable of producing crops of industrial wood and not withdrawn from timber utilization by statute or administrative regulation. Included are areas suitable for management to grow crops of industrial wood generally capable of producing in excess of 25 cu ft per acre of annual growth. *Non-commercial land* is unproductive forest land incapable of yielding crops of industrial wood because of adverse site conditions and productive forest land withdrawn from commercial timber uses through statute or administrative regulation. *Reserved forest land* is non-commercial forest reserved for recreation or other nontimber uses. *Deferred forest land* is land being studied for possible addition to the wilderness system.

species and qualities of timber, but these shortages will have to be worked out by industry.

Since 1960, there has been a 14 percent increase in allowable cut, yet the allocated area increased only 7 percent. Part of the increase derives from new volume data (see Table 2.5). Also, better utilization undoubtedly has been an important contributing factor. While the allowable cut is still in excess of current consumption, it is felt there will be a future shortage. Demand for Canadian wood has been increasing at the rate of 4 and 5 percent per year, and was forecast to exceed the allowable cut by the mid-seventies. (Table 2.6 shows Canadian pulpwood production and consumption from 1966 to 1978.)

The problem of supply is critical in two respects, other than species composition and quality of timber. One is the older, established limits and the other is the use of the northern reserves.

Many of the older, established mills are facing increased costs in delivering wood volumes to the mill. As limit production moves farther and farther back, transportation costs are climbing ever higher. At the same time, the quality of timber is declining. The mills are considering the use of fertilizer for improving stand quality, but this is still a rather new application in Canada, and there are questions of cost and effectiveness. In any case, it appears that the cost of wood will continue to increase.

The enormous Canadian forest reserves are constantly alluded to in literature regarding timber supply. The northern forests do have potential, but they also have problems. Climatic conditions in the northern areas of the country leave much to be desired and are generally inhospitable. Also, there are social

Table 2.5 *Canada's Forest Volume*

Province	Volume (in thousands of cu ft)
Alberta	54,066,499
British Columbia	274,252,400
Manitoba	20,235,208
New Brunswick	23,554,771
Newfoundland	21,894,990
Nova Scotia	7,627,930
Ontario	150,792,910
Quebec	97,538,649
Saskatchewan	16,209,355
Northwest Territories	5,791,578
Yukon Territory	8,934,568
Total[1]	680,898,870

Source: Data from Bowen 1978. (Courtesy Petawawa National Forestry Institute, Environment Canada.)

Note: Wood volume of main stem of trees or stands. It includes the volume of all species and maturity classes in all ownerships on inventoried, productive forest land in 1976.

1. No inventories available for Prince Edward Island.

factors that greatly affect the value of these stands. These lands are rather fragile and subject to rapid change under the influence of commercial pressure.

There are three more factors relating to the northern forests and those forests generally designated as unsuitable for harvest. From an economic point of view, these factors are the most important of all. First, these forests are not easily accessible and have few roads. Second, utilizing these areas would require enormous investments in transportation systems. No small part of the transportation cost would be attributable to their great distance from established markets. And third, because of sparse stocking, species composition, and other factors, the production costs would be higher than those on presently allocated lands.

The fact remains, however, that these timberlands cover about 190 million acres. In an emergency situation, these lands would represent an important source of timber.

One of the major differences between Canada and the United States is the pattern of land ownership. In Canada, the Crown owns the majority of both allocated and reserved forest land. Furthermore, this pattern holds true in all the provinces. This is not the case in the United States. For example, Figure 2.1 illustrates the ownership patterns in the two countries.

In the United States, the federal government owns approximately 20 percent of the commercial forests. In addition, the government owns nearly 40 percent of all the merchantable timber and 57 percent of the softwood

Table 2.6 *Canadian Pulpwood Production (in thousands of cunits)*

Year	Apparent production[1] Roundwood	Residues	Total	Consumption Roundwood	Residues	Total
1966	16,158	4,770	*20,928*	15,090	4,422	*19,512*
1967	15,805	5,414	*21,219*	14,950	4,989	*19,939*
1968	16,418	⋅ 6,304	*22,722*	15,616	5,875	*21,491*
1969	18,080	6,670	*24,750*	17,283	6,395	*23,678*
1970	17,893	6,485	*24,378*	17,070	6,140	*23,210*
1971	16,537	7,706	*24,243*	15,978	7,405	*23,383*
1972	16,765	8,703	*25,468*	16,427	8,437	*24,864*
1973[2]	17,304	10,212	*27,516*	16,879	9,883	*26,762*
1974[3]				17,960	10,701	*28,661*
1975				13,950	8,495	*22,445*
1976				14,852	12,304	*27,156*
1977				14,847	11,976	*26,823*
1978				15,476	13,275	*28,751*
1979				15,860	14,498	*30,358*
1980[4]				15,955	14,756	*30,711*

Source: Canadian Pulp and Paper Association 1981, p.11.
Data from Statistics Canada. Reproduced by permission of the Minister of Supply and Services Canada.

Note: A cunit equals 100 cubic feet of solid wood, excluding bark.

1. Computed by adding exports to and subtracting imports from the consumption figure.
2. Consumption figures are actual; preliminary production figures are best data available.
3. Figures for 1974–1980 apparent production not available.
4. Estimated.

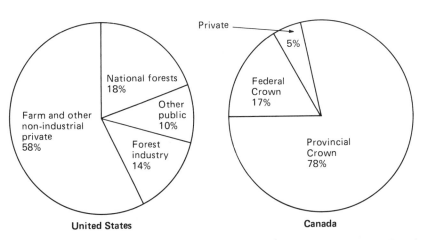

Figure 2.1. *U.S. and Canadian patterns of ownership (from Bowen 1978 and U.S. Department of Agriculture, Forest Service 1977).*

Table 2.7 *Commercial Forest Land in the United States*

Section	Acres
North	170,769,000
South	188,433,000
Rocky Mountains	57,765,000
Pacific Coast	70,758,000
Total	487,726,000

Source: U.S. Department of Agriculture, Forest Service 1977, p. 5.

Note: Totals do not add because of rounding. The definitions for the sections listed are:

North: Connecticut, Maine, Massachusetts, New Hampshire, Rhode Island, Vermont, Delaware, Maryland, New Jersey, New York, Pennsylvania, West Virginia, Michigan, Minnesota, South Dakota (East), North Dakota, Wisconsin, Illinois, Indiana, Ohio, Iowa, Kansas, Kentucky, Missouri, and Nebraska.

South: Virginia, North Carolina, South Carolina, Georgia, Florida, Tennessee, Alabama, Mississippi, Arkansas, Louisiana, Oklahoma, and Texas.

Rocky Mountains: Idaho, Montana, South Dakota (West), Wyoming, Arizona, Colorado, Nevada, New Mexico, and Utah.

Pacific Coast: Alaska, Washington, Oregon, California, and Hawaii.

sawtimber. The balance of the commercial forests are either in non-industrial private ownership or owned by the industry.

The distribution of public and private land in the United States (Table 2.7) leads to individual problems. Nearly 3/4 of the commercial forest land is in the East, where there is a high proportion of non-industrial private ownership. In the South, 72 percent of the commercial land is in the hands of small owners. Only slightly less than 14 million acres belongs to the federal government. The problem in the South is therefore one of procuring timber from a multitude of small owners, many of whom are not interested in timber production as a primary source of income.

On the other hand, 60 percent of Alaska's 375 million acres is under federal control. Passage of the Alaska Lands Bill in 1980 set aside 104 million acres as national parks, wildlife refuges, wild and scenic rivers, and designated wilderness. The bill also reduced the annual allowable timber harvest on national forests in southeast Alaska to 420 million board feet from 520 million board feet. Federal ownership is largely the rule in the entire western United States, where the majority of the federal commercial forests are held. In the

14-state area that constitutes the Pacific Coast and Rocky Mountain regions, the federal government owns approximately 74 million acres of commercial forest land. The total area of federal ownership in the entire country is only slightly less than 100 million acres (U.S. Department of Agriculture, Forest Service 1977).

In the West, because of the large degree of federal control over commercial timber, the industry is extremely dependent upon the government as a supplier of timber. In many cases, federally owned timber is essential to the economy of a community. Furthermore, the government's policies with regard to timber sales can easily result in the life or death of the firms that use federal timber.

Growth versus removals

The western United States is currently looking at a transition from an old-growth to a second-growth economy. The average size of the timber is diminishing, and this is having an effect on the industry in that region. About 1/3 of the forest area there is still classified as old-growth, and a large proportion of this old-growth is on public lands. For instance, in Oregon about 3/4 of the live sawtimber on commercial forest lands is government owned. Most of the large companies with timber ownerships in the West indicate that they have only 10 to 15 years of old-growth left. In anticipation of smaller and more expensive timber for manufacturing they have been building small-log mills, merchandising facilities, or installations that can utilize small timber and investing in new technology to improve product recovery.

The turn into a second-growth economy is hastened by the fact that private firms in the West are cutting the old-growth timber at an accelerated rate. The reasons for this are a combination of management and financial policies, end use requirements, and the desire to obtain the greatest utilization and value from the timber and land.

The timberlands in the West are among the most productive of all commercial forest lands. Large portions of the national forests in the Douglas fir region are capable of producing more than 85 cubic feet per acre annually. The average growth rate on industry lands in this region is 90 cubic feet per acre per year. Because of generally better soils on non-industrial private lands more than 68 percent of this land is capable of producing in excess of 120 cubic feet per acre annually (*Forest Industries* 1969b, p. 35).

The potential for growth is the primary reason that both private owners and industry owners have accelerated the old-growth cut. As a result of this acceleration, they will be able to achieve the goal of a balanced, managed forest capable of achieving high growth rates much sooner. The cutting policy generally involves a larger allowable cut in the early stages of transition, and reduction later on as age classes become balanced and annual growth becomes stabilized.

While the average growth rate on privately held lands in the Douglas fir region is relatively high, the rate on public lands, the national forests, is only about 15 cubic feet per acre annually. The reason for this low growth rate is the preponderance of old-growth timber stands, which have exceptionally high mortality rates. If this deteriorating timber were harvested at a normal rate, rather than held in inventory, the transition could be forestalled as much as 15 years; in other words, there would be up to 30 years of old-growth left for the industry to use.

In the West, net annual growth of softwood growing stock in 1976 was 4.5 billion cubic feet; removals were 4.9 billion cubic feet. Net annual growth of softwood timber is also less than removals. On the Pacific Coast, again with old-growth timber, the removal of softwood sawtimber is nearly double the growth. Between 1952 and 1976, the inventory of softwood sawtimber on the Pacific Coast dropped from 1.43 trillion board feet to 1.17 trillion board feet, a decrease of 18 percent. This decrease is attributed to the rise in the harvesting of old-growth stands, as well as to the low level of growth indicated previously, the heavy mortality in older stands, and the recent cutting on much of the younger stands.

The South and North are in much better shape than the Pacific Coast region as far as growth and removals are concerned. With regard to both softwood sawtimber and softwood growing stock, the two regions appear to be in good condition, with growth exceeding removals in each case. In fact, at the present time, additional cutting could be supported. However, it is the future that must be considered.

Demand for eastern pulpwood and sawtimber is heavy at present. Southern roundwood pulpwood production was more than 32 million cords in 1977, an increase of 5 million cords over the 1967 level (*Forest Industries* 1979, p. 58). During the 1950–1959 decade, pulpwood production nearly doubled in the South. With pulping capacity and demand on the increase, it has been estimated that the South will have to produce 112 million cords of pulpwood by 2000 in order to satisfy demand.

After the late 1940s, the lumber industry in the South declined, but was replaced by a thriving pulp and paper industry. The Southern pine lumber industry has experienced a renaissance in recent years, however, and has taken the biggest share of midwestern and eastern markets away from western lumber producers. At least one large industrial timberland owner has switched management of former pulpwood stands over to sawtimber production.

Additionally, the South has seen the growth of a softwood plywood industry from nothing in 1964 to 73 plants in 1979, accounting for about 30 percent of the nation's production. Between 1976 and 1978, the South recorded a compound annual growth rate of 12.9 percent in plywood production, 1.9 percent in lumber production, and 4.8 percent in paper production, compared with national rates of 3.1 percent, 0.2 percent, and 2.8 percent, respectively.

A substantial amount of the nation's particleboard production also comes from the South, where 31 of the nation's 62 plants are located. Promise also exists for development of a structural, or composite, panel industry.

Now a look at the supply side. In the South, the forest products industry owns about 12 percent of the commercial forest land. Seventy-three percent of the land, or about 145 million acres, belongs to non-industrial private landowners. Seven percent of the forest area is classified simply as "other industry land" and will produce little pulpwood. Public lands constitute about 8 percent of the total and cannot be depended on for any significant volumes.

If the South is to provide the huge volumes needed by 2000, it has only two sources—industry-owned lands and private land. Assuming that industry lands will produce 1.3 cords to the acre per year, total production will be 32.5 million cords in 2000. This is 80 million cords less than we need; the balance must come from private ownership. Estimated present growth on private lands is less than 1/2 cord per acre per year, far less than half its potential.

The industry in the South faces a great challenge. Annual growth on industry lands must be increased from the current 0.7 cords per acre per year to 1.3 cords per acre. Private lands must increase growth from about 0.5 cords per acre to at least 0.66 cords per acre. Millions of acres of bare or poorly stocked land must be planted or converted to more valuable species. An estimated 90 million acres of timberlands will require some sort of stand improvement. Most of this work must be done before 1985 if volume requirements for 2000 are to be filled. The estimated annual cost—$100 million.

It is obvious, simply on the basis of the number of acres, that the real problem lies in the non-industrial private ownerships. This is where the efforts of industry must be placed, and it is unfortunate that the majority of those who own small woodlots do not "think timber." The land is used for cattle raising, farming, investment, or recreation. Few of these small owners are practicing forestry, because of lack of money, interest, knowledge of the potential return, and help. Those who are inclined to sell their timber frown upon and often refuse to allow clearcuts, tree-length logging, or any other logging practice they feel will be detrimental to the land.

The attitudes discussed in the preceding paragraph have a serious effect on the industry in two ways. First, without the help and cooperation of these thousands of private landowners the target volumes required by 2000 will not materialize. Stumpage prices will be higher than necessary because of supply shortages. Second, logging costs, and therefore the cost of wood to the mill, will be higher because the more efficient logging systems cannot be supported economically by woodlot operations.

The problems existing in the South also exist in the North. The difference is in magnitude only, although in some ways, for example species composition, the problems of the North are unique.

The northern forests include some 176 million acres of commercial forest land. Most of this area, about 61 percent by volume and 35 percent by area, is in oak and hickory. Both of these species are major sources of hardwood sawlogs and veneer logs. Only slightly more than 30 percent of the volume is in softwoods, spruce-fir types, white pine, red pine, and jack pine. In the Lake States region, the predominant type is aspen-birch. The aspen-birch type is valuable for pulpwood and has taken over large areas of cutover lands harvested many years ago.

The increase in growing stock and sawtimber in the North has substantially exceeded removals. Another measure of improvement is timber inventory. The sawtimber inventory in the North increased 44 percent between 1952 and

Table 2.8 *U.S. Roundwood Pulpwood Production (in thousands of cords)*

Year	Total	Softwoods	Hardwoods
1954	25,470	20,945	4,525
1955	28,600	23,365	5,235
1956	32,145	26,210	5,935
1957	30,535	24,525	6,010
1958	28,090	22,445	5,645
1959	30,580	23,380	7,200
1960	33,465	25,450	8,015
1961	32,115	23,995	8,120
1962	33,330	24,315	9,015
1963	34,665	25,120	9,545
1964	37,450	26,920	10,530
1965	40,290	29,250	11,040
1966	41,840	29,590	12,250
1967	41,810	30,100	11,710
1968	44,250	32,130	12,120
1969	47,070	33,590	13,480
1970	50,220	36,660	13,560
1971	46,720	33,390	13,330
1972	46,090	31,920	14,270
1973	48,840	32,810	16,020
1974	53,950	37,010	16,930
1975	44,280	31,660	12,610
1976	47,650	32,970	14,680
1977	45,800	31,100	14,700
1978[1]	47,880	31,610	16,270

Source: Ulrich 1978.

Note: Data may not add to totals because of rounding. Figures for 1954–1963 are domestic receipts at pulp mills.

1. Preliminary figures.

1976, from 248.6 billion board feet to 359.0 billion board feet. However, most of the gains in inventories have been in smaller diameter classes, thus limiting the use of such timber to pulpwood. In addition to the increase in small diameter classes, a major portion of the growth has been in less desirable species, such as upland oaks, hickory, and beech. (Table 2.8 gives comparative production data for softwood and hardwood pulpwood on a nationwide basis.)

Each region of the country is faced with its own peculiar problems. In the West, one of the more serious problems is how to achieve maximum utilization of the national forests while at the same time achieving greater growth and utilization on private lands. In the South, the biggest problem is utilization of lands belonging to thousands of small landowners. The North has a problem with undesirable hardwood species and, like the South, an ownership problem.

Decline in commercial forest land

Up to this point, the discussion has been centered on growth. The decline in commercial forest land is also a problem. Between 1952 and 1962, the total area of commercial forest land increased to 509 million acres from 409 million, but has declined since. The area in 1977 was 482 million acres. Much of the decline is the result of withdrawals for recreation, wildlife management, environmental improvements, highways, airports, reservoirs, croplands and pasturelands, wilderness areas, and urban development.

Many of the withdrawals are being legislated into existence. In recent years, a steady procession of land withdrawal proposals has been placed before Congress. The Wilderness Act of 1964 established the National Wilderness Preservation System. By July 1, 1979, Congress had designated 191 federal wilderness areas containing 19 million acres, 80 percent of which came from national forests. At that time, Congress was considering 320 additional proposals covering 26.1 million acres in the contiguous United States. Passage of the Alaska Lands Bill in 1980 tripled the size of the wilderness system.

The current and expected rate of withdrawals is forcing the wood industry in both the United States and Canada to look to other means of bolstering wood production. It should be no surprise that the industry and government are finding there is potential for added production through greater utilization of the lands currently being used, use of mill residuals, timber stand improvement, and technological advances.

Better timber utilization

In the early days of logging, when there was no thought of anything save moving the wood at least cost, only the best logs were taken; the rest were left to rot. This behavior was only natural, since there was no market for low-grade logs. Today, no part of the log is without value, although we have not yet learned to use all of that value.

Actually, there are many products in a single tree, or at least the potential for them is there. For instance, the tree may supply the following primary products: pulpwood bolts, sawlogs, veneer logs, poles, and piling. It may also supply energy fiber, plus a number of secondary products, including sawdust, chips, and bark. Complete utilization means achieving maximum value from the tree by using as much of the tree as is possible. Care must be taken to compute the value of low-grade products such as energy fiber against high skidding and transportation costs.

It is easier to comprehend the contents of a tree if we think in terms of fiber instead of board feet. Indeed, in thinking of more complete utilization, the whole entity, including branches, tops, stumps, and root system, must be considered. All of these parts contain usable fiber of varying quality. The merchantable bole of the tree from stump to 4-inch top contains only 65 percent of the tree's fiber. Twenty-five percent of the cubic volume is in the stump and roots, and the branches make up the difference. Usually, about 35 percent of the usable fiber in a tree is not used.

Until the past 10 years or so, nature has been allowed to take its own course in the growth cycle of the tree. For the most part, the cull trees, the down and the dead, with nothing but fiber volume, were left on the forest floor to rot or were burned in the fall slash fires. There is well in excess of 8 billion cubic feet of salvable dead and cull timber in the states of Washington, Oregon, and California alone. And this backlog increases at a rate of 1.6 billion cubic feet per year. Salvage operations performed at the present time yield about 5,000 board feet to the acre of usable culls plus an equal amount of sawtimber grades. This is 10,000 board feet to the acre more than we harvested when the culls were left on the ground.

The problem of salvaging cull logs, or *utility logs*, as they are sometimes called, is especially serious in the western United States, particularly on national forest lands where there are vast amounts of overmature forests. In these forests, the overaged trees are decaying and dying. This phenomenon accounts, in part, for removals exceeding growth, as was discussed earlier.

At the present time, the industry is expanding its cull removal and stand improvement activities. However, it still has a long way to go. The potential in the area of cull removal is enormous, although the cost of producing the wood must be low enough to realize a profit, given its low value. In spite of the industry's efforts to clean up the land, there is still a great quantity of residues left in the woods after logging and road building. And this does not take into account billions of cubic feet of fiber that were never touched.

Commercial thinning operations are still another source of additional wood volume. Thinnings remove overtopped and excess trees from forest lands before these trees die and decay. This practice spaces the trees for optimum growth or releases the growth on the residual stand. In the national forests of western Oregon and Washington, there is an estimated 25 million acres with a potential of 500 million cubic feet per year that could be gained from thinning operations.

There is a certain amount of thinning done in the West, but it is expensive. However, in spite of the expense, the amount of thinning is and will continue increasing. Commercial thinning in the South is becoming more and more common. This sort of silvicultural treatment is especially applicable to the softwood plantations of the South. Like salvage logging of cull volumes, thinning is an excellent source of additional fiber and is a way of achieving greater utilization of our forests and forest land. It is one alternative to the decreasing land base with which the industry will shortly be confronted.

Commercial thinning, especially in the West, is expensive and produces very small pieces, so logging costs are high. The technology is not yet available to perform commercial thinning economically. Where cable logging is necessary, loggers can compensate partially for the small piece size by keeping the volume per turn as high as possible and presetting chokers.

The management philosophies of the wood industry are changing. The old attitude, which gave greatest consideration to harvesting timber and manufacturing wood products, is no longer primary. It is now understood that the key to the increased growth necessary to provide tomorrow's demand lies not simply in harvesting more trees, but in growing more trees. It is obvious that by

2000 the industry will not have the same acreage in commercial forest land it now has. Yet, the demand will be much greater. While increased utilization and thinning will provide a partial solution, they will not solve the problem entirely. The answer must come from utilizing the full productive capacity of the land.

Forest management

It is estimated that 59 percent of all commercial forest land is capable of increased growth. Public and private agencies are spending millions of dollars per year on forest management programs that include insect and disease protection, pruning, fertilization, planting, and genetics. Private industry has been exceptionally progressive in this type of program, and as a result, some of the best-managed lands are industry lands. There is likewise an opportunity for increasing timber supplies on public lands, but the programs will require sizable investments. Of all ownerships, national forests are growing new wood at the lowest rate, 47 percent of potential. Forest Service budgets are characteristically low, by industry standards, when it comes to appropriating funds for reforestation and intensive management.

As was pointed out, some intensive management programs are already in effect and have proven fruitful. Tree improvement programs have resulted in more and better seedlings selected from superior trees. Seed orchards established from these seedlings will provide industry with genetically superior trees that will be more resistant to certain diseases, have better growth characteristics, grow to merchantable size in a shorter period of time, be of better quality, and substantially increase forest yield.

The geneticists have no plans to grow a square-boled, crownless, rootless, and barkless tree. But they have already produced trees having superior growth characteristics, and have accelerated the production of seed from younger trees. One eminent geneticist has stated that productivity increases on the order of 25 percent can reasonably be expected in what are already considered productive types (Blackerby 1969a, p. 29).

Another silvicultural practice is having a dramatic effect on growth characteristics of timber. In recent years, fertilization has grown out of the experimental stage, and the results are extremely encouraging. The first operational fertilization was done by Buckeye Cellulose Corporation near Carrabelle, Florida, in the early 1960s.

In some areas, the results are not as favorable as the industry would like, but there is much to say for the process. In Douglas fir, an average cubic volume increase of 30 percent is expected over a five- to seven-year period. A slash pine plantation in the South that normally would produce 30 cords per acre at 25 years is expected to produce 36 cords per acre in the same period. In short, through fertilization the plant owner can expect to get the same volume from 20 acres that he normally would have expected from 25 acres.

During the period from 1969 to 1974, well over 100,000 acres of timberland were programmed for fertilization in the United States and Canada.

Private industry is applying thousands of tons of fertilizer annually to commercial forests of the United States and Canada, wherever the yield returns support the investments. The results are faster growth, a revitalization of stagnant areas, and more uniform stands. Most important, the industry is able to grow more timber on less land to meet future demand.

3.

Woods Labor

The woods industry spends millions of dollars a year on new equipment, research and development, and forest management. These are the aspects of the business that are essential if the industry is to progress. Unfortunately, while investing all of these millions of dollars and thousands of hours, the industry often overlooks the biggest cost factor and the most valuable resource it has—people. This creates a problem, not with supervisory people, although they are most important to a logging business, but rather with the people who set the choker, run the skidder, and drive the logging truck. These are the men who get the job done, the men whose attitudes can make or break a business, and these are the men who need the attention.

Logging has always been labor intensive, and despite the current trends in mechanization it is still labor intensive. This is especially true of the many small producers in the South and North. The large producers and contractors also have the same problem, although there is a difference in degree. Twenty years ago, the American Pulpwood Association reported that labor costs made up 60 percent of the total logging costs on a per cord basis (American Pulpwood Association 1969, p. 5). More recent figures verify the high proportion of labor cost and indicate that even in mechanized operations labor constitutes between 40 and 45 percent of total variable cost.

Aside from direct labor costs, there are numerous other ways that labor has an impact on the industry as a whole. Wood product costs can go skyrocketing in anticipation of a labor strike. Stumpage costs, calculated as the difference between selling price of the timber and logging costs plus an allowance for return on investment, have a serious effect on the selling price of wood products. This effect begins with the lumber, plywood or pulp mill manager, who finds his raw material costs may be up to 75 percent of total costs. Of course, many other variables can affect log costs, such as season of the year, current and expected wood products markets, species mix, and the number of firms bidding on a sale, but labor cost is among the most significant.

Labor also has had a dramatic effect on the rate of mechanization. During the early 1970s, more than a few operators mechanized simply because they could not get enough labor to do the job or to improve labor productivity. The hardest hit area was production, which has been curtailed in all producing regions due to labor shortages.

The labor supply

The industry has been faced with the labor problem for many years. And the older and wiser among us always said: "There will be labor as long as there are logs." Nonetheless, the shortage was discussed at many industry conferences, logging training schools were started in an effort to recruit and provide more trained workers, and federal funds have been spent for the same purpose.

In 1951, there were 106,400 employees working in logging. By 1971, the number had dropped to 69,100, a 37,300 reduction in 20 years (Table 3.1).

The trend changed during the 1970s. By the end of the decade, the 69,100 employees had grown to 90,300. There was a dip during the 1975 recession, just as there was a dip during the 1980 recession. But employment, as a trend, kept on growing.

It is difficult to pinpoint employment trends either by region or in the nation as a whole. Some regions have fairly good employment statistics for logging and others do not. For instance, the Northwest has fairly good statistics, but this is to be expected since the region depends heavily on the wood products industry. In many cases logging employment is lumped in with total wood products employment, so it is difficult to separate the two.

Table 3.1 *Logging Camps and Logging Contractors (all employees – in thousands)*

Year	Annual Average	Year	Annual Average
1950	91.5	1965	84.2
1951	106.4	1966	81.1
1952	98.5	1967	81.1
1953	96.9	1968	79.1
1954	90.9	1969	78.4
1955	92.5	1970	70.3
1956	99.8	1971	69.1
1957	86.3	1972	69.0
1958	87.2	1973	75.9
1959	94.4	1974	80.7
1960	91.0	1975	73.5
1961	84.6	1976	81.5
1962	83.6	1977	85.1
1963	83.5	1978	88.3
1964	87.7	1979	90.3

Source: U.S. Department of Labor, Bureau of Labor Statistics.

There are probably several reasons for the growth trend in the woods industry. One major cause obviously is the growth in consumption of timber resulting from increased demand for wood products in the United States and Canada. Over the past 10 years the average annual growth rate for lumber and plywood has been 2.2 percent. Likewise, in the paper industry the rate of growth has been 2.1 percent (Clephane 1980).

A second important factor is the growth in the southern solid wood business, particularly in plywood manufacture. Between 1968 and 1978, the southern plywood industry grew at a 12.8 percent compound annual rate. During the same period, southern lumber production grew at a 1.9 percent compound annual rate and paper production at a 4 percent rate (Clephane 1980).

Finally, the U.S. civilian work force grew 23 percent during the 1970s, from 78.6 million to 96.9 million. This growth obviously made more people available for the woods work force.

Other factors undoubtedly had an impact on logging employment. For instance, piece size has decreased on average, both in absolute terms in the West and as a result of the dramatic increase in southern production. Since the average piece size is smaller, more pieces must be handled to supply a given demand for timber. Small pieces require, on average, more manpower. Wage trends represent another factor that has an impact on labor supply.

Wage levels

The level of wages paid in the industry is probably one of the most important determinants in the labor supply problem. Depending on geographical locations, mobility of the labor force, and level of education, firms and industries paying higher wages tend to experience fewer recruitment problems than do low-wage firms and industries. In addition, there is a positive correlation between quality of labor and wage level. High-wage industries have fewer problems recruiting quality workers than do low-wage industries.

Wages in the logging industry have not always been conducive to resolving either the recruitment or the quality problem. Back during what are often referred to as "the glory days of logging," labor was a cheap commodity. There are stories of logging companies having a crew coming, a crew working, and a crew going—all in the same day. Those were indeed the "glory days."

In 1906, a rigging slinger—a vital man on a highlead logging side—worked for $4.00 per day. Buckers, who are now paid more than $12.00 per hour in the Northwest, union scale, were then paid $3.50 per day. The hours of work were long—often from sunup to sundown—and living conditions left much to be desired. The loggers lived in a camp a long day's hike from the nearest town, ate in community cookhouses, and slept three deep in shotgun bunkhouses amidst sweaty long underwear, smoky wood stoves, and bedbugs.

The wages were low, though not much different from those paid in other industries. But loggers had other expenses, such as room and board, clothing, gloves, and caulked boots, which had to be deducted from their meager wages. There is another side to the coin, however. In spite of the adversity, the low wages, and the terrible working and living conditions, the logger cheerfully,

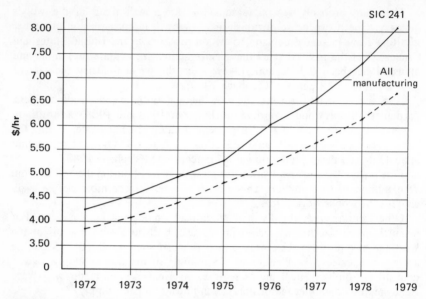

Figure 3.1. *Growth in wage levels, logging camps and logging contractors (SIC 241) versus all U.S. manufacturing (from U.S. Department of Labor, unpublished data).*

or mostly cheerfully, put up with it because he like the independent life. His was a carefree existence, and he rather enjoyed roaming around the country working in logging camps with other men like himself.

In recent years there have been some significant changes in the industry. In 1979 the average hourly wage in Standard Industrial Classification (SIC) 241, Logging Camps and Logging, was $8.06 per hour. This rate was up from $4.24 per hour in 1972, representing a compound growth rate of 8.33 percent. As a matter of comparison the average hourly wage growth rate for all manufacturing in the United States was 7.26 percent, better than 1 percent lower than the logging industry. Of greater interest is the rate of increase between 1975

Table 3.2 *Hourly Rates for Comparable Jobs*

Occupation	Western Canada	Northwest U.S.	Southwest U.S.	Southeast U.S.	Eastern Canada
Skidder operator	$10.28	$ 9.85	$ 5.81	$ 6.58	$10.00-11.00[1]
Loader operator	10.78	10.10	6.48	6.92	8.16
Log truck driver	10.78	9.33	5.68	6.58	8.65
Tractor operator	$10.01	$ 9.95	$ 5.88	$ 6.58	$ 8.16

Note: Experience from selected companies—1980 data.

1. Piece rate: ranges between $10.00 and $11.00 per hour.

and 1979. Logging industry wage rates increased at an 11.2 percent rate during those four years, during which time manufacturing in general increased at an 8.5 percent rate. Figure 3.1 indicates that wage levels are not only higher but, for the past few years, increased at a significantly higher rate.

While the differential between the logging industry and all manufacturing is favorable and is likely to remain so, given the growth rate, there is a significant differential between the southern United States and the Northwest. Table 3.2 illustrates the differences for a few selected jobs that are common across the country.

The quit rate

Hiring the men to run a logging business is just one problem—keeping them is even more of a problem. The quit rate for the wood products industry in general is high, much higher than the rate for all manufacturing industries as can be seen from Figure 3.2. While there are no figures for the logging classifications, it is assumed that the quit rate for logging is higher than for the wood products industry as a whole.

The quit rate, defined as voluntary separations from a firm and measured by the number of quits per 100 employees, is caused by several factors. It may be caused by seasonality, a common problem in the logging industry. It may be caused by workers who seek to upgrade their positions by moving to other industries.

The unemployment rate also has an effect on quit rates. When there is high employment and jobs are plentiful, the quit rate tends to rise, as in 1966, 1969, and 1973. However, when there is high unemployment such as occurred during the recessions of 1967-1968, 1970-1971, and 1974-1975, the quit rate falls. In 1975 when the recession bottomed and unemployment was at its highest the quit rate was at the lowest level since 1961.

Between 1961 and 1969, with the exception of a slight dip in 1967 and 1968, the quit rate climbed dramatically. Even during periods of recession in the industry, when the rate would be expected to decline significantly, it still retained its upward trend. Of course, the severe 1974-1975 recession, with its high unemployment rate, resulted in an extraordinarily low quit rate during that period.

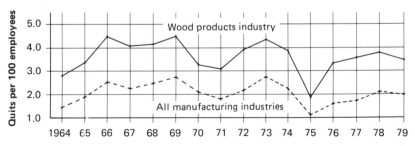

Figure 3.2 *Wood products industry quit rate (data from U.S. Dept. of Labor).*

Following the recession, the rate once again climbed to a peak in 1978. However, the peak reached was not as high as those attained in previous recovery periods. The rate began to drop again in 1979 in the face of recessionary psychology. The rate for 1980 was expected to be very low in the face of massive layoffs in the wood products industry caused by the 1980 recession.

Voluntary quits, or labor turnover, represent a serious problem in any industry where the rate is high.

There is a real cost to hiring and training new employees. Not only are hiring costs high, but a new, untrained employee is not as productive as older, more experienced employees. There are a number of approaches to the reduction of quit rates. New employee selection systems, improved training, better working conditions, increased wages, and greater advancement opportunities are just a few of these.

Fringe benefits

An important factor that will cause quit rates to decline is the expansion of fringe benefits that yield their benefits only with long service. Pension plans and profit-sharing plans are examples of these types of benefits.

Unfortunately, the wood products industry in general does not have much to offer in the area of fringe benefits. Of course, the industry has a legal obligation to provide such benefits as unemployment compensation, workman's compensation, and social security. But there is no law forcing an employer to provide pensions, vacations, or profit sharing. To be sure, the large, integrated companies, especially those where labor is organized, do offer a good package of benefits. There are also associations, like the Alaska Loggers Association and the Associated Oregon Loggers, that make benefits available to their members and to their members' employees.

However, as a matter of economics, a small firm often cannot justify pension plans and the like. Some of these firms are not interested to begin with. Where the crews are small and there is no labor organization the voluntary fringe benefit compensations therefore tend to be small or nonexistent. There are many small producers and contractors that do not even offer paid vacations. This may be due, in part, to the seasonal character of the work.

For whatever reasons, the wood products industry has a poor rating with respect to employee benefits. Though there is no proven correlation between quit rates and fringe benefits, one source indicates that the expansion of these benefits would have a positive effect in reducing the readiness of a workman to leave one employer for another (Williams 1970, p. 97).

Employee motivation

Logging companies, large and small, are continually complaining about the difficulty they experience in hiring and retaining quality people who have the skill and experience necessary to get the job done. It does not seem to make any difference whether the companies are in the West, the North, or the South; whether they produce pulpwood or sawlogs; or whether they are large or small.

Some of the factors that have an impact on employee continuity and job satisfaction have already been discussed: competitive wages, good fringe benefits, and working conditions. Yet there are other factors that management tends to overlook: demotivational factors that cause working people to become unhappy with their jobs, come to work late, become less productive, and look elsewhere for employment.

Many managers correctly assume that one of the prime motivational factors is wages. People need money to satisfy basic physiological needs–food, clothing, and housing. Beyond that they need money to support the affluent lifestyle we have become accustomed to in recent years. There are few occupations today that do not at least satisfy the basic physiological needs. Beyond this point wages are no longer of primary importance. A whole new set of wants and desires takes over and the satisfaction of this new set of requirements depends upon the skill of management and the on-the-job environment. The logging business has changed over the years and management techniques must change with it.

Not too many years ago, a man working in the woods worked with the owner right beside him. The owner was there every day and chances were there was a one-to-one relationship between labor and management. Today, in many firms, that relationship has changed. For one thing some businesses have gotten larger. And when it comes to people and keeping them happy there are no economies of scale. Because of bigness there is now a different relationship between labor and management. The difference pervades industry in general and logging, in this case, specifically. Workers have not changed. They still respond to the same things–strong leadership, a feeling of dignity, and a sense of accomplishment. Management, however, does respond differently.

The personal interaction between labor and management has, in all too many cases, become nonexistent. Where the owner was once a member of the crew, he is now an outsider seldom seen about the job site and when he is around it is a distant relationship. The strong one-on-one relationship once enjoyed between boss and subordinate has now been delegated to a manager or foreman who may not have the interest or the skill needed to develop strong interpersonal relationships with employees. This lack of personal interaction between management and employees reduces the positive interaction between the two parties. The employees, as a result, no longer identify with company goals and objectives; no longer care about company property; no longer have a "can do" attitude about their jobs.

Wages will not change the circumstances described, nor will improved benefits. Wages guarantee only that an employee will show up for work in the morning. Benefits may help keep an employee on the job. What a workman does once he arrives depends on the environment management provides.

A. H. Maslow, writing on the theory of human motivation, speaks of a hierarchy of needs. The first we have already spoken of–the physiological needs. The remaining four needs in the hierarchy are (Maslow 1963, p. 71):

1. Safety and security.   3. Esteem needs.

2. Love needs.   4. Self-actualization.

Assuming that an employee's physiological needs are already fulfilled, his attention is then directed to the next need in the hierarchy—safety and security. This need refers to both safety from injury on the job and the knowledge that the job will continue to exist in the future. As far as safety is concerned, we must acknowledge that working conditions in the woods have been vastly improved over the years. But even considering the many improvements, working in the woods is still a most dangerous occupation. The need for safety is satisfied to the extent that the employee feels that he does not come to work at the risk of his health or life. Most companies have safety programs, and every state, as well as the federal government, has safety regulations, which are generally enforced. However, in most cases the motivating factor for these programs is not the satisfaction of the employee's need for safety but rather management's desire to minimize the costs of accidents. This attitude on the part of management is unfortunate. With a little additional effort the safety program could become a real morale-building tool.

Job security is also of great importance and one of the industry's most serious problems. Like the rest of the workers in today's society the woods worker desires a new automobile, a boat and trailer, a new home, and time for recreation. All of these things require money and steady work. But there are a great many areas in the United States and Canada where this steadiness cannot be guaranteed. In Alaska, Idaho, and portions of the North and South the woods work is seasonal. Labor is always a problem, and especially so where a man can expect to work only eight or nine months out of the year.

Mechanization is solving a part of this problem. As more sophisticated machines are introduced into the woods it will become possible for the worker to remain on the job for longer periods during the year. The cold and rain will become less of a hindrance as the industry becomes more fully mechanized. However, it does not seem likely that the industry will be able to completely control the working environment—at least not for a while.

The seasonal nature of the logging business has always been a problem. Some companies have worked out solutions without the use of mechanization. Some companies manage to retain their key men, who comprise about one-third of the crew, by keeping those men employed during the off-season. They work on the equipment, perhaps operate one logging side, and in general perform those tasks that cannot be done during the regular season. Another solution, applied by one western corporation, is to transfer the woods workers into the sawmill during the winter months when logging is not possible. Of course, this is a solution only if the company has a manufacturing facility. In the future, as wood product companies become further integrated, this may become a more feasible solution.

Once the safety needs are satisfied, when the worker has a safe place to work and a certain degree of job security, he then looks for a meaningful relationship within his work group. He needs to belong to a group and experience some sort of positive interaction and he will strive with some intensity to achieve this goal. There are several problems involved in this area. Because of the increasing availability of a higher level of education, the industry is hiring more and more better-educated people. At the same time, there are a large number of men already working in the industry who have only a minimal education. Thus there is often a problem in establishing common ground between these two groups. If in their actual work activities they were split into groups, the problem might be minimized, but this seldom happens.

When individuals are continually shifted from one crew to another there is no opportunity to establish ties that bind. A new member to a group is often not readily accepted and is considered an outsider. He has to earn membership by conforming to established group goals and establishing himself as a group member in good standing. With a high degree of turnover in woods crews it is difficult to establish the required relationships.

After the need for belonging comes the need for self-esteem or self-respect. The worker desires a stable and firmly based high evaluation of himself. Here the emphasis is on achievement, independence, prestige, recognition from others, and appreciation for a job well done.

The logger has long been described as as brush ape. Of course, this is a severe misconception and management must somehow reduce the impact of this misconception on labor. Logging is not like building airplanes or engineering the construction of skyscrapers and bridges, but it does take a special sort of intelligence to perform the job well. Stupid men, men with subnormal intelligence, are not able to run the high-speed machines used in the logging industry, and they are not able to cope with the high-speed operations resulting from the machine technology currently being used in the industry.

Because of the brush ape stigma attached to logging labor, there is little prestige in the job—thus the self-esteem need is not fulfilled. Achievement, appreciation, and recognition can go along with any job—providing the right environment exists. If management is aware of the need for fulfillment in these areas it can easily communicate recognition and appreciation. Meeting this part of the self-esteem need is management's responsibility.

Supervision in the woods is changing, but it will be a long while before the industry has the type of management that will recognize and react to the needs of labor. Many of the front-line supervisors, the foremen and leadmen, are excellent loggers, but human relations is definitely not their long suit. They often regard labor merely as the extension of a machine. This sort of attitude is not conducive to creating the environment in which a worker can enjoy any great degree of self-respect.

The last need in the hierarchy is the need for self-fulfillment or self-actualization. In the strictest sense, self-actualization refers to the need to fulfill potential—to become what you are capable of becoming. All of the other needs—physiological, safety, relationship, and self-esteem—may be satisfied, but the worker will still be discontented if he cannot realize his full potential.

Of course, if realizing full potential means that a man should be a carpenter or an engineer instead of a logger, management can do little good. However, if the man wants to make a life in the logging business there is much that management can do. One serious problem in this area is the lack of turnover in the better jobs offered in the woods. A man who becomes a hooktender, leadman, or machine operator is not apt to leave the industry for another job, since his pay is good, his job offers a certain amount of prestige, and management will generally attempt to keep him working because of his key-man status. If the key jobs remain filled, however, there is no opportunity for advancement up through the ranks. Therefore, the choker setter, the chaser, and the swampers are faced with the prospect of forever remaining choker setters, chasers, and swampers—not the brightest of prospects for those who are interested in other positions.

In answer to this problem many companies have established policies regarding promotion from within the ranks of the currently employed rather than hiring from the outside. This policy involves upgrading jobs, and promotions into supervisory capacities. Another policy provides for training, either on the job or through adult education, to help the employee upgrade his skills and prepare for better positions within the company. Of course, small firms without sufficient financial resources find it difficult to do any more than pay wages. Perhaps, in this situation the role of the loggers association could be expanded to offer assistance in employee satisfaction through special training schools in much the same way that a large membership in the association makes it possible to provide economical pension plans.

There are many ways to provide job satisfaction and an environment in which a worker can be motivated—good communications, setting achievable goals, and allowing worker participation, to name a few. It is management's job to get labor involved in the work process and to understand its own importance. If management can satisfy all the worker's needs, the result will be improved interest in work, better work quality, and higher productivity.

The Systems Approach

Generally, when thinking about harvest operations, it is in terms of the work performed to prepare the trees and move the logs from the woods to their respective end-use points. The usual mental picture is one of huge, rumbling machines; shiny, taut cables; and whining power saws. For those not familiar with the industry, however, the picture is misleading. There is much more to logging than cutting down trees, bucking them into logs, and moving them to the sawmill. Depending on individual circumstances, harvest operations involve everything from buying the timber and planning the operations to scaling and grading the logs before they arrive at the mill site. All of the numerous functions included in harvesting a timber crop are interrelated and can affect the overall operation.

Strictly speaking, a woods operation is a loose grouping of mobile factories. The manufacturing process taking place in those factories is not unlike the processes taking place in other extractive industries, such as coal mining and oil production. In each case a series of work elements is performed resulting in the production of primary products—products which become the raw material for downstream operations.

From the roundwood produced in logging, the industry's conversion plants manufacture lumber, plywood, plastics, chemicals, medicine, writing paper—literally hundreds of products. And the chain does not end with these products. A contractor uses lumber and plywood to construct homes and places of business. A publishing firm buys newsprint from a paper mill. In other words, the timber industry is but a part of a total system, which includes manufacturing, marketing, transportation, sales, and much else. And all these functions mesh together. Harvesting functions are only part of a dominant system.

The preceding paragraphs constitute a brief introduction to the *systems concept*. While this book is not specifically concerned with manufacturing and marketing, the concept nevertheless provides the reader with a logical framework within which the harvesting operation can be studied and understood.

The systems concept

A *system* comprises a group of components that are interrelated and jointly contribute to some common objective. Furthermore, each system may be regarded as a part of some larger, dominant system. At the same time, the system in question may be composed of a number of smaller systems called *subsystems.* The subsystems, like the components, are also interrelated. The systems concept, then, involves an approach that describes a set of relationships and the interaction between them.

The word *system* suggests plans, methods, and order. This naturally leads to some assumptions about the conditions which must be met before a system can exist. The three primary conditions are:

1. All the components of a system must contribute to the achievement of a common goal or objective.
2. There must be some hierarchy present within a system to assure coordination of activities and allow for specialization of system components.
3. The inputs to a system—energy, information, raw material, labor, etc.—must be introduced according to some plan.

The conditions listed are all contained or implied in the definition of a system. Indeed, they are essential to the definition of any system whether it is a logging system or an information system (Johnson, Kast, and Rosenweig 1963, p. 91). If these conditions are not met, all that exists is a collection of mutually independent parts with few, if any, systematic relationships.

Harvesting system objectives

As we have said harvesting is a subsystem of a large dominant system. However, for the sake of simplicity, in this book harvesting will be regarded as a system. The major functions that make up the harvesting system will be called *components.* It is no easy task to define the objectives and conditions of a harvesting system. An explanation of logging and its objectives depends largely on who is asked.

For example, a forester asked to define logging would probably describe it as the principal forestry operation—the critical end result of many years' labor. The men who operate the sawmill or pulpmill would tend to describe logging as a supply function. For them, the logs represent a raw material input. A logger, in all likelihood, would simply say he is in the business of moving wood—a transportation function.

These examples represent three frames of reference and, therefore, three different descriptions of what logging is. In a manner of speaking all three are correct. In this book, however, the frame of reference is basically that of the logger. However, the overall description developed in this book is necessarily a bit more elaborate than simply "moving the wood."

No matter what sort of logging system referred to, the process always involves log transportation—the movement of trees, logs, or segments of logs from one point to another. By means of various modes of power, the trees, logs, or whatever are dragged, hauled, or carried out of the woods and across the roads and byways to some end concentration point. Thus, two of the first objectives of any logging system are to prepare the trees for transportation and to transport them to the proper conversion facility. How the trees are prepared and transported will depend on the facility or market for which they are intended. Of course, there are other important objectives, such as moving the wood at least cost, maintaining a safe environment for the workmen, and maintaining a good working relationship with public and private agencies.

The actual work of logging involves an aggregation of man-machine components (see Figure 4.1). These components function together to achieve the transportation objective. Within the logging system there are only four major components. They are cutting, skidding or yarding, loading, and log transportation. *Unloading* might be defined as either a component of the logging system or a part of an interfacing system. At any rate, unloading has a direct impact on the log transportation component. Each logging system component can be further reduced to elements that better describe its function. On the following pages the various components and their elements are described.

Timber cutting

Timber cutting includes all work elements leading to the total preparation of a primary product, whether that product is the whole tree, logs, or pulpwood bolts. The elements that describe the timber cutting component are:

Felling: *Felling* describes all steps necessary to sever a standing tree. Most felling is accomplished with either shears or a power saw. In some parts of the world, cutters still use more primitive methods, such as the crosscut saw and the ax.

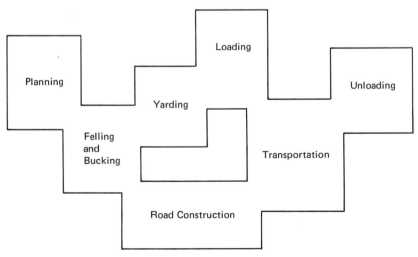

Figure 4.1. *The components of the harvesting system.*

Felling a tree. In this case a power saw is being used.

Bucking: *Bucking* describes the work done to cut a tree into segments. The segments are called *logs, bolts,* or if only the top is removed, *tree-length logs.* Like felling, bucking is accomplished with power saws or shears.

Measuring: Before a tree can be bucked into segments it must be measured to insure that proper log lengths result. Log length depends on end use and ranges from 100-inch bolts to logs over 50 feet in length.

Limbing: In many timber regions all the limbs must be trimmed off the bole of the tree. This is true of species with limbs substantially the full length of the bole. In some cases this function is performed with a power saw or ax. In other cases, limbing is performed by the cutting head on a machine such as a Beloit harvester or a Buschcombine.

Topping: Topping is really a form of bucking except that the only cut made severs the tree at what is called a *merchantable top* or *merch top.* A merch top is the smallest utilizable top.

At least one of the cutting operations is always first in the tree preparation process and is considered the initial step in the harvesting process. The success of all subsequent operations, including final manufacturing at the conversion plant, depends on the quality of workmanship in the cutting operation. In view of its importance, the cutting operation should be very carefully managed.

Primary transportation

Logging is a serial operation; that is, certain steps must be performed in a given order so the objective may be achieved. However, the order of these steps varies from system to system. In most cases, after the cutting is done the logs must be transported to a landing for loading. Any movement from the stump to a landing is called *primary transportation*. This is the second major component. A landing may be either a cleared area where logs are stored for further

Bucking a tree into segments—either logs or pulpwood bolts.

Two ways to skid a log– with a tracked vehicle (left) or cables (right).

handling or a wide spot on the shoulder of the road. Major methods involved in primary transportation are:

Skidding: *Skidding* describes the movement along the ground of trees or tree segments from the stump to a landing area. It may be performed by wheeled or track-type machines. When cables are used it is called *yarding* or *cable skidding.* Airborne vehicles such as helicopters and balloons are also used to move logs to a landing area from the woods.

Bunching: This function is actually an element of skidding. It is sometimes done by hand when bolts are being handled, but mostly it is a mechanized process. *Bunching* refers to piling or stacking small groups of trees or tree segments in the brush, but not necessarily at the stump. The bunches are then moved from the brush to the landing, generally by a wheeled or tracked vehicle.

Forwarding: This form of skidding, which is also called *prehauling*, involves loading the tree segments on a carry vehicle and carrying them to the landing. In some cases large skidders or tracked vehicles will drag several bunches from a single concentration point. Forwarding is considered a component rather than an element.

Loading

The third major component, loading, is not as difficult to explain as the other components. However, there are many variations in this component since it can be accomplished at any number of locations from the landing to stump-side, depending on the logging system being used. It is still not uncommon to observe handloading operations, but most loading is now mechanized. The following is a more precise description of this component.

Loading: This component involves placing some form of tree segment or tree length onto a haul vehicle. Ordinarily, loading refers to woods loading. However, loading also takes place at various transfer points. A transfer point is used when segments are transported part way to a conversion point and unloaded for temporary storage or directly reloaded onto another mode of transportation—truck, rail, or barge—to continue to the conversion point.

Secondary transportation

The final component in a logging system is *secondary transportation* and includes all wood movement from the landing or transfer point. Secondary

Loading logs onto a trailer.   *Photo courtesy Northwest Engineering Co.*

Logs can be transported by truck, rail, or water.   *Photo courtesy* Forest Industries

transportation includes movement by truck, rail, or water. The following is a description of log transportation and unloading.

Log Transportation: Log transportation includes all log movements from a landing to an end-use or transfer point. Movement may be by truck, rail, water, or by a combination of two or more transportation modes. When logs are loaded at the stump, as in some pulpwood operations, this is also considered secondary transportation.

Unloading can be considered either as an element of the log transportation component or as a system component in its own right. Because of its direct impact on log transportation, however, we will consider unloading an element of the log transportation component.

Unloading: The unload or offload element involves removal of trees or tree segments from any transportation mode, such as truck, rail, or barge.

Unloading may be accomplished with cranes, heavy-lift wheeled or track machines, or any other type of machine suitable for the purpose.

Variability of logging systems

Logging systems, defined by the components and elements just described, can and do vary depending on a large number of external conditions, which include geographic location, terrain, weather, and the form of the primary product.

Systems, for example, may be classified by the piece length handled. A tree-length system deals with handling tree-length logs from the stump to a landing, or, perhaps, all the way to the mill where they are bucked into shorter lengths for processing. A second system is the long-log system, with logs about 20 feet in length. A short-wood system describes handling bolts between 4 feet and 100 inches. In each case the order in which the logging system components come into play, the size and type of machine, the skill requirements, and the timber size are different.

Generally, however, a system is described in terms of its components rather than the length of wood handled. Though the description of a long-log or

Unloading logs from a truck.   *Photo courtesy Totem Equipment Co.*

short-wood system implies an ordering of components as well as equipment size, the description is really not specific. There are endless variations of components which can describe a system. The result is an endless number of unique systems, performing under a variety of cost, production, and environmental conditions. Furthermore, the hierarchy of components can also vary. Consider some of the variations that apply to the three systems (Figure 4.2).

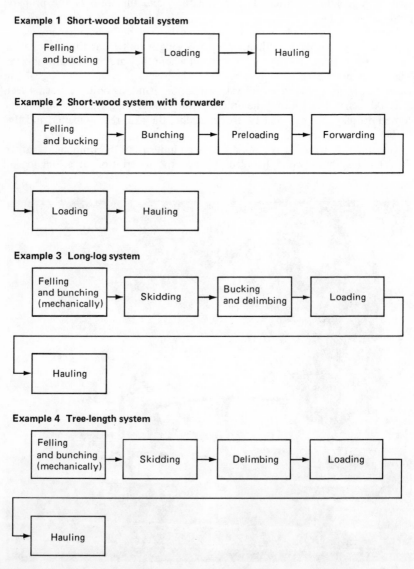

Figure 4.2. *Flow diagrams of representative systems.*

A short-wood system can be a one-man operation. A single workman fells the trees, measures them, bucks them into segments, loads the bolts at stump-side, and delivers the bolts to the concentration yard or mill site. Loading is accomplished either by hand or by use of a winch mounted on the truck. The winch may be hand operated or mechanically driven. The skidding component has been eliminated since the truck hauls the bolts directly from the stump.

Another short-wood system involves felling and bucking the trees into bolts and stacking or bunching the bolts in the woods. A four-wheel vehicle called a *forwarder* then loads the bolts onto an integral bed or trailer and transports the bolts to a landing.* Once at the landing the forwarder either offloads onto a waiting trailer or directly to the ground from where the bolts will be reloaded later.

Other systems have been employed to mechanize various elements of short-wood production. Some machines combine felling, delimbing, and bucking functions, with or without forwarding capability. Or, tree-length stems may be skidded to a landing where a mobile piece of equipment delimbs and bucks to short-wood bolt lengths.

These short-wood systems are not the only possibilities, but they do demonstrate the variety possible in both the components and their order. There is always a basic design and within that design certain components are activated before others. Referring back to the flow diagram (Figure 4.1), the bobtail system applies cutting first, loading second, and hauling third. Each component has a place in the overall system and each component is specialized with respect to the portion of the work to be accomplished.

This ordering of components is one of the basic conditions of any system. A systems analyst would describe this ordering as the definition of hierarchies within the system. The hierarchies exist and allow for a coordination of activities. They also suggest a level of specialization among system components. The concept allows for successively inclusive system levels, each of which can be studied either separately or as a part of the whole system. We will take both approaches in this book.

Proper planning

The existence of a large number of possible combinations within a logging system means planning is essential. The components must be capable of performing together to maximize the system objective. This means the components must be matched for effective performance. The man must be capable of operating the equipment or lifting the bolts. The machine must be equal to hauling the load. If there are too many men or if the equipment used is such that production rates in primary transportation exceed loading capability or hauling capacity, then the entire system is thrown out of balance.

An example of this imbalance occurred in a helicopter logging show observed by the author. The helicopter, a Sikorsky 64, was capable of moving

* Loading from the ground to a forwarder is called *preloading*.

significantly more logs to the landing than the loading machine could handle. Under favorable conditions the helicopter could and did plug the landing with logs, creating inefficiencies in both the loading and hauling components.

A logging system should be designed so that men, machines, and log input are correlated. A great many variables must be considered: labor skills, size of timber, primary product, skidding distances, equipment speed, capital investment required, terrain characteristics, and haul distances, to name just a few. Failure to consider any of the variables will result in operating problems and inefficiencies.

For example, a logging operator was using a cable system designed for small timber. His loader was likewise small and fast, to facilitate handling small logs. For two years the system operated successfully. Then the operator took a contract to log some timber that was of larger average size than his system was designed for. He broke the cable on his yarder, and logging production was very slow. The loader had to handle logs one end at a time rather than picking up the logs and swinging them to the truck. At times the loader simply could not pick up the logs. There was definitely an imbalance in the system caused by the timber size. Because the logging operator had failed to plan properly, production fell to an unacceptable level, maintenance costs increased, and costs rose to offset any possible profits.

Another example is a pulpwood producer who elected to expand his operations and increase production by adding an additional skidder. The increased production was too much for the one truck he owned but too little to support an additional truck. He could not contract a second truck and driver on a full-time basis. The result was that none of his equipment, except the truck, ever operated at maximum effectiveness.

The objective in any system design should be to develop a system within which the components are able to work together at least cost to achieve stated objectives. If the option exists to choose among system components to put the right system in the right place at the right time, costs will be minimized. It is imperative to understand that each system functions best under certain conditions. At the same time, however, the objectives of dominant and co-dominant systems must be considered.

Dominant systems

Harvesting operations are always a part of a larger dominant system. In an integrated corporation that harvests or controls harvesting and manufactures a full line of wood products in its own conversion plants, the dominant objective is to maximize earnings per share over time. A small logging contractor supplying logs to a user mill is or should be strongly influenced by the mill management's objectives. And when dealing with public lands and timber a whole new set of objectives comes into play.

For instance, a great deal of timber, the raw material of a logging operation, comes from public lands. This means a logging contractor, independent or not, must deal with some third parties. There are times when the contractor will have to do things that will not support either the logging system objective or his own firm's objective. He will be forced to operate within the constraints imposed by a dominant system.

For example, there are many areas in the United States where most of the available timber belongs to the government. If an operator is to survive in this environment he must conform to the needs and desires of the public as imposed by the government agency authorized to manage the timber resource. This problem has been especially obvious in recent years during which the U.S. Forest Service, the Bureau of Land Management, and state agencies have imposed contractual constraints requiring buffer strips, clean logging, and special care to leave the forest with an untouched look and to protect water quality. Though these activities all cost additional money because they require more time, the additional cost has not always been included in the logging price. However, logging contractors do the work required, in spite of smaller profit margins, because that is the only way they can stay in business.

Harvesting operations are the first steps in a series resulting in a finished product. The systems that utilize tree segments as raw materials represent interfacing systems and must also be considered. As a result, logging must often be held subordinate to downstream manufacturing systems.

The same rationale holds even among the components of the logging system. The cutting process is the first step in the harvest operation. Not only does the manufacturing process depend on the skill and workmanship of the cutters, but the subsequent logging system components are affected as well. If the logs are cut too short, total skidding time will be increased, as will the unit cost of skidding. It is generally most cost-effective if the cutting function is subordinated to the skidding function. An increase of 50 cents per cunit in cutting costs may save one dollar per cunit in skidding costs. Felling to the appropriate lead (see Chapter 7) and minimizing stump heights are examples where extra care in cutting can improve skidding and yarding efficiency. Likewise, hauling costs will increase because of the additional pieces that must be handled in loading. In the sawmill or plywood mill, short logs can mean excessive waste and a significant decrease in product value.

Results such as these are not premeditated. In most cases logging operators are simply attempting to maximize efficiency and profits in their own business and give little thought to the impact of their actions on interfacing systems.

There are many such examples of this in the logging industry. And what they make clear is the need to understand the interrelationships between components of the logging system and between the logging system and manufacturing. Of course, interrelationships also exist between the logging system and the firm to which the logging system is subordinate.

The firm is composed of a group of systems or subsystems which must all work together (Figure 4.3). Forestry is responsible for planning the long-term growth and use of the timber and land. The logger can have a serious impact on regeneration. Leaving too much debris on the ground after logging makes regeneration activities more costly and less effective. The use of tracked and wheeled equipment on the ground during wet weather can significantly reduce the land's ability to grow trees. Carelessness in logging through what will be a residual stand can result in serious damage to the remaining trees.

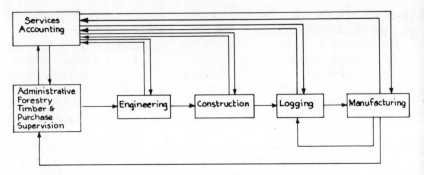

Figure 4.3. *Major systems that make up a total integrated firm.*

Road engineering and construction, to look at another side of the picture, can have a very serious impact on the success of logging systems. Poor road alignment or mislocated roads directly increase hauling costs and the effectiveness of the various logging systems in use. Steep adverse grades, although they are sometimes necessary, always have an unfavorable effect on hauling costs. A setting, the block of timber to be logged, which is poorly laid out by engineering also affects the efficiency of the logging system.

Multiple objectives

Forestry, engineering, and manufacturing all represent major systems with which the harvesting system must interface. Each relies on the others and cannot function well without them. Thus more than one objective is necessary.

While the basic objective of any harvesting system is to operate at minimum cost there may be situations which call for suboptimization of the harvesting system in the interest of serving the objective of a dominant system. One such objective might be to carry out the logging process in such a way as to minimize disturbance of the forest environment. Still another might be to preserve aesthetics of the forest. In any case the logging operation must leave the ground in such a condition that successful regeneration is possible. Obviously, there can be a great many objectives, some of which are conflicting. Objectives can only be determined after examining the systems involved and their internal and external environments.

The discussion in this chapter should help to put the harvesting operation into perspective. Harvesting, specifically logging, represents a system which often has multiple objectives and which satisfies the conditions of a hierarchy for specialization and planned input of raw material and resources. All of the components are closely related, and unless they work together toward common objectives the venture is likely to fail or at least have suboptimum results.

A logging system operates within the internal environment of the firm of which it is a part. Strictly speaking, however, a logging system's internal environment is composed of all the physical and operating variables present in the forest. The external environment is represented by pressures and variables imposed by the business, social, political, and economic systems within which a logging or harvesting system must function. The external environment contains dominant systems to which the objectives of harvesting are subordinate. Even though the objectives of the 'outside' systems sometimes appear to be in conflict with the economic objectives of the logging firm, the firm must nevertheless subordinate itself to those objectives.

5.

Operations Planning

Planning is the most essential function to be performed in a logging business. It is essential because it provides the discipline that welds together all parts of the harvesting system, identifying and resolving conflicts, recognizing constraints, and providing for an orderly input of resources. The plan is a projected course of action which defines a necessary sequence of activities, identifies the techniques to be applied, and determines the timing requirements. In addition, the plan must take into account the needs of the interfacing subsystems—manufacturing, sales, environment, regeneration, etc.—as well as the company's or firm's overall objectives. Without a plan or with an inadequate plan, the result is waste, underutilization of productive resources, and excessive cost.

Understanding objectives

In any planning activity those people responsible should begin with the highest level requirements—the firm's objectives, particularly its marketing objectives. There will always be multiple objectives, at least some of which apply to the activity being planned. For instance, a top priority in any integrated wood products firm is to provide raw material for conversion plants and third party sales in a form defined by volume, quality, and timeliness. A second priority, of equal importance in the long term, is regeneration of the forest so the marketing objective can be satisfied at the end of the rotation period.

There are, of course, other objectives, such as those dealing with social-political problems. One example is preservation of the environment. Streams must be left free of siltation and debris; the forest must be left in an aesthetically pleasing state; the air must be left pollution free. Another example is the use of the forest. The forest has many uses other than timber growing. It is a major source of recreation. People should be able to use the forest for fishing, hunting, and hiking. Wildlife management and watershed management are important, not only for recreation, but also for preserving certain wildlife species and providing clean water for use by the general population. In many sections of the country, livestock grazing on forest lands makes a real contribution to regional economies.

In addition to major corporate or business objectives and social-political objectives, all of which provide direction to the harvesting system, there are also those objectives dealing with the harvesting system itself. This third level of objectives involves such things as utilization of the resource, worker productivity, value improvement, and safety.

At the operating level most of the social-political considerations are already settled. Care must be exercised when logging near streams. Buffer strips are left along well-traveled roads to mask the effects of logging operations. The size of the blocks of timber to be removed is kept to some minimum in order to improve the visual environment. In some cases the forest is only partially cut so that what remains is left in an aesthetically pleasing state. All these procedures are a part of the operating plan in a modern harvest system. Experience has shown that quick recognition of the social-political objectives is a requirement for continued operation. In many cases, however, the same is not true of the marketing function.

The rather simple economics of the marketing objective seem to be the most difficult to understand—perhaps because this objective has not been legislated into existence as have many of the social-political objectives or perhaps because it has not been thought out properly. Yet the laws of economics are basic to any business. If the markets are not satisfied then profits, which are necessary for staying in business, are not forthcoming.

The basic question is that of raw material supply. But the question has less to do with the volume of wood delivered for conversion than it does with whether that volume of raw material is supplied in a timely fashion and in the correct form—species, grade, and size—to yield marketable products.

In both the long and the short term the firm's marketing objective is the driving force behind all its other activities. In the short term, one year at a time, the raw material must be supplied through the harvest function. In the long term the marketing objective causes the forest management objective to shift from harvesting to timber growing. Once again the market place, where profits are actually earned, is the driving force.

The demands of the market place must be satisfied both today and in the future. Nobody can deny that logging is certainly an expensive business. But logging expense is not the only consideration. Forest products are the raw material of the sawmill, plywood mill, and pulpmill. These conversion facilities also require an enormous capital investment. The huge capital investment in land and timber, logging equipment, and conversion facilities requires a financial return throughout the economic life of the investment. And that return depends on a continuous supply of timber from which marketable products can be manufactured.

Whether the timberland is public or private, owned or not owned, the basic considerations are the same. The woods manager must consider both the long and the short term needs of the firm, the present and future values of the land and timber resource, and the noneconomic as well as economic uses of both the land and timber. All this must be balanced against the harvest systems, the downstream conversion requirements, and, finally, the market place. Clearly, planning the harvest involves much more than preparing for timber removal.

The complexity of harvesting

Even if harvesting were the only consideration there is more to harvesting than cutting the timber and removing it. To gain access to the timber, roads must be engineered. A single mainline road may well open up a large area to logging. However, the timber to which access is gained by a single mainline will not be completely logged in one year but, rather, over a number of years. After the mainline is built, spur roads must be engineered and constructed for more direct access to the timber. The logging area, called a block or tract, must be engineered and laid out for maximum efficiency.

Since the forest products must be used in some conversion process downstream from logging, it is also necessary to have fairly accurate information, by species and grade, regarding volumes to be removed. A fair amount of planning is required simply to minimize damage to the timber through falling breakage where that is a consideration. The main point here is value extraction—removal of the maximum volume of marketable material.

There are other considerations as well. Such things as seasonality, wetland logging, and land drainage are all important depending on the geographic area. Finally the harvest system must consider the regeneration objectives of the firm.

Harvesting functions can have a serious effect on regeneration. On wetlands, logging can and often does lower the site index. (Site index is stated as a ratio of height over age and is a measure of the land's capacity for growing trees.) A poor job of timber removal or poor utilization leaves an unnecessary accumulation of logging debris, chunks, and logs to be cleaned up or burned in preparation for planting trees. In every case, the logging manager must do what he can to reduce site degradation and in general reduce the cost of downstream forestry activities. One example is logging cleanup. The more fiber that can be removed during logging the easier and less costly it will be to prepare the site for planting. In short, logging planning must look closely at all activities which are necessary before, during, and after the harvest.

Planning must also take into account the many variables that affect harvest operations. Forests grow under a wide range of conditions. Topography may be gentle, steep, or broken. The land may be wet and capable of being logged only during dry seasons or it may be all-weather land. Climate, of course, varies with the seasons. In Canada and the northern sections of the United States the winter brings snow. In the southern and southeastern United States late summer and fall often bring hurricanes, creating serious harvesting problems. The timber may be young and vigorous or old and decadent. Timber types change from one side of a ridge to another. Soil types differ from area to area and from region to region. To further complicate the situation, the logging manager has a variety of logging systems to choose from. There are several types of cable systems, wheeled skidders, and track-type machines. Transportation modes and networks must be considered. Another key consideration is labor skills, along with the amount of labor available. All of the variables mentioned, from topography to labor, interact in a systematic manner to define logging systems.

Operating variables must be identified and their impact on operations must be estimated before logging begins. Some information can be developed from topographic maps, aerial photographs, survey notes, and timber-type maps. These tools provide much of the data required for logging planning. However, any plan requires a certain amount of ground reconnaissance to give the planner firsthand familiarity with the timber stand in question and to verify and supplement data originating from maps, photos, and survey notes.

The extent of the reconnaissance will depend upon the size of the ground area involved. A plan for logging a single setting will be developed only after the entire setting has been walked over. A logging plan for an entire drainage, including haul road location, setting and landing location, and spur road location, will probably include both ground reconnaissance and helicopter survey. When ownerships are mixed a complete land survey is necessary.

The data derived from reconnaissance, maps, and other data sources provide the basis for a total logging plan. To these data must be added the operating constraints imposed by the timber-owning agency. For instance, policies have been made regarding types of logging systems that may be used on public land. Logging distances and road spacing are specified. Standards for road building are also specified, along with volume to be removed, water-quality protection plans, and fire-prevention plans. The final plan, whether established by contract or company policy, becomes the blueprint of the logging operation.

The specific plan for removal of timber from a setting depends on the logging systems available. Setting layout will be quite different for a highlead cable system than for a skyline system, which can remove logs over a greater distance. Ground skidding may be possible, depending on the operating variables. Roads built with skyline systems in mind will result in inefficiencies if highlead logging is used instead.

The balance of this chapter will deal generally with the operating variables and their effect on operations. Specific relationships between the variables and the logging systems will be described in subsequent chapters dealing with the various systems.

Road layout

Most commercial forest areas have fairly well-developed road systems. However, in some areas, including several drainages, road systems have yet to be developed. Road layout—location and construction—is a prime consideration in logging. Road spacing and location can restrict the use of certain logging systems. Several factors, such as gradient, surface, alignment, number of lanes, and availability of turnouts, have a direct impact on the efficiency of the transportation function.

The logging operator is always considering the productive capability of his equipment. The haul can easily create an expensive bottleneck. No matter what the unit of measure—cords, board feet, or cubic feet—the amount of wood delivered to the conversion plant or other end-use point is the ultimate measure of production. Many operators estimate cost based on loads per day produced. However, no profit is earned until the wood is delivered. The capacity of the trucks is a given quantity, but the number of loads hauled varies with distance and road conditions.

Road conditions are at least partially specified by a definition of road classes. The following are the road classes recognized by the State of Washington Utilities and Transportation Commission (Olympia, Washington):

Class A: Paved or macadamized, reasonably free from chuck holes, ruts, "washboard" conditions and other hazards, not exceeding grades of 6%.

Class B: Paved or macadamized, or graveled, other than the specifications in Class A as applicable, not exceeding grades of 12%; also permanently and continuously maintained fine gravel, smooth surface, free from chuck holes, ruts, "washboard" conditions and other hazards, with grades exceeding 6% but not exceeding 12%; also good plank not exceeding grades of 12%. Good plank road shall be defined as at least 10 feet wide with side guards at least 6 inches high on cross planking, or center guard at least 6 inches high on longitudinal planking; constructed of planks 3 inches by 10 inches, firmly spiked down and with sufficient turn out space for truck passing at least every 200 yards; supporting timbers to be at least 10 inches by 10 inches.

Class C: All roads with grades exceeding 12% but not exceeding 18%; also all dirt, rock or plank other than good plank specified under Class B, not exceeding grades of 18%.

Class D: All roads with grades exceeding 18% but not exceeding 22 percent.

Class E: All roads with grades exceeding 22%; also, roads consisting of mud or water to a depth of 8 or more inches, or any road that cannot be negotiated by the truck under its own motive power.

There are other useful road descriptions. For instance, one author describes woods roads, graded dirt roads, and gravel roads for southern operation. A *woods road* is an undrained road pushed out by a bulldozer or a road that is simply a brushed-out trail. A *graded dirt road* is constructed with ditches on either side but cannot usually be used after prolonged rain. *Gravel roads* are all-weather roads which cannot be traveled as fast as paved roads (Tufts).

Forest road standards vary depending on the forest land owner, whether public or private. Roads standards for private land depend on the size and horsepower of the trucks; the road construction problem (that is, elevation to be gained within a given distance); the volume to be hauled; and who will use the road. In some states, for instance Washington, rates for contract hauling are regulated and depend on the road class as described above. The rate per mile per thousand board feet is contingent upon the grade and relative roughness of the road. The steeper the grade and the rougher the road, the higher the rate.

The actual standards for public or private roads specify width of road, width of subgrade, maximum adverse and favorable grades allowed, and minimum curve radius. In each case the owner must consider the equipment that will travel on the road. For instance, if the road is going to be used for transporting a mobile yarder with a steel tower, the minimum curve radius depends on the radius the equipment can turn without scraping the side slopes. The maximum adverse grade depends on the horsepower of the truck and the payload it will carry. Trucks with high horsepower ratings can naturally negotiate a steeper grade than those with low horsepower. The road width depends on the width of the trailer to be used. Off-highway trucks with 10-foot bunks need a greater road width than trucks with 8-foot bunks.

Main roads or primary roads with high traffic density are usually built to the highest standard. The total volume of timber to be transported is also a determining factor. Spur roads and branches of the main road are called *secondary roads* and are built to lower standards than primary roads. Branch roads are built to intermediate standards, while spur roads (short roads leading to a landing) are built to the lowest standards.

Forest roads are classified as either permanent or temporary, all-weather or dry-weather. Permanent roads are constructed and maintained for traffic for many years. Temporary roads are abandoned after logging is completed. In Alaska, for example, roads built in many logging areas are not only abandoned, but the bridges and culverts are removed as well. The aim is to return the land to a natural state. The roads are often seeded or planted to bring them back into production.

All-weather roads, those surfaced with rock, gravel, or asphalt, can be used winter or summer. Unsurfaced roads break up during wet weather and are usable only during the summer or during dry periods.

Road variables

A logging operator must pay special attention to the condition of the road gradient, alignment, and haul distance. A very rough road with poor alignment and steep adverse grades will slow down the haul cycle and limit the loads that can be delivered. If the road is well maintained and surfaced the number of loads hauled will increase.

Roads broken up with chuckholes and ruts force a driver to slow down. Even then, traveling fully loaded on such a road will result in higher truck maintenance costs and less production. Poorly aligned roads also increase round-trip time. A road with many curves forces the driver to maintain an exceedingly high level of alertness, resulting in physical strain and fatigue. In addition, the constant maneuvering causes wear on the steering and braking mechanism of the truck. Hills on any type of road slow down logging truck traffic. With long adverse grades, at a high percent, either the truck must be geared down or trucks with greater horsepower must be used. Continuous, steep, favorable grades require greater braking capacity.

All the variables mentioned can have an adverse effect on log hauling. However, haul distance is probably the most critical variable an operator will look for. Many log haulers figure their costs on the basis of the loaded mile. A single-trip haul of 250 miles one-way will result in a higher total cost than four 50-mile hauls one-way. On a per unit basis, with one haul versus four, it should be obvious which the hauler would elect to take. From the logger's point of view, the difference is simply the revenue from one load as compared with four. He will either have to settle for less production in terms of delivered wood, or he will have to hire more trucks, if possible, which it often is not.

Our discussion thus far gives just an inkling of the importance of road layout and location as it relates to secondary transportation. The problems and considerations become more complex if additional transportation modes such as rail or water are used. The objective then must be to minimize travel time so as to maximize the number of loads delivered, given the constraints of the other system components. The other modes of transportation represent a number of new variables not yet discussed. Further discussion on secondary transportation will be found in later chapters.

Before continuing with a discussion of operating variables, it is important to consider the conditions and time cycles imposed by constraints at the delivery point. Specifically, the operator must concern himself with the time required for log measurement at or en route to the delivery point. Forest products generally must be measured or scaled for payment to the logger, the seller, and the buyer. Also of importance is the time required to unload. When many loads are being delivered per day to one concentration point, there will be surge periods when the trucks are delayed. This may also occur at the scaling ramp or scales if the logs are being weighed. The number of loads per day that must be handled, the distances traveled by various haulers to the unload point, and the log-handling facilities available are vital to the success of the operation. This information will help the operator schedule hauling to his best advantage.

Logging variables

Thus far the discussion has centered on roads and transportation. But transportation is just one part of the total system being considered. Before the logs can be hauled they must be logged, and in this area there is a wide range of factors, all of which are critical to an operator and which must therefore enter into the planning process.

No matter what type of logging method or logging system is being used it will be more sensitive to some variables than others. On the other hand, certain variables can have an adverse effect on logging operations no matter what the method—the effect can only be minimized, not eliminated. Such variables include topographic conditions; stand characteristics such as total volume, volume per acre, and volume per piece; setting size; setting layout; and yarding and skidding distance. The following explanation of key operating variables makes clear how these variables affect logging operations in general. The two most important variables are volume per stem and yarding/skidding distance.

Volume per stem

There is a definite correlation between tree size and volume per acre. In general, logging costs will be higher as tree size decreases. This is true because more trees must be handled per unit of output—a 5-inch stem has about half the volume of a 7-inch piece. This inverse relationship holds true for all logging system components. Timber-cutting costs will be high with small stems. The faller, for instance, must spend nearly as much time felling a small tree as he does a large one. The same is true of skidding or yarding. The movement of a group of logs (called a *turn* or *drag*) to the landing takes nearly as much time for small logs as for large ones, but the volume per turn is smaller with small timber. Figure 5.1 illustrates the impact of small pieces on yarding costs. In the loading component, more stems must be handled and the loading time will be longer than for large timber.

Multiple-length and tree-length logging, where applicable, have emerged as responses to the declining piece size trend.

Thus far small stems may seem to be the problem. But this is not necessarily true. With proper equipment a logger can handle a log of nearly any size. Of course, every machine or system has its limitations. One northwestern logger, who operated a very efficient thinning or small-log system, found himself in trouble, cost-wise, when he tried logging in old-growth Douglas fir. The thinning system was taxed beyond its capacity. The result was increased cycle times, higher line wear, higher maintenance costs, and unnecessary destruction of the residual timber (Conway 1971d, p. 53).

Another consideration related to stem size is the number of stems per acre. This is a critical factor when the stems are small. Couple small stems with few stems per acre and the costs are bound to be higher than normal. Once again, this factor affects all system components.

One eastern Canadian firm, which is operating in low volume per acre and small stems (i.e., few stems per acre) has experienced extremely high logging costs—nearly two and a half times what would be expected with average stems. Of course, other factors besides small stems and low volume per acre contribute to those high costs, but the relationship is still obvious. The cutters must walk long distances between trees, and then only for a small volume. The skidders have to travel farther for a full drag or return to the landing with less

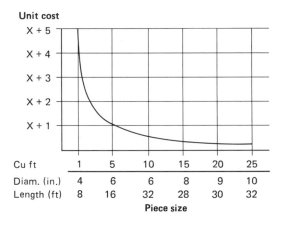

Figure 5.1. *Piece size as a factor in highlead yarding costs (from Conway 1977b).*

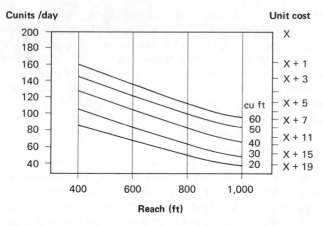

Figure 5.2. *Highlead yarding and loading production and cost by yarding distance and piece size (from Conway 1977b).*

than a full load. The impact is felt right up to the landing, where loading takes up to an hour per truck and trailer.

Yarding and skidding distance

The cost impact of extending ground skidding and cable yarding distances is important to consider, and must be balanced carefully against road and landing construction costs. Figure 5.2 illustrates typical highlead yarding and loading costs for reaches between 400 and 1,000 feet, and for piece sizes between 20 and 60 cubic feet.

Linked with the increased yarding costs are decreases in productivity encountered with longer reaches, due to a loss of production. A highlead system with a seven-man crew will produce about 80 cunits a day, yarding 40-cubic-foot logs a distance of 800 feet. At 1,000 feet, production falls to about 65 cunits.

This suggests a critical analysis of the yarding method to be employed. For example, a highlead system reaching 800 feet for 40-cubic-foot logs will require 0.68 manhour per cunit, versus 0.46 manhour per cunit for a gravity system operating under the same conditions. With a yarding distance of 1,000 feet, and the same piece size, the gravity system turned in a productivity figure of 0.48 manhour per cunit, while highlead manhours rose to 0.86 per cunit. Clearly, as distance increases, gravity is the more efficient system of the two, providing the needed deflection is available.

Volume per acre

Timber varies in size, species, and density from location to location. Furthermore, in many cases it varies even within a given tract. Only in rare

74   Logging practices

instances will it be uniformly distributed and of the same size over extensive areas. It is therefore necessary to determine what volumes are available in each new case.

In some instances the seller will give an estimate of volumes in the sale prospectus. For example, the Bureau of Land Management performs a 100-percent cruise (it estimates the volume and grade of every tree in the setting), and the buyer makes his purchase and pays for the volume said to be present. In other cases—a small southern woodlot, for instance—the volume available may be unknown. In either case, it is in the operator's best interest either to rough-cruise the timber himself or to hire a qualified cruiser to make the measurements for him.

If the volume per acre is relatively high, and if all other variables are favorable, then costs will also be favorable. There is an inverse relationship between volume per acre and unit logging costs. If volume is low, the fixed costs of moving equipment, landing construction, and setup will have a small divisor and the fixed portion of total costs will be high. On the other hand, high volume per acre leads to lower unit costs on the fixed portion. And it does not really make any difference what type of logging system is used—the impact of small volumes per acre is felt whether the operator is using wheeled skidders, tractors, or a cable system.

Volume per acre refers to the stand density per acre. Low stand density requires more work for each unit of volume produced. If the operator is using wheeled skidders he will have to build more skid roads to log the timber. With a cable system, low stand density means more road-changing costs. Using the best system for the job will help if a low-density stand must be logged, but average unit costs will still be higher.

Just what constitutes high (or low) volume per acre depends on several factors, such as the type of cut (clearcut, partial cut, or salvage cut), the logging system being used, and the geographic location. There will obviously be a difference between what a thinning contractor regards as high volume per acre and what is regarded as high by a logger working in a clearcut. On the West Coast, a clearcut running above 30 or 35 thousand board feet per acre (Scribner) is considered respectable. In contrast, one southern company, which operated almost exclusively in selective or partial cuts, would move into a stand for as little as 800 or 1,000 board feet per acre. A selective cut on the West Coast, where equipment investments are generally higher than elsewhere in the country, would be considered marginal at less than 8 to 10 thousand board feet (MBF) per acre.

Pulpwood Production, an American Pulpwood Association publication, cautions that a timbered tract averaging less than 4 cords per acre should only be purchased at a reduced price, indicating that at the going market price such a tract would be marginal (Bromley 1968, p. 78). On the other hand, one study involving 10 eastern companies indicated that the average number of cords per acre was between 10 and 20 (Jarck 1967, p. 25). The implication is that such a range is at least considered to be acceptable. Most sources indicate that any volume below 5 cords per acre is marginal and will require careful planning and price estimation if the operator is to make a profit.

Limbiness

Still a third problem associated with small timber, but not only with small timber, is limbiness. Certain species of timber (like hemlock and spruce) and small timber have an abundance of limbs. The first major problem is encountered in the cutting operation. Since the sawmill does not want limby logs the trees must be trimmed—generally on three sides. This is an additional operation and therefore adds to the cost. When dealing with bushy or limby timber the problem shifts to the skidding operation. Usually the cutters will sever the tops, but the skidding crew still has to wade through the brush. If the timber is thick the problem is aggravated, since there will be logs lying under the brush and limbs. The net result is longer cycle time in the skidding operation and, therefore, higher costs.

A change in techniques can often produce favorable results. For instance, grapple yarding in hardwoods, typically delimbed at the landing, brought a sharp decline in turns per hour. This is because limited landing room forced yarding to stop periodically while limbs were cleared from the plugged landing. Solutions, depending upon conditions, would be to delimb in the brush or to yard ahead of the delimbing and loading operation.

The problem of limbiness follows the timber right into the mill. Invariably some of the limbs will be missed. At the very least there will be some stubs left on the side of the log that was facing the ground. These stubs raise havoc with the conveyors and roll-cases, therefore causing both expensive hang-ups and operational delays at the conversion plant.

Defect

Like limbs, defect affects all operations from the stump to the mill. The main problem occurs because loggers are generally paid on a net scale. In dealing with any of the commonly used log rules (Doyle, International 1/4, or Scribner) the logger will find the difference between net and gross scale to be in direct proportion to the defect in the timber. The exception to this rule is when cubic scale is used and the defect is sound. In this case gross and net cubic scale will be the same.

However, if a logger fells a tree with 1,000 feet of scale and that tree is deducted 30 percent (300 feet) his unit costs are going to increase, because the divisor, which is based on net scale, will be smaller. If the logger spent $25.00 moving this tree to the mill his unit costs would be $35/M*(25/0.7M), rather than $25/M if there had been no deduct.

The amount of defect can cause physical problems in certain components of the logging system. For instance, the presence of defect not only will increase the cost per unit of the felling operation, as just described, but will also increase that cost because the fallers will have to exercise more care in doing their work.

* M equals 1,000 feet.

Defect affects the integrated firm from the woods right into the conversion plant. In a sawmill, highly defective timber lowers the yield of the timber and increases operating costs of breaking down the logs.

Underbrush

In a previous work the author showed that the amount of brush or groundcover on a setting has a noticeable effect on the efficiency of the operation (Conway 1978, p. 32). The author was concerned mainly with felling and bucking, but impact is also felt in other logging components, generally when very heavy brush is encountered.

Brush does not decrease the quality of the work done on a setting—but it does affect the time required for each unit of volume harvested. In the cutting operation, less volume will be cut in a given period of time. One cutting foreman remarked that with an average crew heavy brush increased his costs up to 10 percent. The impact is greater in nonmechanical cutting operations than, for instance, in skidding, since the man-machine equation is balanced heavily in favor of the man.

In a cable skidding operation, brush has an adverse effect since the rigging crew (the men who work in the brush) have to fight the brush to travel to the logs and then back again once a turn is prepared. However, since a good bit of time in the cycle is spent with the cables either returning to the brush or pulling logs in from the brush, the effect is diluted. The same is true of wheeled or tractor skidding operations.

In general, excessive brush has a greater effect when the skid or yarding distance is short than when it is long. This is because on short skids the time spent in preparing the logs for skidding or yarding is proportionately greater. In longer skids, the balance falls to log movement rather than preparation.

Topography

Topography or terrain is probably one of the most important operational variables. It affects both manpower and machines and limits the type of logging method that can be used.

The impact of terrain on log transportation has already been mentioned. In steep, cut-up country, road location is difficult and often expensive. Furthermore, the resulting roads are sometimes "slow" roads. That is, when the terrain is steep switchbacks are built to keep the road grade to an acceptable percentage. Curves in the road increase log truck cycle time because trucks must slow down to negotiate them. Greater brake wear also increases costs. On steep favorable grades the driver must use more braking power to keep his vehicle under control. On adverse grades (uphill hauls) the lower gears must be used, and the haul is therefore slower. One Oregon logger, who inadvertently built a spur road to his landing on a 20 percent grade, discovered that his haul costs increased by 7 percent. This was an extreme case, since the logger had to use a tractor to push the loaded trucks up the grade during wet weather.

Terrain also affects the cutting operation. On excessively steep slopes—from 65 percent up—it is difficult to hold the timber on the hill. The result is that the trees, after being felled, in some cases run all the way to the bottom. Slope also dictates the number of trees that can be felled before bucking begins. Recovery will be decreased because breakage is high on steep slopes. If steepness is combined with short, broken ground, the recovery factor is lowered substantially (Conway 1978, pp. 29-32).

Steep ground seriously impairs the efficiency of both wheeled skidders and tractors, as we will see in subsequent chapters. If the ground is too steep these means of conveyance cannot be used at all. In some southern regions, slopes of 35 to 40 percent are considered inoperable. A highlead cable system could be used, but in many areas where volume per acre is small, using a cable system is not a feasible alternative.

Slope, length of ground, and general topography can also have a positive effect on cable systems. With extremely short, broken terrain and many side drainages, extra moves are necessary to achieve maximum efficiency from the system. On steep and extremely long slopes the problem is one of deflection, or rather the lack of it. The ideal situation is to have a certain amount of lift on the back end of the setting. This allows the rigging to hang over the ground and reduces the number of hang-ups caused by dragging the turn of logs directly on the ground. Instead, the logs will travel with only the trailing end on the ground. When, for whatever reason, no deflection is available, the entire operation is more costly—there is more equipment breakage (cables, chokers, etc.), the work is slower as the slope becomes steeper, and, of no small importance, the work is more dangerous.

Terrain is extremely important to the success of a logging operation. The limitations of terrain vary among forest regions; what constitutes a problem in one region may not be a problem in other regions. For instance, slope is not considered a serious problem in the Northwest, but in the South it can mean the difference between taking a job or not. The conditions described by terrain—slope, length of ground, location of drainages—must be regarded within the context of individual regions and capabilities of individual systems.

Soils

The soil factor, as an operating variable, is directly related to terrain. Its impact on the efficiency of a logging operation varies depending on topography and geographic area. Aside from its impact on logging, the soil factor is important from a purely environmental point of view. Finally, the condition in which the soil is left after logging is critical to the regeneration effort.

One consideration is whether a particular logging system can work under given soil conditions. In extremely wet areas the soil will simply not support some types of logging equipment. For instance, there are swampy areas on the coastal plains of the southeastern United States which cannot be worked on at any time. The ground skidding systems used in that region simply cannot be used in these areas. Even if these systems could be used damage to the site would be irreparable. Other swampy areas can be drained to permit equipment access. In certain wetlands in the same region, ground skidding equipment can be used efficiently only during the dry season. Logging can be accomplished at other times but only at a much higher cost than usual.

Another example of difficult soil conditions is the Alaskan muskegs. Neither tracked machines nor wheeled skidders can operate under those conditions. In addition, roads must be carefully built or they will be lost after only a little use. Alaskan loggers do, however, use cable systems successfully on the muskegs. Of course, timber stand conditions, topography, and values there are favorable to these high capital cost systems. In other areas, the same ground conditions might very well preclude logging because of excessive costs.

Two considerations related to erosion can't be separated:

1. *Water quality.* It is essential to exercise care in selecting the appropriate skidding or yarding system for a given piece of ground and time of year, to prevent heavy particulate runoff into streams.

2. *The ability of the soil to produce timber.* Site damage can be severe on primary skid roads, and these cover 17 percent of the total area of a setting. If soils are fragile, the site can be damaged seriously through compaction and through the churning action that raises the infertile soil layer to the top. If compaction is the only problem, the site can sometimes be rehabilitated by loosening the soil mechanically.

The best solution—although one not always feasible—is not to log wet ground. Otherwise, the impact of skidding on soil can be reduced in the following ways:

1. Use low-ground-pressure skidding tractors, such as FMC and Bombardier machines, when timber size permits.

2. Establish more landings to avoid concentration of activity on one skid road, but take care that the increased number of skid roads does not contribute to a greater total soil damage area.

3. Exercise care in locating primary skid roads.

The question that soil factors really pose is not one of logging costs or even of the adaptability of currently available logging systems, for the industry has proved that it can operate despite problems caused by soil. The question is, rather, how to harvest the present crop of timber and leave the land environmentally sound and in such a condition that a future crop can be grown. When skidders and tractors are operated on wetland with shallow topsoils, they often do irreparable damage to the site. The areas so affected will simply not support tree growth and are lost as productive land.

It is the logger's obligation to use common sense in this regard, although soil considerations are also dictated by regulations in many cases. Bad or marginal practices can only bring more regulations and environmental pressures.

Weather conditions can and do affect logging operations anywhere in the country. The impact may be subtle, but it still exists. The effect varies from lowering productivity to outright work stoppage. In general, the basic weather variables are precipitation in the form of rain or snow, extreme temperature conditions (heat or cold), humidity, and wind.

The impact of these variables depends on the region being studied. However, there is some effect on productivity, costs, and work-ability in all regions. In the Northwest, heavy rains do not, as a rule, stop operations, but they certainly have an effect on productivity. Producers in the southern and eastern United States, however, normally do not require their employees to work in the rain.

Snow is another factor that can put an end to operations. The seriousness of the impact seems to vary with snow depth. On steep ground a skiff of snow can make working very hazardous. As the depth increases work can still be done, but productivity will be hindered. Not only does deep snow make it awkward to work, but it also makes it difficult to find the logs.

A great deal of the effect of precipitation on operations seems to depend on the level of wood supply at the time the precipitation occurs. In times of extreme shortage workers will be asked to work regardless of the weather and despite the increase in costs. If a company has failed to build an adequate log inventory during the most productive seasons, it is necessarily faced with the prospect of working during less-productive months. The alternative is to run the conversion plants out of wood.

At any rate, with regard to snow, an operator must consider when it will occur and be sure to have logging shows that can be profitably operated during the winter months. In the northeastern United States and eastern Canada there is no way to work around the snow short of shutting down, which in fact happens every season. On the West Coast, in the Douglas fir region, the loggers merely move out of the higher elevations during the winter periods. However, as the lower-elevation timber is cut out that alternative will not be available and the loggers may be faced with an off-season of several months duration.

All regions seem to have a fire season when high winds, high temperature and low humidity make fire conditions extreme. In the western United States such extreme conditions cause operations to stop. In other areas of the country this is not the case; the work continues in spite of the fire hazard.

During the fire season the loggers may continue to work, but often under some legal or extralegal constraint. The workday, both in terms of logging and log transportation, sometimes becomes shorter, as is the case on the West Coast. Loggers start working hoot-owl shifts, beginning at dawn and ending at noon or sooner depending on conditions. The results of this curtailment are lower production and higher costs.

Except for fire weather, loggers will or can work in almost any sort of weather, providing the need exists. One other exception is high wind. A logging crew understandably gets nervous when the winds reach gale force. In many states the safety laws do not allow loggers to work in the woods when the winds exceed certain levels.

In summary, the weather has a very real and often adverse effect on logging operations. Excessive rain will stop summer logging; low humidity will accomplish the same thing. Even if operations are not shut down they will cost more, since rain has a negative impact on logging road trafficability and productivity of logging operations. The spring breakup in the intermountain region is expected each year, but it is not really predictable, and sometimes comes early, causing extra problems. Freezing conditions also affect operations. In eastern Canada men are expected to work in the snow. However, the freeze must come and stay before the roads can be used regularly without causing them to break up. Continued freezes and thaws only contribute to increased costs.

Silvicultural considerations

A major consideration in logging planning is the type of cut required. This is not an operating variable and should be known to the persons responsible for operations. However, it is important to understand the relationship between the type of cut and the logging method used.

Basically, there are two types of cutting recognized by loggers, although a forester could easily enumerate others. A logger either cuts all the timber on a tract (a clearcut) or he takes only selected trees (a partial cut or selective cut). Also falling into the selective cut category is *salvage logging*, which refers to salvaging timber damaged by wind, insects, or ice. Of course, such a project could easily turn into a full harvest situation, such as that which occurred in the wake of Hurricane Camille in the South, in 1969. The ice storms that hit the national forest around White Salmon, Washington, on the Columbia River in 1970 also required full harvesting. In each case the salvage operation amounted to much more than a selective cut, although less than a clearcut in some areas.

The 1980 eruption of Mount St. Helens in Washington State also created a salvage job of epic proportions, involving both private and national forest timber. Trees close to the mountain were felled by the blast, and standing timber beyond the blast zone was killed by the heat, requiring prompt salvage to prevent insect infestation and fire. Much of this timber was young-growth not ready for harvest under normal circumstances.

Logging after a catastrophe almost always costs more money, since there is generally a certain amount of cleanup involved. In addition, there is also more work to perform. In the ice-damaged timber resulting from the Washington storms a great deal more cutting had to be done than ordinarily. All breaks had to be bucked. Since some of the trees were uprooted by the weight of the ice there were also root wads to contend with. Bucking off root wads is more costly because of the dirt found at the base of the tree. The dirt must be cleaned away before cutting to prevent dulling of the saw chain. This, of course takes more time and therefore costs more. Salvage of timber around Mount St. Helens was complicated by heavy deposits of ash and powdered rock, which accelerated wear on chain saws and other equipment. Loggers were protected by filter masks in dry weather.

In general, clearcut operations are the most economical in terms of logging, but they are not always the most economical in terms of forest management objectives. Clearcutting in the Northwest, however, creates the best seedbed conditions for natural reforestation and on slopes of more than 30 to 35 percent generally causes less soil movement (Wackerman, Hagenstein, and Michell 1966, p. 83).

A paper published by Weyerhaeuser Company (1971) pointed out that in a selective cutting system the logging costs are 166 percent of clearcutting costs for each unit of wood produced. The combination of higher logging and road costs plus reduced yield means that selective harvest of old-growth timber in the Northwest would reduce the value of the crop to 39 percent of its present value.

Selective cuts are more expensive because the timber is scattered. In a sense, this scattering has the same effect as a low number of stems per acre. If the timber is both small and scattered you have the combined effect of low volume per acre and low volume per stem.

Falling timber in a selective cut is difficult and dangerous, since the cutters are always falling through standing timber. Such a setting will also be more difficult to skid, either with tractor or cable, because more skid road is required and the logs must be skidded through standing timber, which is always difficult, especially if the logs are long.

Basically, the same problems occur in cable skidding through standing timber, although some cable systems are more adaptable than others. In cable skidding there will be more road changes, and the same problems will be encountered in yarding through standing timber as in skidding with respect to long logs and hangups. Yarding tree length is possible and has been done. When attempted, it is imperative that the timber be felled with the lead to the skid road or yarding road, whichever is the case. Tree length with cable skidding is, like wheeled skidding, apt to cause damage to the residual stand and excessive breakage as trees are pulled around standing timber onto the skid road.

The setting

Earlier in this chapter a block or tract was loosely defined as a logging area. More specifically, a *block* or *tract* refers to a timbered area in which one or more logging operations can take place. A block can be further subdivided into settings. A *setting* generally refers to the area where logs are being delivered to one landing, whether by cable, wheeled, or tracked equipment. There are times, however, when a setting may have more than one landing. The setting may be a 5- or 10-acre woodlot, or it may involve a much larger area within a 400- or 500-acre harvest block. In any case the setting is the smallest planning unit that can be dealt with.

Each operating setting is a factory where forest products are manufactured. A great deal of work is necessary if the factory is to operate efficiently. Just as weather, topography, or volume per acre can adversely affect logging costs so can setting size. Whether a setting is large or small has a direct effect on fixed unit costs. Even if volume per acre and stem size are favorable, the total volume is limited on a small setting. The costs of spur road construction, skid road construction, and landing construction are fixed and must be absorbed regardless of the volume available. When the volume is small the unit cost for fixed expenses will be high.

Large settings are preferable to small ones because of their lower fixed unit costs and higher operating efficiencies. However, a setting can be too large. How large is large and how small is small? It is not a matter of definition, but of economics. There are also environmental aspects to consider—in this case not damage to soil but the visible impact. Forest aesthetics are of high concern today, and this has produced the effect of reducing the size of settings on an industry-wide basis.

On a very small setting it may be impossible to produce wood and deliver it to a conversion plant at a competitive price—the fixed unit cost may simply be too high. As a result a highly mechanized logger may find it unprofitable to move into a small woodlot, because the small volume involved (in relation to his investment in costly equipment) will not support him competitively in the market place. This means a smaller operator, with his lower fixed cost of moving, will get the timber. Because the smaller operator is likely to be less efficient, however, the seller or buyer of the timber may be placed at a disadvantage. The lower limit of setting size is determined by moving and setup costs plus whatever other fixed costs are incurred. This assumes, and it may be a bad assumption, that variable costs are approximately the same regardless of setting size. The fixed and variable unit cost in addition to stumpage cost and a margin for profit and risk must be less than or equal to the market price of logs delivered to the conversion plant.

Many loggers will say "the bigger the better" when speaking of setting size. This is generally true when a large setting can be laid out to offer maximum efficiency. With optimum skidding or yarding distances, good landing positions, and proper road location a large setting offers many advantages. If any of the operating variables are unfavorable, however, the large setting may prove to be a disadvantage. The fixed unit cost may be low, but higher variable costs may well offset the gains.

Setting layout and operational planning

The layout of a setting must take into account all the variables mentioned thus far if the operation is to be efficient. For this reason the person responsible for the layout must first walk the setting, taking note of terrain, timber size, heavy timber concentrations, drainage, and road and landing locations. The purpose of the reconnaissance is to determine how best to log the setting. As was pointed out earlier, a setting—with its complement of equipment, the timber, and the capital investment in roads and landings—is a factory. It is little different from any other factory except there are no walls surrounding it. Material flow and handling, equipment size and location, and distance from market are just as important in the forest factory as they are in the automobile factory or the bottling plant.

It is important to remember that the man responsible for the operation should be the one to make the plans. These plans can later be revised, but whoever is to be held accountable should have the opportunity to control the variables—at least those that are controllable. The foreman, superintendent, or whoever is responsible for the operation, must be aware of the timber sale requirements, make the production estimates, and finally lay out the entire setting for production.

Road building costs are the first consideration, and must be weighed in comparison with skidding and yarding costs. Figure 5.3 compares skidding costs per cunit, including spur road construction costs of $47, $65, and $100 per station, indicating that on the site specified, optimum average skid distance is 700 feet. If road construction costs per station are excessive, it may be more economical to extend skidding distances.

After the initial reconnaissance, equipment allocation needs to be considered. The planner must determine what system will best apply to the particular logging situation. If the planner has no choice in systems then the setting should be laid out to take advantage of whatever system is available, even though it may not be the most efficient. If he does have a choice, care should be taken to select the proper equipment for the job.

For example, if a logger is running three skidders and two tractors, a loader, and four trucks, and they are all available, can they all be used or will there be excess equipment? Perhaps there is an opportunity to use the available equipment on more than one job. If so, the planner must balance the equipment needs, and much of what he determines will depend therefore on the results of the reconnaissance.

If a setting is too large it can perhaps be broken up, using several landings to complete the job. This may mean *colddecking*, that is, stacking the logs on the landing or at the roadside for future loading. Or it may mean loading out of

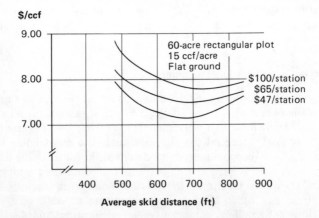

Figure 5.3. *Skidding plus spur road costs at various road construction costs per station (from Conway 1977a, p. 35).*

two landings at once—loading out of one in the morning and the other in the afternoon. The final decision on if and how the setting should be broken up will depend on how the greatest number of loads can be obtained. A small setting might indicate that only two of the available skidders should be used. However, if the land is too wet for the skidders, then only the tractors can be used.

Using the right equipment on the right setting is very important—providing there is a choice. A man with only one logging system will have to make do. An operator with two or more logging systems available has to make a choice. If the right choice is not made costs may be higher than necesary.

For instance, in cable logging several different systems are available. Each system features equipment that will perform better under certain conditions than under others. If the logger has a skyline system and a conventional yarder/spar, and if the yarding distances are extreme, he would probably elect to use the skyline system since it works more efficiently under these conditions, all other things being equal.

Much thought must be given road and landing locations since they can have a serious impact on costs. Landings, roads, and skid trails should be laid out with an eye toward minimizing skidding distances and, in the case of landings, optimizing work areas.

When tractors or wheeled skidders are being used the location of the skid roads must be carefully chosen. It is always preferable to skid downhill rather than up. If possible, the skid roads should slope to the landing. But this depends on haul road location, landing location, and general topography. There are times when road location is not optional. In this case the logger may find himself skidding uphill or extremely long distances. If he has planned his work sufficiently he can at least adjust the logging price to take these conditions into account. Thus there will be no unpleasant surprises.

Once the logger determines the type of logging system to be used he will have to start thinking about manpower. When should the cutting crew be moved in and in what sequence should the equipment arrive? In some cases the equipment will arrive all at once and some of it will remain idle waiting for the work to catch up with it. If mechanical feller-bunchers are being used, several days' leadtime may be required before the skidders arrive so that felling and skidding functions do not interfere with each other. On the other hand, with a small operation the skidding may be done hot—that is, the trees will be skidded to the landing as soon as they are felled and the faller may indeed help the skidder operator by setting chokers. In highlead logging the cutting should always be completed before the rigging crew moves in.

A time interval is also required between skidding and loading. If there is no other need for the loader machine it is probably a good idea not to move the machine and its operator into the area until there are sufficient logs decked to keep them busy. One Montana logger, for example, always attempts to build up a million-board-foot surge deck for the loader to start on. This inventory, which is in the form of several small decks, creates a lag between skidder activity and subsequent loading activity. The lag provides for minimal operational delay in either skidding or loading (Conway 1971b, p. 1).

Interrelationship of functions

The main objective of the harvesting system should be to supply at least cost to the entire system the raw material required, in the form required. But the elements of the harvesting system are sequential in nature. Roads must be engineered before they can be built, and construction must be complete before logging can begin. Timber cutting comes before skidding, and skidding before loading. If the company or firm is also in the tree growing business, plans must be made for regeneration subsequent to logging. Planning begins with engineering and ends with regeneration. A total plan must take into account all the harvesting functions and variables and their interrelationships--and this must occur before the job begins. Each function in the operational sequence must be carefully considered, especially with regard to its impact on the cost of subsequent functions. Obviously costs must be minimized.

Of course, there are many aspects of planning that cannot be discussed without a knowledge of the principles outlined in the following chapters. Further reference will be made to planning as these principles are explained.

Section Two

Felling and Bucking: Introduction

The various parts of the harvesting system have been described as part of a manufacturing process. Each timbered setting, with its complement of men and machines, is a factory. The raw material is the timber and the finished products range from logs to bolts. With each factory, of whatever kind, there is always a first and last step.

In harvesting, the first steps in preparing trees for the market involve one or more of the cutting components—felling, measuring, bucking, limbing, and topping. These steps have a critical effect on all other operations, including downstream manufacturing. If trees are cut without regard to their end use, value is lost. If trees are cut without regard to subsequent harvest operations, costs can be increased. High timber breakage incurred while felling means a loss of volume and the generation of short pieces, which are more costly to handle. Bucking long logs may result in unnecessary breakage during skidding, yarding, or loading, or in logs too heavy for the equipment to handle. Once again, costs are increased because of operation delays, slower cycle times, and increased maintenance.

Timber, standing timber, has only potential value. The actual value is determined by the end products manufactured in conversion plants. Realization of value is contingent on recovering all of the volume available from the stand and, if bucking is necessary, segmenting the trees so that value can be maximized at the end-use points. The latter means bucking for grade if logs are to be sold or bucking to mill specifications for internal conversion.

The bulk of the value lost in all woods operations is lost in felling and bucking. About 40 percent of the loss occurs in timber felling alone, either as breakage incurred during felling or as a result of high stumps. The remedy for high stumps is a matter of discipline and management. With modern, mechanized equipment and increased emphasis on fiber recovery, high stumps are not the problem they once were. But timber breakage is another story.

Breakage is the result of several variables, some controllable and some uncontrollable. Some species of timber—western red cedar and larch, for instance—are relatively brittle and therefore break easily. Large, heavy trees or highly defective trees also tend to suffer more damage than smaller, sound trees. When dealing with fairly large timber on rough, broken ground, breakage will be high. In an uneven-aged stand, with a mix of small and large trees, breakage will be high if both the small and the large trees are felled in one cutting. Finally, more breakage occurs in clearcut settings than in partial or selective cuts.

The balance of the value loss attributable to cutting occurs during the bucking operation. *Bucking* reduces a tree into marketable products usable in veneer plants, sawmills, and pulpmills. The value of timber products is almost entirely set by the market. When there is a strong demand for wood products— lumber or veneer, for instance—there is an increase in demand for logs. Along with this increase in demand there is generally an increase in value. This basic economic fact makes the end use of the logs a prime consideration during the bucking process.

In the past (and even to a certain extent today), logs were cut without regard to end use. There were certain prime lengths for veneer and sawlogs, and it became a habit for the cutters to manufacture those lengths almost exclusively and without regard to log quality. All things considered, the objective of a cutter or an entire cutting crew was to fell and buck the greatest volume possible at least cost. Least cost was, and unfortunately still is in all too many cases, the main objective.

Neither cutting for volume nor cutting for prime length, if length is the only consideration, will result in obtaining the maximum value from the timber. The job of the bucker is not to make little logs out of big ones, but rather to produce a product to market specification. The market, log-length requirements for specific end uses, log-grade requirements, trim allowance, and certain other special requirements are key considerations if a bucker is to perform a conscientious job and obtain the greatest value from each tree. Failure to cut for end use can result in the loss of millions of dollars to the industry every year.

In a sawmill, for instance, the profit margin earned depends on manufacturing a volume of quality products. There is, of course, a wide range of products. Differences in quality and value depend on defects such as knots and on lumber lengths. The ability of the mill to recover all the value possible depends on how the trees are bucked in the first place. Trees must be segmented into units of like quality and must be cut to lengths which will yield desirable lumber lengths at the mill. Improper lengths in certain products may reduce the value of the lumber by as much as $10 to $20 or more per thousand board feet. Failure to buck logs to set quality standards may result in either downgrading of the lumber or significant loss of lumber to trimming and upgrading. When one considers the hundreds of millions of board feet produced each year, the opportunity for increasing value through good bucking practices becomes quite significant.

On the one hand, cost must always be considered. On the other hand, cutting operations should maximize the value of the raw material. These two objectives may appear to be in conflict, but this is not so. There are some

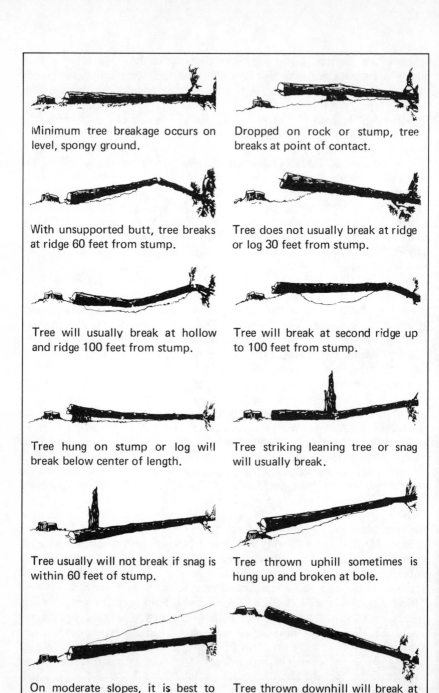

Figure 6.1. *Timber breakage in Douglas fir (Conway 1978, p. 28).*

trade-offs to examine, with cost and productivity on one side of the balance and value on the other side. Good management, intensive training, and proper planning allow the logging operator to enjoy the best of both worlds.

Planning

Good management and planning are nearly inseparable. In a cutting operation, planning time is the most valuable time spent. One day or even a few hours spent reconnoitering a new setting or block will pay dividends in terms of reducing breakage, increasing crew efficiency, and helping to reduce costs in subsequent operations.

The reconnaissance trip should be performed by the cutting foreman or whoever is responsible for the cutting operation. During the trip the foreman will look for areas where breakage is likely to be high and attempt to devise alternative methods resulting in less breakage. He will try to determine what felling pattern best suits the terrain, maximizes the recovery of fiber, and serves subsequent operations. In addition, he will attempt to resolve questions regarding safety, resource distribution (men and equipment), and time requirements.

The operating variables that interest a cutting foreman are not much different from those described in Chapter 5. However, some new variables specifically applicable to cutting operations must be scrutinized.

Terrain

The general terrain features are perhaps the most critical variable when planning for cutting operations (Figure 6.1). If slopes are steep the use of certain mechanized cutting systems are precluded and the efficiency of those used may be impaired. Mechanized systems, such as the Drott feller-buncher, lose efficiency to the point of making an operation infeasible on slopes over 35 percent (see Table 6.1). Human beings, of course, can operate on virtually any slope, with the possible exception of straight up and down. On the West Coast the author has seen cutters working on slopes so steep that ropes were required for traveling up and down the hill and between trees.

Slope has its greatest effect on timber breakage. On slopes over 65 percent it is difficult, if not impossible, to hold timber on the hillside. The felled trees

Table 6.1 *Effect of Slope on Drott Harvester*

Operational description	Slope (%)	Productivity (as a % of max. achievable)
Good	0-10	95-100
Moderate	11-20	80-94
Poor	21-35	50-79
Impractical	35 plus	less than 50

Source: Conway 1978, p. 29.

run, the distance depending on steepness and the presence of obstacles such as stumps and rocks. If the timber ran *straight* down the hill, the damage would at least be minimized. However, timber is quite likely to twist and roll *sideways* down the hill, especially if the faller was trying to hold the timber on the sidehill. The effect of such slopes is excessive breakage. The larger the timber, the greater the breakage will be. Also, if the trees are to be bucked, additional time is required to climb up and down the slope.

Aside from additional travel time, bucking on a slope takes longer and is often more dangerous than bucking on level ground. Trees running down a hill end up in piles or jackpots. Since it is more difficult to buck under these conditions only a few trees at a time can be felled.

On slopes combined with rough ground—that is, ground which is broken

The breakage shown here was the result of free falling on steep ground.

with less than a tree's length between the breaks—the breakage is even higher. Rough ground accounts for between 15 and 20 percent of timber breakage (see Table 6.2). Since minimization of breakage is a primary objective, the faller will have to take more time and exercise more care in achieving this objective. *Broken ground* refers to what the cutters call *length of ground*. This factor also affects cutting efficiency, work safety, and breakage control. The term describes the extent to which the ground is cut up by ridges, gulleys or swales,

Table 6.2 *Breakage in Felling Douglas Fir Trees*

Dia breast high class (in)	Avg dia breast high (in)	Avg height (ft)	Breakage vol		Trees (basis) number
			Brd ft (%)	Cu ft (%)	
Smooth ground					
16	16.4	130	2.3	3.2	179
24	24.3	155	5.3	7.1	254
32	32.2	184	7.5	10.2	356
40	39.8	204	8.1	11.0	324
48	48.0	216	8.9	12.5	222
56	55.8	224	10.2	13.0	137
64	63.4	232	13.0	15.6	46
72	72.7	240	13.9	16.8	29
80	79.8	240	12.5	12.9	23
88	87.2	213	12.5	13.1	8
96	95.5	240	12.6	13.7	4
104	103.5	234	9.7	11.2	4
120	120.0	225	4.2	6.0	1
Avg of Total	38.0	189	7.5	10.0	1,587
Rough ground					
16	17.8	117	6.8	7.1	65
24	23.8	137	5.6	7.4	140
32	31.8	164	7.3	9.9	118
40	40.2	192	10.5	13.1	90
48	48.0	213	14.3	16.5	66
56	55.4	226	17.0	18.7	48
64	63.7	241	13.2	17.3	22
72	72.0	244	14.6	17.0	13
80	76.7	239	17.4	19.2	3
88	90.0	268	42.1	44.3	1
96	95.2	252	18.9	21.9	1
104	–	–	–	–	–
120	–	–	–	–	–
Avg of Total	36.0	173	9.5	11.6	567

Source: J. R. Dilworth, *Log Scaling and Timber Cruising*, (Corvallis, Oregon: O.S.U. Book Stores, Inc., 1964), p. 179.

rock outcroppings, and sharp slope changes. *Short* or *rough ground* means ground that is less than a tree length. *Moderate ground* is equal to a tree length. And *gentle ground* is more than a tree length. With short ground, whether it is steep or not, the direction in which the timber is felled must be constantly changed to follow ground contours. When machines such as the Drott harvester or a tractor-mounted shear, are used on short ground, more time is spent maneuvering the machine into position than when they are used on moderate or gentle ground. And when felling is done with a power saw, short ground causes the workman to spend more time than usual with each tree. This is because wedging or jacking (discussed in the next chapter) is necesary in order to get the tree into a lay that will minimize breakage.

Brush density

Heavy brush concentrations on a strip can have a dramatic effect on productivity. Since a good deal of time is spent walking between trees for felling and bucking, the entire operation is slowed down if brush is thick enough to make walking difficult. In small timber the effect on production is more serious than in large timber. This is true because it takes less time to fell and buck a smaller tree, but the same distances must be traveled.

Travel time is not the only consideration. Good safety practices dictate that thick or heavy brush around trees be cleared before the trees are felled. In addition, an escape route from the tree must also be cleared for the faller before the tree is felled. Bucking time also increases in heavy brush, perhaps at a greater rate than felling time. The bucker necessarily travels farther than the faller does between cuts. As in felling, the effect is more noticeable in small, sparse timber than in large, heavy stands.

Mechanized cutting operations are not affected as seriously by brush as are hand operations. Although brush may slow down the operation slightly, the effect is generally discounted completely. Even with manual operations the effect varies and usually has an impact on productivity only when density is very heavy—where it is more difficult to move around, and it takes more time to clear working areas and develop escape paths. The degree of impact depends on the quality of the workmen. A good crew will experience some production loss; but the real slowdown shows up with a green or less than conscientious crew.

Lean of timber

All timber leans in one direction or another. The timber on a strip will generally have a more or less uniform lean. This lean will determine how the fallers will lead the timber when felling it. As a rule the lean varies with slope. Trees growing on steep slopes will generally lean downhill. A condition called *windlean* may be found in an area affected by a constant prevailing wind.

Whatever the cause, a faller must adjust his lead or felling pattern to compensate for lean. The felling pattern is important, especially in large timber. If felling is not planned carefully in advance, excessive breakage results when timber is felled across rough terrain or across other felled timber. The

Even though this mixed-age stand is on good ground, the bushy tops make bucking difficult and slow the yarding operation, because the rigging crew will have a hard time moving around in the brush:

latter is called *crossing the lead* and is a major cause of breakage. Breakage is not the only consideration, either. A consistent felling pattern will result in more efficient skidding operations as well.

In heavy timber where breakage is estimated to be high it may be necessary to fell the timber against the natural lean. In some cases, providing the timber is not too large, this can be accomplished with wedges or jacks. In other cases, the trees may have to be pulled over and up the hill with the use of cables and powered drums. Pulling is a costly process, but experience has shown that the ends justify the means. The recovery of sound fiber in log form and the values related to this recovery far exceed the costs involved.

Lean is less of a problem with mechanical felling. Most mechanized feller-bunchers or felling equipment is adapted to directional felling. *Directional felling* means predetermining the lay of the tree when it hits the ground. When shears are used, the wedge-shaped blade provides a lever that directs the tree into its lay. Other machines hold the tree upright while severing it from the stump, and then let it drop into a bunch of felled trees. Still others utilize a bar to push the tree in the desired direction.

Defect

Defect affects cutting cost because it lowers the net volume of sound wood that can be harvested. It also dictates what can physically be done with a tree. In addition, cutting timber in a highly defective stand can be extremely hazardous.

For example, a tree with severe stump rot cannot be pulled around into lead and must be felled in the direction it is leaning. *Snags*—dead trees which are still standing—are especially troublesome, for there is no way of determining how rotten they really are. The vibration caused by the saw or the jarring caused by wedging may well be enough to knock the top out of the tree. Worse, it may simply fall apart. In either case, the condition is dangerous and must be approached with care.

Breakage is always higher in defective timber than in sound timber. One reason, of course, is the difficulty in directing the fall of a rotten tree. Secondly, there is simply no way a defective tree can be felled to ensure that it will not break up. Because of the breakage, lost volume, and extra time required, the cost of cutting a defective stand is higher than average.

When the foreman is looking at a defective stand he must assume that cutting time will increase and net volume will decrease. Scale deductions will directly affect cutting costs, since the divisor used to calculate unit costs will be reduced. A set of cutters working in a stand with 25 percent scale deduction for rot, and cutting 250 MBF gross volume, will be paid for only 188 MBF. If the rot is not accounted for in planning, the results can be serious.

Windfalls

The presence of windfall timber presents a major problem in cutting operations as well as in subsequent operations. *Windfalls* are trees blown down by strong winds. Windthrown trees are generally found in random lays, like jackstraws, and are usually found in relatively high volumes. The volume, of course, depends on the severity of the storm. A large volume of windthrow timber makes it difficult to use mechanical cutting systems, slows down subsequent logging operations, creates hazardous working conditions, and results in excessive breakage.

The random pattern of windfalls is what slows down logging operations and makes mechanical cutting difficult. The trees lie in all directions over the ground, in crisscrossed fashion. Heavy equipment simply cannot maneuver through the windthrown trees easily even when they are bucked. Since there is no uniform lead many of the trees will have to be turned before the skidder can begin pulling the logs to a landing. Additional maneuvering is necessary before a skidder operator can accumulate a full turn.

An untrained observer might conclude that in a windfall patch nature has done part of the job for the cutters—felled the trees. This is true—the trees are on the ground. But nature is a little haphazard. First of all, the trees blown down generally follow no lead at all. They are, as mentioned earlier, in a random lay—some up the hill, some down the hill, and some across the hill. To further complicate the matter, some trees are left standing. In any case, a man or group of men must clean up the mess, which is not a simple task.

For example, if a windfall is lying downhill with the roots exposed, there will be a residue of dirt on the bole of the tree close to the root wad. Dirt is also likely to be present on timber lying across a hill. Dirt dulls saw chain and so must be cleaned off the bole before cutting, even though this takes time.

A windfall patch. The large trees left standing will be broken if they are felled over the down timber. Note the random lay of the windthrown timber.

If a tree is lying up the hill, the root wad still must be removed. Dealing with root wads is always hazardous, but on steep ground, with the wad above the tree bole, the danger is multiplied. If the wad is free of the ground it is apt to roll when it is released from the bole of the tree. The direction it will roll is unpredictable. Many timber cutters are injured when working in windfall patches, and the majority of these accidents are caused by the root wad.

Breakage is always a serious problem in windfall timber, especially when the timber is large. Of course, a great many trees split or are otherwise broken when the timber is blown over. However, the only breakage that can be reduced is that caused by felling the remaining standing trees over the windfall timber.

Because the windfall timber falls into random lays it is almost impossible to avoid felling standing trees across them. The windfalls become obstructions like the rocks, stumps, or ridges described in Figure 6.1. Stumps, rocks, and terrain characteristics cannot be easily changed, but windthrown timber can be removed before the standing timber is felled, thus reducing breakage significantly. If removing the timber is not possible for physical reasons, it is still possible to buck the windthrows and swing them into lead with the standing timber so that trees will not have to be felled across the windthrows. The first method is *windthrow removal*; the second method is *lead correction*.

Both windthrow removal and lead correction can be accomplished with cable machines, tractors, or skidders. Another alternative is to buck the windthrow into short logs when there is any possibility of felling standing timber across them. The short logs are more apt to give and move as they are struck by a falling tree and thus breakage will be reduced. This method is not as effective as the first two methods suggested and results, in many cases, in a loss of value to the windthrow.

Regardless of the method used, breakage will be reduced—it is just a matter of degree. A cost is involved, of course—either the cost of removing or moving the windthrows, or the value lost in bucking short logs. However, the breakage is nearly doubled in areas that have a great many windthrows, so the trade-off is generally worth the extra cost.

Stand composition

Stand composition is a variable which, combined with type of cut, can affect breakage. Clearcutting an uneven-aged or a mixed-species stand in a single clearcutting will almost always result in higher than normal breakage. If a stand has an understory of small timber that is felled along with the larger

This tree was broken (splintered) when it was felled across another tree.

trees, the smaller timber will suffer damage. Likewise, with an understory of western red cedar and hemlock and an overstory of Douglas fir there will be excessive breakage in the cedar and hemlock if it is felled at the same time as the fir.

The remedy for the problem is simple and, although it appears costly, results in a value gain that offsets the additional cost. The remedy is stage cutting, which means the understory is cut first and removed. Then the larger overstory is cut and removed. The additional cost stems from the fact that the men and equipment must be moved into the setting twice, the first time for the understory and the second time for the overstory. One logger on the West Coast reported that his recovery increased 25 percent as a result of stage cutting in an old growth redwood stand

Stand density

Stand density is a relative term referring to the number of merchantable stems per acre. Stand density depends on the type of cut, the logging system, and the geographic location, among other things. Low stand density (few merchantable trees) means more work must be done for each unit of production. With low-density stands the cut is scattered and more traveling must be done from tree to tree, whether cutting is performed by man or machine.

There is obviously a difference between what a southern producer and a western producer regard as high or low density. A southern producer working in a selective cut might remove as little as 800 to 1,000 board feet per acre. On the West Coast a clearcut with 30 to 35 thousand board feet per acre is acceptable although higher volumes are much preferred.

Low-density stands, all other things equal, mean lower productivity and

An example of stage cutting. The understory was felled and bucked first. After the understory is removed, the remaining trees will be cut and removed. The result is less breakage and a higher value realization.

higher costs, while high-density stands will generally net higher productivity and lower costs. A Drott feller-buncher working in a stand containing 40 stems per acre will fell and bunch about 60 trees per hour. With 280 trees per acre production is doubled to 120 trees per hour. The same relationship also exists between stand density and manual felling.

In high-density stands there is one main disadvantage. In large timber, all other things being equal, breakage will be higher in a high-density stand than in a low-density stand. The reason is that when a tree is felled it creates obstacles which affect the felling of the remaining trees in close proximity to it. (Bucked logs and stumps are both potential obstacles.) High-density stands with large timber are best cut in stages so as to reduce breakage.

Average volume per stem

Volume per stem and density are obviously related. Volume is also related to average log diameter at breast height (DBH). All costs in a logging operation, whether cutting or subsequent operations, are expressed in a cost per unit basis. Units may be cords, cunits, or board feet. When average volume per stem is low, costs increase; when the opposite is true, costs decrease. When stand density is low, along with low volume per stem, the cost increase is even more serious.

Figure 6.2 illustrates the relationship between cutting time and diameters in Douglas fir timber. The time required to fell a four-foot-diameter tree is little

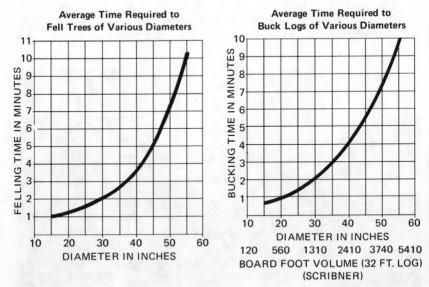

Figure 6.2. *Diameter-time relationships for felling and bucking in Douglas fir timber; no variables (Conway 1978, p. 38). Actual cutting time is included. All cutting is done with a 36-inch saw bar.*

Note the cutting pattern. The timber is all felled to one lead, which not only looks neat but also reduces breakage.

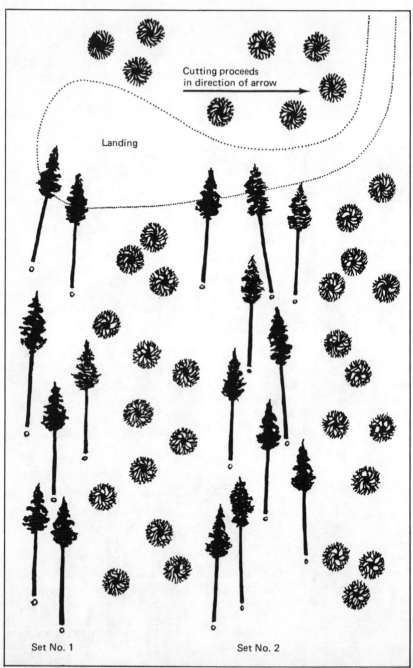

Cutting proceeds
in direction of arrow →

Landing

Set No. 1   Set No. 2

Figure 6.3. *Plan for opening up a strip on flat ground with two sets working (Conway 1978, p. 41). Corridors are cut from front to back.*

more than that required to fell a 2-foot tree. However, the volume of the 2-foot tree is much less. As the diameter increases beyond 4 or 5 feet, the time involved also increases, but so does the volume. The greatest cost increase is found in small-diameter, low-volume timber. The time required is not significantly reduced, but volume per piece drops. Cost per unit therefore increases.

Nor is the volume/time/cost relationship applicable only to manual felling. The same relationships hold for mechanical felling. Small timber results in low volume per hour cut, and larger timber—up to the capacity of the machine—results in higher cuts per hour. Shears slice through wood at a rate of 3 to 5 inches per second. More slices have to be made per unit of production as tree volume decreases.

Cutting pattern

When the cutting foreman is conducting his investigations of the strip or block he will determine what cutting pattern will be used and where cutting will begin. The pattern will be determined by the natural lean of the timber, the logging method to be used, the terrain, and the alternative cutting systems to be applied. One of the first considerations is where the strip is to be opened up. This decision can have a very great effect on the pattern that is developed and may well constrain the entire cutting operation.

The opening will generally be at the lowest point on the strip when cutting is manual. Cutting always begins at the bottom or lowest point and works upward, so the men can avoid working under felled and bucked timber. This precaution is especially important on steep ground. On flat or nearly flat ground, cutting alternatives are greater. One example is shown in Figure 6.3. The cutting begins on the road right-of-way next to the landing. Two corridors are cut, one by each of two sets. The corridors are cut in the same manner as in cutting road right-of-way. Once the corridors are cut both sets work in the same direction indicated by the arrow. The two sets can work the strip with no danger since there is a man-made break between them.

Opening a strip is simply a matter of good judgment and logical planning. When it is impossible to open at the lowest point the cutters are forced up the hill to a point where a good opening is possible. They will then work downhill as soon as possible. When strips are large two or more sets are often started at once, which means several opening points. This is done on large strips with natural topographic breaks which make it possible to safely work more than one set at a time.

The lead of timber or cutting pattern to be used is often set when the strip is opened. Opening without giving some consideration to lead can cause difficulties in terms of breakage, cost, and time.

No strip can be planned as a single unit unless it is very small. A strip is really composed of several smaller units, each of which may have unique problems. An opening along the bottom of a strip may result in excessive breakage if the timber has to be felled across contours to avoid dropping timber into a stream or out of the cutting area. If cutting begins at the bottom of a rather small drainage and is parallel to a stream, one must consider what is going to happen as work progresses up the hill. In Figure 6.4, for example, while the cutters have kept the timber out of the stream, the end result was more breakage than necessary, because work progressed up the drainage rather than down.

Figure 6.4A shows a situation in which cutting was opened at the stream and continued parallel across the contour to avoid felling trees into the stream. As cutting commences up the hill the cutters must change the lead and fell timber across the felled and bucked timber below or continue the same lead all the way to the top. In the latter case breakage would occur because the timber would break across the ground. A better alternative, shown in Figure 6.4B, might be to pull the timber that is next to the creek uphill. In fact, if the timber is large, the entire area in the drainage should be pulled.

In the situation just described, when pulling is used in an uneven-aged stand, the large, heavy trees are pulled uphill. Then the cutters go back down and pick up the smaller trees, felling and bucking the small timber and bucking the large timber at the same time. In this manner the cutters avoid having to work under felled and bucked timber.

It may be necessary to change the cutting pattern or lead several times after the strip is opened. This, of course, is part of the planning process. Sharp lead changes should be avoided. By planning ahead it may well be possible to make the changes smoothly and avoid excessive breakage.

Another factor to be considered when establishing a cutting pattern is the impact of this pattern on subsequent operations. On steep ground, with large timber, the timber will be felled to minimize breakage. On the West Coast such ground will be logged with cable systems, most of which are not adversely affected by timber lead. On moderate to flat ground a ground skidding system—either track-type or wheeled machines—will perhaps be used. In this case, the lead should be to the skid road if possible, to facilitate bunching and reduce cycle times. If a residual stand is to be left, the cutters should fell the

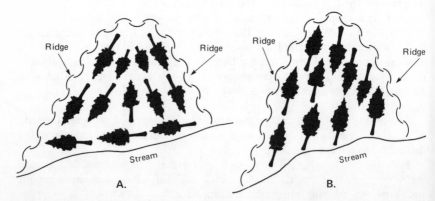

Figure 6.4. *Two ways to open cutting. In A, cutting is opened at the stream and continues parallel across the contour. In B, the timber that is next to the stream is pulled uphill.*

timber to minimize damage to the trees that will be left. Also, cutting should be done with an eye toward subsequent operations. If cables are to be used, then the trees must lead to the landing. For ground skidding, as mentioned, the lead is to the skid road. Lead in this case affects not only the skidding or yarding cost, but also the value of the residual stand.

Lead to terrain

When cutting large timber such as the old-growth Douglas fir in the Pacific Northwest the only lead acceptable is to the terrain. Leading to the terrain means the trees are felled along the contours of the ground. When confronted with a physical barrier, such as a short ridge, a shallow swale, or a sharp change in ground slope, the lead must be changed. The objective of this type of lead is to achieve maximum recovery in terms of volume and value.

Leading to the terrain is almost exclusively the result of logging on steep, rugged ground. Under these conditions, downstream logging operations are not considered. The only objective is to save the tree. With modern equipment and cable systems it does not make much difference where the logs are lying. An exception is grapple yarding systems. In these systems the lead should be such that the trees are felled perpendicular to the yarding direction or lead to the yarding machine. This method exposes more of each tree or log, making grappling easier.

Cutting for ground skidding

When tractors or wheeled skidders are to be used, it is essential to lead the timber toward the probable skidding route or skid road. This is especially important in tree-length operations if excessive breakage is to be avoided. Felling timber for ground skidding requires both skill and experience. Not only must the timber be saved, but the workman must have a working knowledge of skidding operations and machine capabilities. The cutter must be able to visualize where the skid roads will be located and lead his timber accordingly.

One of the primary justifications for investment in some of the sophisticated timber-cutting equipment is the savings in skidding costs resulting from the reduction in choker-setting time and bunching time. The timber cutter, with nothing more than a power saw, some wedges, and an ax can do a great deal to help make the skidding operation more efficient. Because of the cost relationship between cutting and skidding, it is possible to justify a small incremental cost increase in cutting if it will decrease the cost of skidding or yarding.

Besides leading the timber correctly, the faller can also provide partial bunching and tree orientation, which will help speed up the skidding operation. Trees can be felled into a fan-shaped pattern, leaving either the butts or tops of several trees in position to be picked up by the skidders in one operation rather than one tree at a time. Generally, it is best to have a butt orientation, since this will result in less breakage and in fewer logs being lost during the skidding cycle. Group felling is one way of laying out felled trees to assist the skidding operation.

Figure 6.5. *Group felling pattern (Conway 1978, p. 46). Group felling is one method of orienting butts for skidding when logging small-diameter timber.*

In group felling, which is illustrated in Figure 6.5, the first tree is felled with the butt pointing in the direction in which it will be skidded. The second and third trees are felled across the first so their tops are raised slightly off the ground. The idea is to fell the trees into groups of two or more. When skidding begins the first choker is set on the third tree and the last choker on the first tree. Choker-setting time is reduced and the skidding cycle is reduced. This system of felling is especially helpful in tree-length operations such as those commonly found in the southern United States and in Canada.

Cutting patterns and mechanized equipment

There are several patterns used when feller-bunchers or other mechanized cutting equipment is used. (Mechanized cutting is described in Chapter 8.) The pattern depends on road location, topography, and the number of machines being used. As with manual felling, the relationship between cutting and skidding costs must be kept in mind, and if more than one machine is being used, the machines must be separated for safety and maximum efficiency. Lead time is also a consideration. In the South, for instance, where blue stain is a problem, cutting operations are separated from skidding by only a few days. In the West the lead time is much greater. In large operations, cutting may be as much as three months ahead of yarding or skidding.

Mechanized equipment ordinarily works along a face of timber. Several faces may be worked out in sequence by one machine, or simultaneously by two or more machines. A *T* pattern can be used and is illustrated in Figure 6.6.

The *T* pattern is so named because of its shape. First a corridor is opened near the road or landing. The corridor is then cut back to the far end of the block. The corridor will be anywhere up to 50 feet wide, depending on the machine being used. Once the first corridor is cut a second corridor or face is started at right angles to the first, at the back end, thus making the *T* shape. The side areas are cut last. Skidding is done in the same sequence in which the block is cut. The first corridor is skidded, often while the back is still being cut. Finally, the logs from the two sides are removed.

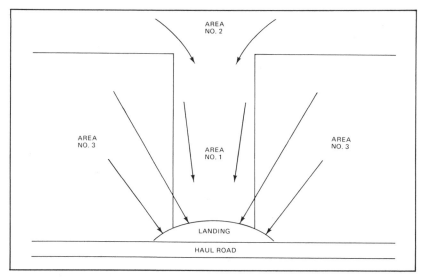

Figure 6.6. *T cutting pattern (Conway 1978, p. 152). Area 1 is cut first, followed by areas 2 and 3. This pattern allows several shear units to work together. As an additional advantage, skidding production is balanced, with skidding performed in the same sequence as cutting.*

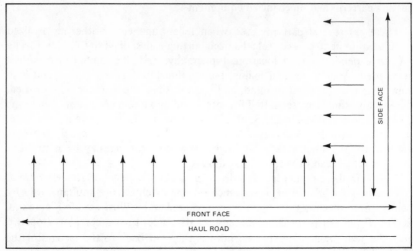

Figure 6.7. *L or block cutting pattern (Conway 1978, p. 154). This block pattern allows two machines to cut at the same time. If the side face were modified or two sides were worked along with the front face, then more than two cutting machines could work the strip. As in the T pattern, skidding distances in the L pattern are balanced between long and short skids.*

This pattern allows for the use of two or more cutting machines. When the first corridor has been cut, additional machines can be started at the back end or along the remaining sides.

The *L* pattern illustrated in Figure 6.7 is a second pattern common where mechanized equipment is used. The *L*, or block, pattern is started along the front face of the timber next to the roadside. The machine works back and forth along the face felling the trees with tops toward the road. When sufficient timber has been felled to allow skidding, the cutting machine begins a second face along another side of the block more or less perpendicular to the original face. This face is worked up in the same manner as the first.

If skidding has not commenced or if the face is long enough, a second machine may be started so that both faces are brought up at the same time. Sufficient room must be available to insure a safety factor either for skidders or for a second cutting machine. The *L* pattern allows either for simultaneous skidding and cutting or for the use of two or more cutting machines. Like the *T* pattern, the *L* pattern allows a balance of long and short skidding distances as well as maximum utilization of cutting equipment.

There are many factors to consider when laying out a strip and selecting the lead, regardless of whether the cutting is done manually or mechanically. The most important factors are control of breakage and ease of subsequent operations. Planning is the key issue. The entire setting must be considered as the sum of many parts, the smallest being the individual trees. The work must progress so as to maximize the overall value of the stand. This way a pattern can be selected and sustained with due consideration to all variables.

7.

Felling and Bucking: Manual

Several steps must be performed in order to fell a tree. Before the cutting begins the faller must check the lean of the tree and the lay into which the tree is to fall. Actually severing the tree from its stump involves three steps: (1) making the undercut, (2) making the backcut, and (3) directing the fall if necessary. Directing the fall may not be necessary if the lean is right. However, if the lean is in a direction other than the desired direction of fall, then wedging or jacking is required. A careful faller will generally put a wedge in the backcut of any tree with a doubtful lean—just for insurance.

The cutting to be discussed in this chapter is manual cutting; that is, cutting performed by a man with a power saw. As a rule, power saws used by professional cutters are powered by a five- to seven-horsepower motor and can be operated by one man. The cutting is done by an endless cutting chain which rides on a bar. The bar length (and therefore the chain length) varies depending on the size of timber to be cut. Motor size varies for the same reason.

In the large timber found in the Pacific Northwest, large saws with bars from 30 to 54 inches or longer are used. The smallest saws used in professional cutting probably utilize a minimum of an 18-inch bar. In addition to a fully equipped power saw the cutter also carries an ax (both for cutting and for driving wedges), wedges, saw fuel, and tools for making minor repairs.

Choosing the lay

Every tree on a strip both stands and falls in relation to every other tree. Broken timber can be caused by failure to carefully plan the lay of a single tree. The *lay* is simply the position on the ground where a tree will fall when severed from the stump. In choosing the lay, consider several conditions:

1. The presence of solid obstacles such as stumps, rocks, and windfalls.
2. The standing timber surrounding the tree to be felled.

3. The contour of the ground.
4. The condition of the tree with respect to lean and defect.

Several of these variables have already been described in sufficient detail. Others require more explanation.

Lean is a key variable. Although the trees in a single stand generally lean uniformly there will be some variation in degree. *Lean* refers to the degree and the direction which a tree leans from the perpendicular. A tree seldom stands straight up and down; more than likely it will have a visible lean in two directions. The most pronounced lean is called the *head lean*, while the least pronounced lean is called the *side lean.*

Weight distribution and lean must be considered together. The weight distribution in the top of a tree varies, depending on limbs and leaves, and their location on the tree's bole. A very large limb or many limbs on one side of a tree definitely affects the faller's ability to direct the tree's fall without the help of wedges or some other device. A faller must be able to estimate the total effect of lean and weight if he is to fell a tree into a specific lay (Figure 7.1).

The lean is determined in some cases simply by sighting up the tree on all four sides. Many men can easily see the variance from perpendicular; others have difficulty. The best way to measure the lean is to use a plumb bob of some sort and to stand several feet away from the tree. By placing the plumb bob on a line-of-sight between the eye and the tree's butt and by then sighting up the plumb bob string, any variance from perpendicular can be spotted.

The lean of the tree indicates the natural lay for a tree. Most trees with a relatively light head lean can be felled into a lay between 30 and 45 degrees from the head lean. This depends on the location of the side lean and how much lean the tree has. A large tree with a very heavy lean, say 15 feet or more, is going to fall in the direction of the lean, and there is little that can be done about it without mechanical help such as jacks or pulling. In some cases nothing can be done and breakage occurs.

With the lean determined the cutter knows the general area into which he can fell the tree. Exactly where to fell the tree depends on the other variables and on the lead of the timber already felled. The timber faller will not want to fell a tree across the established lead as this is a major cause of breakage. So the tree will have to be felled into lead. This narrows the choices somewhat.

Next the faller will look at the ground. Specifically, he will look for the longest ground that is in lead. *Longest ground* means a distance into which the tree can fall without breaking over ridges or in depressions. He will also make sure there are no stumps or other solid obstacles in the lay which might cause the tree to break (see Figure 6.1). If such obstacles exist the cutter will have to fell the tree in such a way that he causes the tree to miss them.

The ideal lay results in the least amount of breakage and is consistent with the established cutting pattern. In California, when cutting valuable old-growth redwoods, lays are sometimes built for a tree by rippling the ground with a tractor blade to cushion the tree's impact. In other cases, when it is necessary to save a valuable tree, it may be pulled into a more desirable lay. Tree pulling will be described later on in this chapter.

Figure 7.1. *When felling a tree with a side lean it it important to compensate for that lean (Conway 1978, p. 63). For example, if the tree has eight feet of side lean, it will fall eight feet to the side of the point it was gunned at. If an obstacle is closer to the tree than the tree length, then a proportionate distance must be calculated. If the tree in the figure were gunned for the stump as shown, it would fall eight feet from the stump. If it were gunned eight feet to the left of the stump, it would break over the top of the stump.*

Felling and bucking: manual   111

Finally, with the lean and lay determined the faller is ready to begin working on the tree. The first task, if it is necessary, is to clear all heavy brush away from the butt of the tree so there is room to work. Also, an escape path from the tree is chosen. The path should lead about 20 feet back and to the side of the tree so the set can have a safe exit when the tree begins to fall. After this has all been done the tree is ready to be felled.

Stump height

In bygone days it was not unusual to find stumps 10 to 15 feet high. And there were good reasons for this. The old-growth trees tapered heavily from the butt upward for about 20 feet or so. This meant that close to the ground the butt was very large. Spiral grain in these butts made it very difficult to saw through the wood with the hand saws used in those days. So the stumps were cut high to avoid the swell and spiral grain. Today the same conditions exist in some old-growth trees, but cutters have the power saw, which makes it relatively easier and safer to cut shorter stumps. In addition, fiber values make short stumps or low stumps mandatory.

Most timber companies set maximum stump heights at 12 inches (or less if possible) on the high side. One of the advantages of mechanized cutting, which will be discussed in Chapter 8, is that the stumps are cut nearly at ground level, allowing fuller utilization of the tree.

One need only look at the values involved to understand the rationale for low stumps. For illustrative purposes imagine a set is working in a 35-acre setting in the Douglas fir region. (The values will be different elsewhere, but the rationale will be the same.) The 35-acre setting has 80 trees to the acre for a total of 2,800 trees. Assume that half the stumps were left 1 foot higher than necessary and that the average diameter of the trees at normal cutting height is 24 inches. The loss is 3.14 cubic feet or approximately $4.71 per tree if the average market value is $150 per cunit. The total dollar loss on the 1,400 trees with high stumps is nearly $6,600. This is revenue the timber owner would have realized had the stumps been cut 1 foot lower. Had the butt diameter been larger, say 36 inches instead of 24 inches, the loss would have been nearer $15,000, adjusting for increased volume and value. This should be sufficient reason for maintaining low stumps; however, there are other reasons.

Statistics show that approximately 15 percent of the total breakage occurring during the felling process is the result of felling trees across stumps (called *stumping a tree*). Therefore, lower stumps will result in less breakage. In addition, high stumps are a source of difficulty in subsequent logging operations. High stumps cause hang-ups in the yarding and skidding cycles. Hang-ups mean delays, which are costly in any operation. Low stumps, however, add value to the timber, reduce breakage, and help eliminate hang-ups in logging.

The undercut

When the faller begins his work he should be sure he is working close enough to the ground. The undercut is the first step and a conscientious cutter

A
B

C
CORRECT

D
INCORRECT

Figure 7.2. *Types of undercuts (Conway 1978, p. 69). The conventional undercut, A, is used in smaller timber and will allow the stump to be cut closer to the ground. The Humboldt undercut, B, the one most widely used in the Northwest, allows a square cut on the butt log. Modern power saws make the Humboldt undercut feasible. When making the Humboldt undercut, it is important to match the corners where the bottom cut intersects the top cut, as shown in C. When the face is cut in this manner, the resulting wedge can be shaken out easily. When the cuts fail to match, as shown in D, the tree will have a tendency to swing away from the tight side, which comes closed first. In addition to pushing the tree away from the desired lay, an incorrect undercut may make it necessary to chop the face out with an ax.*

will start his undercut at or very near to ground level if possible.

Cutting an undercut, or face, in a tree involves sawing out a pie-shaped piece of wood facing in the direction the tree is supposed to fall. Figure 7.2 illustrates the types of undercuts and the correct ways to apply them.

The face has three functions: (1) it helps direct the fall of the tree; (2) it controls the fall, allowing the tree to slip rather than jump off the stump; and (3) it serves as a means of breaking the holding wood while it prevents the tree

This tree was split as a result of felling it across a stump.

from kicking back over the stump.

Two basic types of undercuts are used—the conventional and the Humboldt. The conventional cut is the older of the two and was the most common when timber was cut with the handsaw. A horizontal cut is made in the tree butt, followed by a slanting cut that intersects the horizontal cut. In the days of the handsaw the slanting cut was made with an ax. Today the conventional cut is still used and allows for extremely low stumps since the *snipe*, or slanting cut, is taken out of the tree rather than the stump.

The Humboldt cut takes the snipe from the stump rather than from the tree. On a large tree the snipe is quite large and this volume is thus saved, since it does not come out of the butt log. A Humboldt cut involves first cutting horizontally to a depth equal to between 1/4 and 1/3 the diameter of the tree. Then, with the saw bar slanted upward, a second cut is made intersecting with the first horizontal cut. The resulting opening should have a height about equal to 1/5 the tree's diameter.

A modification of both the Humboldt and the conventional cut utilizes two slanting cuts—one downward and one upward. The purpose is ease of cutting, achieved by diminishing the angle of the upward or downward strokes used in either of the two types of faces. The same depth and width are accomplished with a smoother and somewhat faster cut.

The backcut

With the face cut out, the wood left holding the tree must also be cut. The backcut should be made 1 to 3 inches above the facecut, depending on the size of the timber. Normally, the backcut is sawed leaving only a narrow strip of wood between the backcut and the face. This narrow strip is called *holding wood* and serves as a hinge for the falling tree. Cutting this hinge on one side or

The steps involved in cutting a tree are shown in these five numbered photographs: (1) Cutting the top cut in the face. (2) "Gunning" or aiming the tree into a lay. (3) Starting the backcut. The backcut is 1 to 2 inches above the top cut in the face. (4) Finishing the backcut. Note the wedge—one should always be placed in the backcut if there is any question about the tree setting back. (5) Watch it! The tree is falling into its lay.

the other will cause the tree to pull around to the side opposite that which has been cut off.

If all the wood is cut, leaving no hinge, control is lost. The outer edge of the hinge, at the surface of the tree bole, is called a *corner*. Since the hinge extends across the tree each cut has two corners. Cutting the corner can be planned to affect the fall of the tree as was suggested. However, careless fallers can lose control of the tree by cutting corners. If the faller cuts the corner opposite the side lean the result might well be the tree falling over sideways. This is especially true with a heavy side lean and is one reason a faller should not work under the lean of the tree if it can be avoided.

Felling against the lean

Unfortunately, it is not always possible or desirable to fell a tree in the direction it is leaning. There may be a poor lay ahead of the lean or some obstruction that would break the tree or throw it across lead. For whatever reason, it is often necessary to fell a tree either against or partially away from its lean. It all depends on the size of the tree and how far out of lean the tree must be pulled.

In a tree being felled with its lean, rather than against it, the wood on the face side will be under compression while the wood on the back side will be under tension. When the tree is felled naturally, it lifts off the stump in the back and closes against the face. When felling against the lean the situation is reversed. The tension and compression are in the same place and their effects must be nullified.

If a tree does not have a heavy lean and the faller is simply trying to swing it less than 90 degrees away from its head lean, then holding wood might accomplish the desired result. Figure 7.3 illustrates what happens. The face side is under compression under the lean while it is under tension opposite the lean. If the faller cuts his corner (or nearly cuts his corner) under the lean and then cuts up the back, leaving a wide strip of holding wood opposite the lean, the tree will pull around. The taper of the wedge of holding wood will determine how far around the tree will pull. The heavier the lean the more wood that must be cut under it.

A word of caution. Cutting should be done from the side opposite the lean, which means that the saw bar, if fully into the cut, will have its tip in wood that is under compression. If the bar is left there it may be pinched and hung up. However, if the faller is watching the top of the tree he will be able to see it begin to settle and will remove the saw bar before it becomes pinched.

A second alternative is the use of wedges. In the preceding example, a wedge or two in the backcut would probably be enough to keep the tree from settling on the saw bar. If the lean is very heavy, wedge use is strongly recommended.

As soon as there is sufficient room in the backcut the wedge or wedges are placed in the kerf. Then, as the kerf is completed, the wedges continue to be tapped in slowly, using the flat end of the ax. The wedge actually picks up the tree and keeps the pressure off the saw bar until the cut is finished. Once the cut is completed and sufficient holding wood is left the sawbar is removed and

Figure 7.3. *Top view of a stump illustrates how the corner is cut in order to pull a tree away from its lean (Conway 1978, p. 73). A wedge-shaped strip, with the majority of the holding wood on the side to which the tree is to be pulled, is left. By this means the tree can be pulled around into the desired lay. This is necessary only with a relatively heavy lean. Otherwise the tree can be felled up to 60 degrees either side of its lean (depending on the amount of side lean) without any measures other than facing the tree in the desired direction. Note: Care should be taken when cutting up the* corner in this manner. In the diagram, the corner is not quite cut off. This is for two reasons: first, if the side lean is in the direction of dashed line A and the corner is completely cut off before it has had a chance to start swinging around, there is a good possibility that it will set down on the stump and the saw bar. In this situation the faller will have to wedge himself free. Second, if the faller has misjudged the side lean and it is in the direction of dashed line B, then the tree could very well fall off the stump sideways in the direction of B. Since the faller will probably be cutting from this side, it could mean a wrecked saw, or, in an extreme case, injury to the faller. This situation relates to an important safety rule: never, unless it is impossible to do otherwise, stand under the lean of a tree to do the cutting. Always stand opposite the lean.

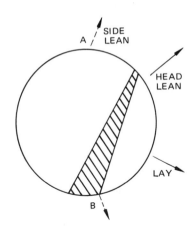

the wedges are driven into the kerf until the tree is forced over.

For very large conifers, however, which have very thick bark, the bark is generally chopped away before placing wedges into the backcut. The wedge must be placed in the woody portion of the tree to have any effect. On trees with very heavy lean several wedges may be necessary. In fact, it may be necessary to stack two or more wedges in order to pick up the tree enough to offset the lean. When using more than one wedge, the wedges are driven in alternately so they achieve the same depth in the cut.

Most of the wedges used today are plastic and work very well. However, on very large trees, with heavy lean, it may not be possible to drive the plastic wedges in without breaking them. Some cutters still carry steel wedges for such an eventuality. Steel wedges are generally longer than plastic wedges and are used along with flat steel plates, two plates per wedge. The plates allow the steel to be driven in more easily and help the wedges lift the wood rather than simply cut it. A good set of steel can lift a very large tree, but driving wedges is hard work and takes time.

There are times when wedges simply cannot offset the lean. *Jacking* is a third alternative and is very effective and relatively faster than wedging on large

The second faller is driving wedges to lift the tree and allow it to fall into its lay. The wedges are driven alternately.

timber. A hole is cut in the back of the tree and the jack is placed in the hole. Along with the jack a couple of wedges are also used as a precaution. The jack is placed into the tree before the back is cut up. Once the jack is in, sufficient pressure is developed to keep the saw bar running free. When the back is finished the saw bar is removed and the tree is jacked over. If necessary more than one jack may be used. Jacks commonly used in woods operations are either 25-ton hydraulic or 50-ton jacks. While some jacking is done by hand, automatic systems in general use allow the cutters to operate the jacks electrically or from a power takeoff on a specially modified power saw.

At times, in spite of best efforts to prevent it, a tree sets back. This occurs when the faller has attempted to fell a tree directly opposite its lean, or has misjudged the lean. In this situation the backcut is under the lean and the tree simply sets down on the backcut—at times trapping the saw bar as well. This is one reason for watching the top of the tree while felling. The faller can see the tree setting back in plenty of time to remove the saw.

A set-back tree can be wedged over most of the time. Or it can be jacked if a jack is available. There are times when the cutters have no alternative but to cut a small face in the back, cut the hinge, and let the tree fall over backwards on the stump. However, if it is imperative the tree fall in the desired direction it may be possible to push or drive it over with another tree.

Using a second tree to drive a set-back tree involves selecting a tree behind the one that is set back and felling into the forward tree, which is already cut up. The momentum of the second tree is sometimes sufficient to carry both trees over.

If the tree used for driving is less than 40 feet from the tree to be driven, the back tree should be felled directly into the forward tree. If the back tree is farther away than 40 feet, it should be felled past the leaner to one side or the other, depending on the side lean, so that the tops of the two trees will brush.

When driving, the faller takes care to watch for limbs and tops that are broken loose and thrown by the impact.

Small trees present another problem since the saw bar and wedges will not both fit in the backcut. This occurs in timber with a butt diameter of less than 15 to 17 inches. In this case the backcut is sawed first and the wedges are placed. Then the undercut is made and the tree wedged over. Another alternative is to push the tree over. This is a common practice in small timber when directional felling is required. The faller uses a light but sturdy pole with a sharpened end. The sharp end is placed as high as possible against the bole of the tree. The man then attempts to lever the tree over with the pole. On small timber this method works well.

The dutchman

A dutchman is another method of felling a tree away from its natural lean. It is a modification of using holding wood, which was explained earlier, and is considered by many to be an unsafe practice. At any rate it is not recommended to any but the most experienced cutter. In fact, some companies will not allow the use of a dutchman. However, when correctly applied the dutchman can pull a tree around to a desired lay by redistributing the forces and weight of the tree. Figure 7.4 illustrates how it works.

The dutchman, in effect, changes the center of gravity of the tree. The danger is in the possibility of losing control of the tree and in the fact that the cutter has to keep cutting the holding wood until the tree is in the right position to fall into the desired lay. There is always the possibility the tree will barberchair or split, that the roots will pull on the side opposite the lean where the cutter must stand, or that it will pull around too far and roll off the stump—once again, right where the cutter must stand.

Dangers not withstanding, a tree with a moderate lean can be felled or pulled 90 degrees or more from its lean using a dutchman.

Tree pulling

So there are several basic ways to make a tree fall against the natural lean, and they have been described briefly in the preceding pages. However, there are times when it seems impossible or too time-consuming to use any of these methods. When a tree weighs a hundred tons or more, is 200 feet high, and is leaning heavily, it is difficult to do anything but fell with the natural lean. Pulling trees is one method that has been used to fell heavy trees uphill successfully and greatly reduce breakage.

Tree pulling can be used to pull trees uphill against the lean or simply to maintain the lead when heavy leaners are found on steep ground. In some cases, on the West Coast, the loggers use tree-pulling techniques to keep the timber out of the streams.

The equipment required for tree pulling ranges from logging tractors and line trucks to old triple-drum yarders and especially designed pulling machines. The procedure is quite simple.

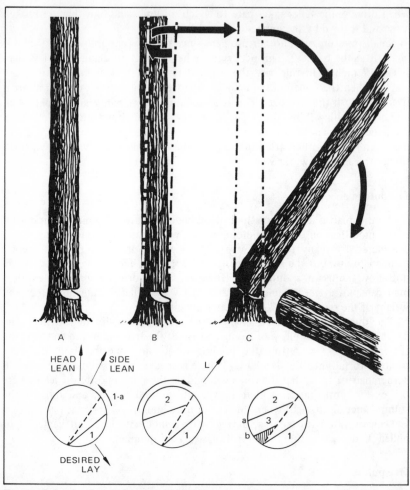

Figure 7.4. *Applying the dutchman (Conway 1978, p. 79). A. The face is cut first, facing the direction in which the tree must fall. Before the face is removed, cut 1-a is made a single kerf wide, this being the dutchman. B. The saw dogs are placed at point a, leaving sufficient holding wood, and cut 2 is made in the direction of the arrow. At this time the tree's center of gravity will change as the tree settles on the dutchman, and its lean will be in the direction of arrow L. As soon as the tree settles on the dutchman, the saw dogs will be moved to point b and the holding wood, indicated by the hatched area, will be cut to the point indicated in 3. The faller will continue to cut off the holding wood until the tree is at a point where it will continue to fall into its desired lay unassisted. If it is impossible to cut the dutchman under the center of gravity, then it may be necessary to use a wedge as soon as the cut is deep enough to permit it. The fall of the tree, once it has started, should be in a continuous spiral movement into the chosen lay.*

A climber ascends the tree to be pulled and attaches a cable between 40 and 80 feet up the tree. The cable is between 5/8 and 9/16 inches in diameter and is attached to a strap that encircles the tree and is fastened with a shackle. The cable, or pulling line, is in turn attached to a drum on a yarder or logging tractor.

With the cable attached in the tree, the faller can go to work. First the slack is taken out of the cable. Then the faller faces the tree uphill into a desired lay. A slight strain is then put on the pulling line so the backcut can be made. Communication between the faller and the yarder engineer or tractor operator is by radio whistle or hand signal. As the backcut is completed enough strain is kept on the cable to keep the tree from setting back. However, care must be taken to avoid too much strain, which would pull the tree prematurely. Wedges may be used to hold the tree up until the backcut is completed.

When the backcut is completed everyone involved retreats to a safe position well away from the tree. The machine operator is then signaled and goes ahead on the pulling line to pull the tree over. Because everyone is clear of the tree, the pulling system is basically safer than normal cutting. Also, in some cases it may be faster than normal operations, especially if the faller is trying to fell his timber uphill with small jacks or wedges.

Once in a while a tree will run when it has been pulled, but most of the time its butt comes to rest against the stump. When a tree does start running

A mixed-age stand. The larger trees were pulled uphill and the smaller timber was then felled with a normal lead. Had the larger trees been felled along with the smaller timber, both would have suffered high breakage.

downhill the butt will dig into the earth and gradually bring it to a halt. If a tree does begin running, let it go. Any attempt to hold it will result in a broken cable.

Some machines work better than others in tree pulling. All that is really required is a drum that can transfer power to a pulling line. The drums on a tractor can be used, but the line speed is generally such that the tree falls faster than the line can be pulled. It is desirable to wind the line in at the same speed the tree is falling since this allows for greater control and, if the line is kept tight, minimizes the chances of the tree running back down the hill.

Tree pulling is more expensive than conventional cutting because of the equipment and manpower requirements. However, the value recovered through reduced breakage far outweighs any additional costs involved.

Sidenotching

Heavy lean, especially in large trees, presents another problem. The excessive weight and constant lean over many years makes the sapwood opposite the lean brittle. However, moving around the tree, away from the tension caused by the lean, the sapwood retains its toughness. If the tree is cut conventionally it will most likely split or barberchair from the stump upward and kick back over the stump. To prevent barberchairing, the tree is sidenotched, which means the sapwood is cut on both sides of the tree before the back is cut up. Then when the brittle wood opposite the lean breaks, the nonbrittle wood on the sides of the tree will also break, allowing the tree to fall rather than split. Figure 7.5 illustrates sidenotching techniques.

Many other techniques can be applied in felling timber. It is possible to fell a tree without any undercut at all, although this is recommended for only very small timber, since without an undercut there is little or no control. Also, pieces of limbs can be used in the undercut to make a tree jump off the stump. Safety notches are sometimes cut on one side of the face to allow the hinge to hold longer on that side and thus to help direct the fall of the tree. However, these methods, while they do work and are used extensively, will not con-

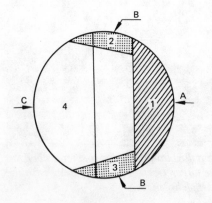

Figure 7.5. *Sidenotching (Conway 1978, p. 81). A. The face, 1, is cut first and should be as deep as possible. B. The sides are cut out next as indicated by 2 and 3. C. The backcut is made last, allowing the tree to fall without splitting.*

Felling birch. The cutter will have to sidenotch this tree to prevent barber-chairing, a lateral split in a tree.

tribute to the reader's basic understanding of timber felling. More important is to see what happens once the tree is on the ground—hopefully in one piece.

Bucking techniques

Bucking is the activity resulting in a felled tree being cut into segments. Though this is a simple definition, it may be misleading. Bucking should really involve cutting trees into log lengths acceptable for a specific end use. In addition to knowing how to cut the logs the bucker must know log descriptions by grade or log type, the range of acceptable lengths, and the correct trim allowance required. In essence, the bucker's job is manufacturing products that will result in the greatest value being recovered from the tree. The specifications for lengths, log types, and trim are the responsibility of management, as is the development of a rationale for those specifications. Once committed to bucking instructions such as those shown in Figure 7.6, it is the cutter's responsibility to perform in accordance with those instructions.

Felling and bucking: manual   123

Diameter (inches)	Specifications	Lengths (feet)
	Douglas fir	
30+	No knots or indicators.	40, 36, 32, 20, 16
24+	Scattered indicators (2 visible knots).	40, 36, 32, 20, 16
16+	Well scattered knots or indicators to 1.5".	In multiples of 2 (20 min)
12+	Scattered knots to 2.5".	In multiples of 2 (20 min)
5+	Scattered knots to 3.0".	In multiples of 2 (12 min)
	Plywood logs	
11+	All conky trees and rough, knotty trees, knots to 3.5".	17, 26, 34 (May include up to 7' or 8' of higher grade wood.)

Figure 7.6. *Bucking instructions for Douglas fir and plywood logs.*

Measuring

After a tree is felled, the faller or the bucker will measure the tree so it can be cut into segments. While walking the log for measuring, the cutter should carefully examine the log for changes in surface characteristics, such as knots and knot size, that indicate a change in grade or log type. If he is cutting for grades, diameter is critical. For instance, according to the Douglas fir log grades used on the West Coast, grade breaks occur in logs at 12-inch, 16-inch, 24-inch, and 30-inch diameters. A log with 1 1/2-inch knot indicators and a 24-inch diameter is more valuable in the market place than a log with the same surface characteristics and a 23-inch diameter. The first log is a #3 peeler while the second, simply because the diameter is too small, is only a special mill-grade log. A good set of bucking instructions will spell out surface characteristics and minimum diameter requirements.

Log lengths are also determined by end use. Minimum lengths may be set by the requirements of the logging equipment. Trucks, for instance, may require at least a 34-foot log for the bunks. Mill requirements should also be a prime consideration and will vary from mill to mill. At any rate, lengths should be such that they allow the mill to recover the highest value per log.

Each log length, except for pulpwood bolts and certain special products, must be cut to a range of specific lengths plus some overtrim which compensates for yarding damage and unsquare bucking, and insures proper-size lumber or veneer after trimming in the mill. Trim allowance varies. It may be from 3 to 6 inches per 16 feet of log length, or 12 inches of trim may be required regardless of log length. The exact allowance depends on local conditions, log-grading rules being used, contractual agreement, timber size, or mill requirements. Table 7.1 shows certain trim requirements specified in the official log-scaling and grading rules generally used on the West Coast.

Trim allowance will naturally vary even if measurement is very accurate. A power saw is not a precision tool and human beings are not perfect either. However, with a reasonable amount of care trims can be quite close, at least close enough to suit the purpose. Because insufficient trim can, both in terms of scaling and end use, result in loss of volume and excessive waste to the log, checking trim length is considered extremely worthwhile.

On the West Coast, a log measuring 37 feet will be scaled and sold as a 36-foot log. If the same log were cut to 36 feet 5 inches, with insufficient trim, the log would be scaled as a 34-foot log—resulting in a 2-foot loss in scale. If, on the other hand, the bucker arbitrarily added 4 inches, cutting a foot of trim where only 8 inches was required, he would be giving away a 20-foot log for every 60 logs bucked. Valued as #2 sawlogs as per the Columbia River rules, with an average diameter of 15 inches, the cost of the overtrim would be nearly $2,300 on a 35-acre setting, with a net of 2,800 logs. Accurate log measurement is important.

The measuring devices used by cutters range from a metal rod or wooden measuring stick to a 50-foot automatic tape. The accuracy depends on the measuring device used. With pulpwood bolts, measurement need not be very accurate. For sawlogs or plywood, however, measurement must be quite accurate.

Measuring rods are quite common wherever pulpwood is cut. In some areas of the United States and Canada measuring rods are used even for sawlogs. In the Rocky Mountain region and the Pacific Coast region 50- or 75-foot tapes are used almost exclusively. Measuring is done by the faller if a set system is used although the bucker may also do the work. In some areas, for instance northern California, a trained scaler may accompany the cutters and perform all the measurements. Because of the high values of old-growth redwood timber the marker system results in greater value recovery.

Table 7.1 *Minimum Trim Allowance*

Logs 40' and shorter: 8" trim.

Logs over 40': add 2" trim for each additional 10' or part thereof.

Exceptions when required:
    34' logs with trim in excess of 12" shall be scaled as 35'.
    42' logs with trim in excess of 14" shall be scaled as 43'.

Source: Official Log Scaling and Grading Rules for the Columbia River, Grays Harbor, northern California, Puget Sound, southern Oregon, and Willapa Harbor Log Scaling and Grading Bureaus (July 1, 1972), p. 12.

Note: Both the 35' and 43' logs are primarily plywood lengths.

Measuring the tree for log lengths. The cutter will measure the entire tree, make any corrections necessary, and then do the bucking.

When using a tape, which is usually attached to his belt loop, the person doing the measuring walks on or beside the tree, allowing the tape to pay itself out. When he arrives at the proper log length he holds the tape as close to the log as possible and marks the spot with an ax. In some cases a marking keel is used; in still other cases the cutter may simply scratch a mark on the tree with his caulks. A marking keel or ax is best for accurate measure.

Measuring should begin at the butt of the tree and proceed to the top. If maximum value is the objective the entire tree should be examined and measured before the bucking runs are made. Another consideration is safety. If the mark falls in an awkward or unsafe position it should be moved in 2-foot increments one way or the other until a safe position is found. Trees should be bucked to the smallest top diameter consistent with good utilization practices. The merchantable top varies from company to company and from agency to agency—probably from a minimum of 2 inches to a maximum of 6 inches.

Wherever bucking is done log lengths are in even feet of sawmill lengths. This is not true for plywood use where the lengths are calculated for peeler block lengths. Acceptable plywood log lengths are 17, 35, and 43 feet. On the Pacific Coast the most common sawlog lengths are cut in 2-foot multiples from 20 to 40 feet. However, some logs are cut shorter and some longer, depending on end use, the logging systems to be used, and transportation constraints. In some areas, northern Ontario for instance, the trees may be too short-bodied or too limby to allow any log over 25 or 30 feet. In many cases the only bucking done is to sever the top at a merchantable diameter.

Bucking the logs

Bucking, like felling, is a matter of leverage—the application of forces and counterforces to achieve the desired results. Each tree must be considered as a separate entity, and the consequences of each run must be carefully thought out and planned to make each succeeding run easier and safer.

In considering the bucking problem the cutter will consider at least four variables:

1. The immediate terrain and its effect on the tree to be bucked.
2. Surrounding trees and logs and how they will act once bucking commences.
3. How the tree to be worked on will act when it is cut.
4. The safeness or lack thereof resulting from the interaction of all the variables.

In no case should a bucker try to make a run his experience tells him has a high probability of causing injury. If a cutter is afraid of a run he should not attempt it. Also, no bucking should take place below felled and bucked timber or where felled trees are likely to shift downward and endanger the bucker.

The basic lays

The *lay of a tree* refers to the position of the tree relative to the ground and other timber. In choosing the lay one must consider other trees and bucked logs, saplings, slope, and soil conditions, among other things—anything that can have an effect on what a tree will do when it is being bucked. The lay will lead to five basic conditions the bucker must contend with when bucking a tree. These conditions are top bind, bottom bind, side bind, drop, and end pressure.

Top bind describes a condition where there is a downward pressure on the tree as a result of its being suspended either partially or completely between two points. The top of the tree is under compression while the bottom is under tension (Figure 7.7).

As a bucking run is made in a tree with top bind the tree will settle downward. If the bucker attempts to cut straight through the tree from the top the saw will become pinched and hung up. Even when a tree is seemingly lying flat on the ground there is likely to be some bind, which will result in a hang-up. The way to cut a tree with top bind is illustrated in Figure 7.8. The actual depths of the various cuts will depend in every case on the amount of bind, the size, and the tree species.

A tree with a severe top bind will not allow much wood to be cut before the saw begins to bind. This tree is treated like a very heavy leaner. The sapwood is cut on all sides. The bucker will cut as deep as possible on the lower side of the

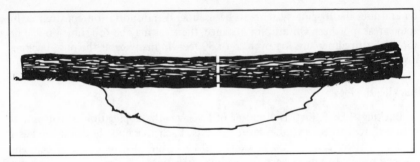

Figure 7.7. *Illustration of top bind (Conway 1978, p. 124). The weight of the tree that is suspended over the dip causes the top portion of the log to be compressed. If cutting is started on the top, the saw bar will be pinched out before the run is finished.*

log, which can be reached safely, and will then cut a pie-shaped piece out of the top. Then, when the bottom is cut out the tree will break off rather than splitting which sometimes happens when a heavy tree settles with some sapwood still holding.

Bottom bind occurs when the tree is lying over some solid object or when one end or the other is hanging unsupported. Figure 7.9 illustrates the method used for making a run when there is bottom bind.

Bucking a tree with top bind on a steep hillside. The bucker stands on the upper side, cutting the far side first, then the top, and finally the bottom.

Figure 7.8. *Top bind run (Conway 1978, p. 125). The vector line above the diagram represents the direction in which the pressure is being applied to the log, causing top bind. The saw is first dogged at point A. The cut is made in a downward arc ending at C, cutting out all of area 1. The saw is then pulled back to point B, cutting area 2. The sap along area 3 is cut next by partially withdrawing the saw bar and cutting in a downward arc. The saw bar is then pushed into the bole at point D and, in a downward motion, area 4 is cut out. As soon as this is finished, the saw bar is pulled back up*

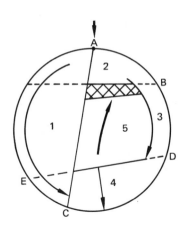

to line D-E, where the run is completed by pulling the saw bar up through area 5. Note: In the last step, the bucker gradually pulls the saw toward himself, cutting wood as he comes. He should always try to get all the wood on the off side first (area 1). In this manner, when the run is finally completed only a minimal amount of the saw bar will be in the tree, thus decreasing the chances of hang-up if it does not break away cleanly. If the bar is hung up and the bucker cannot wedge himself out, he will have very little chopping to do as only a few inches of saw will be bound.

Figure 7.9. *Bottom bind run (Conway 1978, p. 126). The vector line below the diagram represents the direction in which pressure is being applied to the log. The saw is first dogged at point A and, with a downward arc, area 1 is cut out. Area 2 is cut out next, starting from point C and cutting upward to point D. The saw is then partially withdrawn and the sap is cut up as far as E, indicated by area 3. Areas 4 and 5 are cut out next by cutting straight down in the direction of the arrow in area 5. The crosshatched area indicates the approximate amount of wood that will break away at completion.*

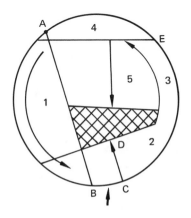

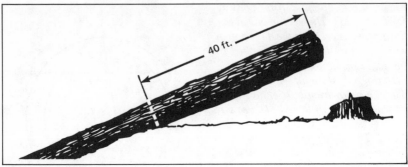

Figure 7.10. *Illustration of bottom bind (Conway 1978, p. 126). The top of the log is under tension while the overhang causes the bottom to be under compression. If the supporting wood on the top is cut first, the saw bar will pinch out before the compressed wood is cut.*

With bottom bind the tree is under tension on the top side, while the bottom is under compression (see Figure 7.10). The run is started on the bottom, where the bar would pinch out first, and it is finished from the top. On large timber it is necessary, in most cases, to cut the off side of the log first. In this case the bucker might stand or kneel on top of the log and reach over the side and as far down as possible to make sure the bottom is well cut out.

Figure 7.11. *Side bind (Conway 1978, p. 127). The bole of the tree is being bowed outward, and when the run is completed it will swing in the direction indicated. The portion of the tree closest to the sapling is under compression; the side in the foreground is under tension.*

Using a wedge in bucking. This is sometimes done in order to prevent the saw bar from getting bound in the cut.

Side bind occurs when a tree is felled in such a position that the bole is sprung to one side or another, so that when a bucking run is completed and the tension released, the tree springs sideways. Figure 7.11 illustrates side bind, and Figure 7.12 illustrates how a cut should be made under side bind conditions.

The bucker should always stand in a protected position, preferably on the side of the tree which is under compression. This side needs to be cut off first, thereby increasing the tension on the off side. The side under tension is cut last. If the bind is extreme it may be necessary to use the pie cut so that a clean break will result.

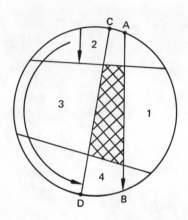

Figure 7.12. *Side bind run (Conway 1978, p. 128). The vector arrow indicates the side of the log that will be compressed. With the tip of the bar, the area indicated by 1 is cut out. The run will start by placing the tip of the bar at A and pulling it downward, ending the cut at B. Area 2 is cut next. The saw bar is placed on the top of the log, and the cut is made straight down in the direction of the arrow in area 2. The saw is then dogged at point C and the cut is made, taking all of the wood in area 3 as indicated by the arc. This cut ends at point D. The run is finished by cutting out area 4. The bucker will continue to cut in an upward arc until the wood indicated by the crosshatched area breaks away. In this manner, the bucker will be cutting from behind the area that will be compressed, allowing the run to be completed without hanging up.*

Drop exists when one end of a log to be cut is supported but the other end is not, as illustrated in Figure 7.13. The run, when completed, will allow the unsupported end to drop while the supported end will remain stationary. The amount of drop may be small or the log may drop completely to the ground. At any rate, if the run is not made correctly there is a good chance the saw bar

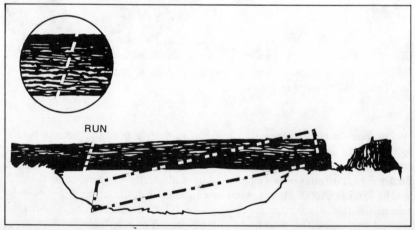

Figure 7.13. *Drop (Conway 1978, p. 129). The run is made at a slight angle, slanting away from end that will drop.*

will become hung up. Avoiding the hang-up is a simple matter of slanting the run in the direction of the unsupported end, as shown in Figure 7.13.

End pressure is the last of the five conditions to be discussed. It is simply the result of gravity. When a tree is lying straight up and down a hill, and a run is made, the part of the tree above where the run is made will slip downward, putting pressure on the bottom segment. This end pressure will either pinch out the bar or cause the bar to be hung up.

Avoiding a hang-up in this situation is simple. As soon as there is room, a wedge is placed in the kerf and driven in as the run is completed. The wedge will keep the kerf open.

The five basic conditions are quite simple to explain, but often difficult to recognize. Further, they often occur in combination, which further complicates the problem. Only experience allows the bucker to 'read' the tree and to recognize what is going to happen before he gets into trouble. Basically, however, the foregoing pages describe the operation of making logs out of trees.

In the next chapter, the subject will be mechanized cutting. Emphasis will be on shearing logs, the variables that affect machine performance, and operation of mechanized cutting systems.

Felling and Bucking: Mechanized

Most people, when thinking about mechanized cutting, think only about shears and felling. While shears are probably the most common cutting tools and felling the most common function performed, there are other important tools and functions. There are any number of machines that perform two or more of the cutting functions, including felling, limbing, bucking—and even, in a manner of speaking, measuring. In addition, some of the new equipment is capable of bunching and of actually forwarding the timber products to a landing.

Cutting is done with a variety of cutting tools including not only shears but also hydraulically driven chain saws, circular saws, and augers. In this chapter a number of different types of mechanized cutting equipment will be discussed and illustrated. Three specific types will be described in more detail, including the functions they perform and the degree of sophistication that differentiates them.

The *single-function machine* is the simplest of the felling machine types since it can only fell timber. A good example is the carrier-mounted shear, which is mounted on either a crawler tractor or a wheeled skidder. The *dual-function machine* is more sophisticated since it not only fells trees but also bunches them. A good example is the Drott feller-buncher. A third type of machine is the *multi-function machine*, which fells, limbs, bucks, accumulates, forwards, and loads pulpwood bolts. The first operational multi-function machine was probably the Bushcombine, no longer produced. A more contemporary machine, the Koehring feller-forwarder, will be used in this chapter as an example of the multi-function machine.

It will probably be a long while before mechanized cutting equipment replaces power saws in the Pacific Northwest or in the Rocky Mountain region, where rugged terrain restricts the use of mobile equipment. However, in small timber, with fairly uniform ground and slight slopes, mechanized cutting can be or already is a feasible alternative. Mechanized cutters are much faster than men with saws and allow continuous operations day or night, under a range of climatic conditions not always practical for men.

A shear, no matter what it is mounted on, can fell a tree in between 3 and 6 seconds. A Drott feller-buncher is capable of felling and bunching 100 trees per hour in deep snow and up to 200 trees per hour under more favorable conditions. In timber of the same size as that handled by mechanized cutters (as a rule, from 6 inches to a little over 24 inches) a man with a power saw could not cut anywhere near that amount. In timber up to 14 inches it would take a man between 30 and 90 seconds to fell a tree. And a man cannot work as steadily as a well-maintained machine.

Drott feller-buncher at work. The tree is sheared, swung around upright, and laid in a pile. The machine is capable of a two-way sort to the left and right of the machine.   Photo courtesy Howard Cooper Corp.

The examples just cited involve the use of shears. While there are other cutting devices, shears are the most common ones used. Most of the discussion in this chapter will therefore deal with shears—their applications, advantages, and disadvantages.

Types of cutting tools

There are many different types of knives or shears used in the timber industry. They are usually differentiated by knife design. Basically, they can be divided into two types—single-action and double-action. *Single-action shears* operate like ordinary pruning shears; one movable blade works against an anvil.

Double-action shears work like a pair of scissors; that is, there are two, offset blades working against each other. Figure 8.1 illustrates the two basic types of shears.

There are three types of single-action shears in general use. The first type is the same as pictured in Figure 8.1 The second type is the draw shear and the third is the guillotine shear.

The *draw shear* operates by drawing a single knife through the tree much as a carpenter would use a draw knife. The draw shear requires two hydraulic cylinders as compared to the one cylinder shown in Figure 8.1. One cylinder is located on each side of the knife. The cylinders are positioned on either side of

John Deere 743 tree harvester delimbs and tops stem in carriage while operator fells another with shear boom. Electronic control determines automatic delimbing sequence while topping is automatic at preset stem diameter.

Photo courtesy Forest Industries

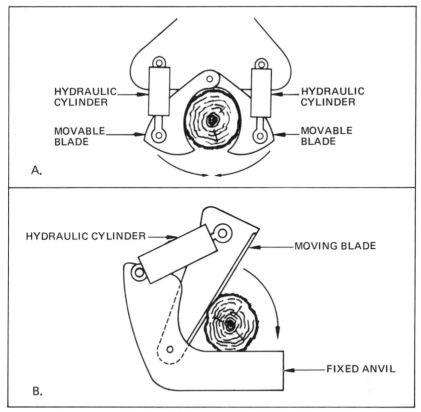

Figure 8.1. *Two basic types of shears (Conway 1978, p. 135). A shows a double-action shear; B shows a single-action shear.*

a tree; the blade is moved down into cutting position; and, finally, it is pulled through the tree very close to ground level. When the blade is not in use it is held in a vertical position. The draw shear is used with a carrier that can carry the severed tree a short distance and bunch it with other felled trees. For carrying, the severed tree is held in a vertical or near vertical position by a V-shaped plate and two hydraulic holding arms.

The *guillotine shear* is also an anvil-type shear, but the action is the reverse of the draw shear; that is, the knife is pushed rather than pulled through the tree.

Shearing knives vary in thickness, cutting bevel, and taper.

The single-action shear generally has a thicker knife than the double-action shear. Both types of shears are capable of directional felling. In the single-action shear a wedge-shaped knife throws the tree opposite the direction of knife travel. The double-action shear relies on some type of holding arm to carry the tree. The double-action shear has two knives that are parallel-edged

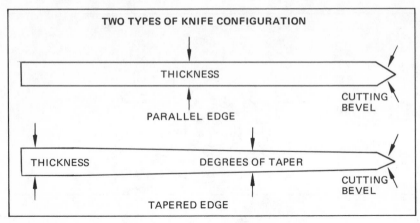

Figure 8.2. *Parallel- and tapered-edge knives (Conway 1978, p. 137).*

rather than tapered, and do not wedge the tree over. The parallel- and tapered-edge knives are illustrated in Figure 8.2.

There is another type of cutting head that does not use a shear or a knife. The auger head contains a rotating tool that is thrust laterally through the stem, reducing the amount of damage sustained by the butt.

Feller-buncher drops one load of stems beside another to form a turn for a grapple skidder. This operator in South Carolina paired a Franklin 170 XLN carrier with a Morbark 16-inch feller-buncher head.

Photo courtesy Forest Industries

138   Logging practices

Typical of many such attachments, Morbark feller-buncher head can be mounted on rubber-tired or crawler carriers. Note how clamp arms accumulate stems. Photo courtesy Forest Industries

Cutting-force requirements

Force requirements vary with the two knife types. A single-action knife employs an increasing amount of force as it is pushed through a stem up to about two-thirds the stem's diameter. Beyond the two-thirds point, force declines at about the same rate it builds up until the cut is completed. The force in double-action knives increases steadily throughout the cut. Tree diameter and knife thickness are the most critical variables affecting force requirements.

As a knife's thickness increases, so does the cutting force required for shearing. In general, doubling the knife thickness will increase force requirements by 50 percent. Tests performed on eastern hemlock by the U.S. Forest Service indicate that cutting-force requirements increased nearly 55 percent when knife thickness increased from 3/8 inch to 3/4 inch.

Cutting bevel, the bevel of the knife's cutting edge, does not have a significant effect on cutting-force requirements. However, a tapered knife has the advantage of requiring somewhat less force than a parallel-edged knife.

Anvil width does significantly affect cutting-force requirements. A wide anvil has higher cutting-force requirements than a narrow anvil. The anvil opposes the bending tendency of the fiber as the knife slices the wood. The cut opens more easily with a narrow anvil, reducing the pressure between knife and wood. Less pressure means less cutting force. With a wide knife, pressure between knife and wood increases; thus more cutting force is required.

Tree diameter, wood density, and temperature all have a significant effect on cutting force. However, density seems to have the greatest effect. Temperature does not become critical until freezing levels are reached; then force requirements increase substantially. Tree diameter is the most predictable variable and directly affects cutting-force requirements. As tree diameter increases, more of the knife's surface is needed to shear the tree. The increase in cutting force in relation to tree diameter is linear with a double-action blade and somewhat more than proportionate to diameter increase with a single-action blade. The variables described will be discussed again later on in the chapter, with reference to wood damage caused by shearing.

Types of machines

The use of shears is widespread, especially in Canada and the southern and eastern United States. They are also used in the western states, but that use is somewhat restricted by tree size and terrain characteristics. The cutting tools themselves are just as varied as the companies that manufacture them and the applications to which they are put. Table 8.1 offers some examples of the three types of machines available and the functions performed.

As we have already pointed out, the single-function machine, of which the carrier-mounted shear is an example, is the least sophisticated type of these

Table 8.1 *Mechanical Cutting Devices and Functions*

Name	Functions performed
Koehring feller-forwarder	F, B, Fw
Drott feller-buncher	F, B
John Deere 743A tree harvester	F, L
Timberjack TJ-30 harvester	F, L, Bk
Carrier-mounted shear[1]	F
Carrier-mounted bunching shear[1]	F, B

Note: F=felling, B=bunching, Bk=bucking, L=limbing, Fw=forwarding.

1. Any one of several makes of shears, or feller-buncher heads, may be mounted on the carrier. Examples are Esco, Morbark, Fleco. The bunching shear, or feller-buncher head, shears the tree and grabs the butt with a grapple-like attachment, which also serves as the anvil for the shear. The machine hauls it to a point where the tree is left with several others. In other words, the felled trees are bunched together.

machines since it can only directionally fell timber. The dual-function machine, of which the Drott feller-buncher is an example, both fells and bunches. And the multi-function machine performs more than two functions. The Koehring feller-forwarder is an example of this type of machine.

The economic justification of these various machines, from single-function to multi-function, is based on increasing productivity, while at the same time reducing overall harvesting costs by mechanizing all or part of the logging functions, from stump to landing, and accommodating a perceived problem in securing woods workers. Multi-function machines replace labor expenditures with capital outlay. Maintenance expense is also an important factor with multi-function machines, in some cases exceeding the capital cost. In dealing with small timber, an argument can be made for multi-stem handling machines rather than multi-function machines. The objective, in any event, is to develop combinations of equipment which work together to minimize handling and costs. Thus feller-bunchers are valuable not only because they can fell trees faster than a man with a power saw, but also because they can bunch trees, which makes the skidding operation more efficient and productive. In fact, the effect on skidding operations is the major justification for most single- and dual-function machines.

Between 40 and 45 percent of the time involved in a skidding cycle is spent setting chokers and bunching loads. Man-machine requirements are high in conventional skidding and in all too many cases load size, even after all the work, is suboptimum. Bunching allows for optimum load size and less choker-setting time. If grapples are used the efficiency is even greater.

The bunching ability of dual-function machines and the directional felling ability of single-function machines both enhance skidding operations. The bunching ability dramatically reduces time required to gather sufficient logs for a skidder load or full drag. In addition, because the trees are placed in piles rather than being spread out, less time is required for limbing. Under normal conditions 60 percent of a faller's time is spent limbing where limbing is necessary. Bunching allows the cutters to work from bunch to bunch rather than from stem to stem. This saving, along with an approximate 30 percent saving in skidding operations, makes the dual-function machines very valuable.

Single-function machines

Directional felling, the contribution of the single-function machine, means that all the butts are oriented in the same direction and in a uniform lay. This makes the skidder function more efficient. According to one Florida producer, the directional felling ability of his felling equipment increased his production by three tree-length truck loads per day.

The shears on most single-function machines are attached to the front end (despite the fact the machines are engineered for back-end loading). The cutting cycle is simple. The shears are raised slightly above the ground and the blades are opened during the approach to a tree. The operator maneuvers the machine into position by using the open shears as a sort of gun sight. Once in position the blades are lowered and allowed to float along the ground. Shears are not forced against the tree, but gently pushed. The trees are severed at ground level or as close to ground level as possible.

Adjustments to obtain a better position on the tree are accomplished by backing off and starting again if the carrier is a crawler tractor. Adjusting an articulated skidder is simpler; the machine is shifted right or left depending on the adjustment required. Once shearing has begun, however, the machine should not be moved. Movements while shearing is taking place put unnecessary stresses on the shears. With the shears in position the carrier is placed in neutral and the blade activated. During the shearing cycle the engine is run at full throttle.

Single-action shears ordinarily throw the tree to the right. When a grab arm or an accumulator is used the tree is not thrown—it is carried and bunched.

In the procedure used for directional felling, the tree should always be thrown away from the standing timber. There are three reasons for this. First, the driver can work into the trees rather than around them. Second, felled trees are not hung up in standing timber. Finally, felling through standing timber, even if the felled trees do not get hung up in the limbs of standing trees, creates safety hazards. When timber is felled through standing trees, limbs and broken tops are often thrown back toward the machine or carrier. A sturdy, protective canopy is especially important on carriers. Figure 8.3 illustrates the cutting pattern that results in timber being thrown into the open.

When operating on slopes it is best to run the machine up and down the hill rather than along the contours. Remember, skidder operation is recommended on slopes in excess of 30 percent. The maximum slope recommended for the operation of crawler tractors is 50 percent.

The Drott feller-buncher

There are some differences between operating with a carrier-mounted shear and with a dual-function machine such as a Drott. The Drott feller-buncher is a track-mounted, hydraulically powered machine with a knuckle boom equipped with a cutting head. The machine can cut trees up to 24 inches in diameter. Sixteen inches is the approximate limit in hardwood.

The Drott's operation is similar to that of the carrier-mounted shear. It works in swaths, cutting the timber and bunching it in the open. However, the swaths are wider and accomplished with much less maneuvering since the Drott can reach several trees from a single position—anywhere from two to six trees depending on stand density. The boom is 26 feet long and can be swung a full 360 degrees. It can cut a 50-foot swath (25 feet on either side of the machine) working in either a clockwise or a counterclockwise direction.

A full cycle includes positioning the shears, shearing, lifting, swinging, and bunching. Between 14 and 28 percent of the Drott's available time is spent moving or traveling. Service and personal time together constitute about 12 percent of available time.

Travel time involves moving from one setting to another, usually between 10 and 15 feet. Processing time involves the time that the machine is actually cutting and bunching timber (see Table 8.2).

In operation, with the machine in position for cutting, the cutting head is placed against the tree with the shears open. The shears are lowered to ground level. In deep snow the head is lowered at a slight angle so the shears can penetrate the snow. After the shears are in position, the grab arms are closed about the tree and the stem is cut. The whole tree is then lifted upright and

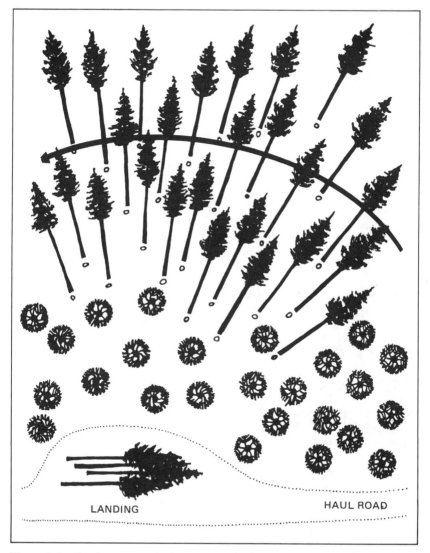

Figure 8.3. *Cutting proceeds in counterclockwise direction. The trees are thrown away from the standing timber and the butts are oriented toward the skid road or landing (Conway 1978, p. 147).*

Table 8.2 *Time Distribution for Drott Feller-Buncher*

Processing time	60 to 74 percent
Traveling time	28 to 14 percent
Service and personal time	12 percent

Source: Conway 1968a, p. 148.

swung to one side or the other for bunching. During the swing the tree is resting on the knife and secured upright by the grab arms.

Depending on stand density, there may be from 10 to 15 trees in a bunch. Each bunch is laid about 70 degrees from the line of travel and is slightly fanned out to facilitate limbing and topping. In a very dense stand the machine may be moved frequently to keep bunch size manageable. In very sparse stands it may be necessary to carry trees from position to position in order to accumulate full bunches. When trees are carried, they should be carried at an angle, leaning back toward the cab but not over it.

Optimum conditions for a Drott-type machine are slopes up to 10 percent, 300 or more stems per acre, and stem volumes ranging from 8 to 10 cubic feet. Heavy brush and a high percentage of unmerchantable stems lower efficiency.

Heavy brush will increase processing time by as much as 50 percent of that experienced in an ideal stand, since it lowers operator visibility and makes it more difficult to position the shears. When there is thick brush the shears are sometimes used to tamp the brush down in order to improve visibility. This

Drott 40 in operation in Ontario. The trees are laid in windrows. Limbing and topping is done with a power saw. The bundles are then transported to the landing for loading.   Photo courtesy *Graphics Division, Environment Canada*

should be done with the boom slightly retracted to prevent damage to the machine. Nonmerchantable stems, which are a hindrance to cutting, must be cut, knocked over, or avoided.

As slope increases, productivity decreases. A Drott can work on slopes up to 35 percent but only with a severe effect on productivity. On slopes up to 10 percent the machine works best following the contours of the ground. On slopes exceeding 10 percent the machine must work up and down the slopes. On steep slopes the trees are laid at a 45-degree angle to the left and right of the machine rather than at the 70-degree angle mentioned earlier.

Working on steep slopes can be dangerous, and extreme care should be taken to avoid injury to the operator or damage to the machine. When working on steep ground the boom should not be fully extended, but should be fairly close to the machine. This is because the boom, with the cutting head, weighs more than the counterweight on the rear of the machine.

The operator of a dual-function or single-function machine works in the timber just as a farmer plows his field. Work begins on the outside of the block (probably near the road), and commences in a counterclockwise direction along the face of the timber. Working in a counterclockwise direction is not always necessary, but seems to be most common. There are several patterns used with mechanized cutters, two of which were described in Chapter 6.

The Koehring feller-forwarder

This tree harvesting machine can fell trees and forward them to roadside in one operation. The boom-mounted felling head is an accumulating 24-inch shear capable of holding up to six 6-inch (DBH) stems. Once the felling head has harvested as many stems as it can hold, the stems are placed in the forwarding bunk capable of holding approximately 9 cunits (50,000 pounds).

The early model of the feller-forwarder, the KF2, had a fixed cab and the engine at the rear. The current version, the KF3, has the engine mounted at the front. The cab and boom swing together, allowing for a 380-degree rotation. These two improvements allow for a better loaded weight distribution between the front and rear axle and an effective cutting width of 50 feet.

The KF3 is hydrostatically driven. Forward and reverse are selected using switches mounted on the boom controls. This allows the operator to move the machine while cutting and bunking stems.

Productivity is affected by stem size, terrain, and cutting pattern. Highest productivity is achieved on level, dry ground with road spacings such that the machine can cut while it moves away from the road and on its return, thus eliminating the time to travel empty or loaded without cutting.

Typical speeds are 120 feet per minute traveling empty, and 80 feet per minute while loaded. Productivity has been reported as high as 142 stems per machine-hour in 4-cubic-foot piece sizes and 87 stems per machine-hour in 9.2-cubic-foot piece sizes (FERIC 1978).

An important drawback of the feller-forwarder concept is the fact that the stems must be mechanically delimbed as they are placed at roadside in piles of up to 10 cunits. In spite of higher costs usually associated with mechanical delimbing, Ontario Paper reported that the wood delivered by the KF2 and delimbed with a boom-mounted roll delimber was costing 60 percent of its conventional felling, delimbing, and skidding costs (Schabas 1979). The above costs include labor, depreciation, repair and maintenance, and supplies and expenses.

This dramatic cost reduction is the result of the combination of two concepts: the accumulating felling head and the forwarding capability of the machine, which eliminates a secondary skidding operation. In areas where road costs are expensed as part of logging costs, the forwarding capability of the machine can be used to advantage to reduce the amount of road required.

The multi-function aspect of the feller-forwarder is accomplished with only slightly more complexity than a feller-buncher. This small increase in complexity, plus the low speed and ease of mechanical access due to the large size of the machine, results in acceptable mechanical availability.

As this is written, there are more than 50 feller-forwarders in operation, most of them in northeastern Canada cutting small-stem spruce and pine.

Advantages of mechanized cutting

There are many advantages to mechanized cutting, but not all of them apply equally to single-, dual-, and multi-function machines. Some of the main advantages are:

1. Cutting production is improved with most mechanized cutting devices.
2. Skidding and limbing costs are reduced since the machines and men can work from pile to pile rather than from stem to stem.
3. Skidding production is improved because the skidders can work on bunched timber and because of butt orientation in single-function machines.
4. Mechanized cutting allows for a separation of cutting functions from the skidding and loading functions.
5. Wood utilization is improved generally because of the low stump height.
6. Mechanized cutting can be used on a multiple-shift basis, providing for better equipment utilization.
7. Mechanized cutting is safer because it takes man off the ground.

Disadvantages of mechanized cutting

Although the advantages of mechanized cutting outweigh the disadvantages, it is important to understand these disadvantages. The main ones are:

1. Limited range of stem sizes.
2. High cost.
3. Wood damage.

Stem size

Most shears are limited to cutting a specific range of stem sizes. The lower limit is set by the economics of handling and marketing. The shears can cut the smallest stem required. The upper limit is set by the size of the hydraulic ram operating the shears. Theoretically, shears could be designed to cut a tree of any diameter. While there are machines equipped to cut trees over 24 inches in diameter (the Beaver clamp is one such machine and is capable of felling trees up to 36 inches), most shears attached to the equipment discussed thus far are limited to severing trees up to 24 inches in diameter.

Cost

Even the single-function machines (skidders or tractors equipped with shears or some other cutting device, and the least sophisticated of the mechanized cutters) are expensive. Exclusive of the carrier cost, which can extend past the $90,000 level, a shear attachment with bunching clamp can run between $13,000 and $29,000.

The dual-function and multi-function machines are more sophisticated, require more engineering, and use more working parts—they are just plain more expensive to build. The Drott Model 40 costs about $150,000. Other entries of similar configuration reach beyond $200,000. The cost of the supersophisticated multi-function machines ranges up to and beyond $400,000.

The cost of the dual-function and multi-function machines would or could be a deterrent to many potential users—especially the small producers. With limited capital such producers would find it all but impossible to purchase such a machine no matter what the advantages.

Wood damage

As shears pass through a stem the fibers are bent, compressed, and fractured. The damage results in a loss of lumber recovery and in degradation of butts used for plywood blocks. The butt logs damaged by shearing must be trimmed to remove the damaged parts. At best, the recoverable product is lowered from veneer classification to chips. One southern mill reported a 50 percent loss in overrun when sheared logs were used. Logs scaled at 7,280 board feet Scribner, which should have yielded 8,820 board feet of lumber, only produced slightly over 8,000 board feet. Mill tests also showed that between 2 and 4 feet of trimming would be required, depending on the butt diameters, to remove the splits caused by shearing (Koch 1971, pp. 21-26).

The damage cited in these tests appears to be quite high. However, a test made in British Columbia at summer temperatures indicated the damage was negligible (Letkeman 1972, p. 41). Indeed, there seems to be no normal level of damage. It runs between extremely high and negligible, apparently depending on such variables as temperature, species, wood density, blade thickness, and blade configuration.

Shear damage is caused by knife action as the knife passes through a stem. The wood fibers are bent ahead of the cutting edge as the knife is pushed or pulled through the tree. The knife action, as it passes through the tree, concentrates stresses, causing fibers to fracture or split. The bending action caused by the knife is illustrated in Figure 8.4.

The greatest lateral depth of failure is caused by splitting. This is the most serious type of damage since it affects the most wood and causes the largest

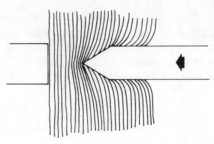

Figure 8.4. *Knife action bends fibers ahead of edge (Conway 1978, p. 158).*

amount of degradation. In the example just cited it appears the splits may extend up the bole of the tree from 2 to 4 feet. Other tests performed using single-action shears placed the damage site much lower, at approximately 10 inches. Although the estimates of damage severity often vary, the facts seem to indicate that damage certainly does occur.

Aside from species, the major variables affecting splitting are cutting bevel, knife thickness, and anvil width. The thinner the knives and the smaller the bevel, the less splitting damage will result. Also, damage is more severe in frozen wood.

Fiber compression is another shear-related problem. The wood on each side of the blade may be crushed as the knife passes through the tree. The damage is relatively slight, extending up the butt to a distance of about half the thickness of the blade. Knife thickness is the major factor causing compression. The thinner the knife, the less the damage.

Pullouts, another type of shear damage, occur when the fibers are not allowed to bend; they are similar to stump pull, which was mentioned in an earlier chapter. Anvil width is the key factor here. When the anvil is narrow the fibers bend, opening in a V as the knife cuts. The bending action does not take place when a wide anvil is used. Instead, the wedging action of the knife creates a tensile force resulting in a ragged fracture. Fingers of wood ranging in length from 1 to 3 inches or more are pulled from the stump and the butt. Pullouts are common when double-action shears are used.

Barberchairs, such as those described earlier, also occur in shearing. While barberchairing does not happen often, it does occur and is caused by slow knife speeds or incomplete cuts. The tree begins to fall before the cut is completed. Barberchairing is associated with large trees, when single-action shears are used.

Although there are several disadvantages to the use of shears, it should be pointed out that these disadvantages are decreasing in severity. This decrease is the result of better research by manufacturers, improved technology, and increased industry knowledge of the equipment and its capabilities. The auger-type cutting head, for example, has proved useful in reducing fiber compression and pullouts, especially in frozen wood. Though capital costs remain high, shears are being developed which can harvest larger diameter trees on a volume, operational basis. Thus mechanized cutting will continue to be an important factor in harvesting timber crops.

Section Three

Ground Skidding: Introduction

All log movement is divided into two categories—primary and secondary transportation. Primary transportation involves all movement of logs or trees, after cutting, from the stump to the landing. Primary transportation may be performed by tracked machines (crawler tractors), four-wheeled tractors (wheeled skidders), any one of several cable systems, or aerial logging systems. *Skidding* generally refers to the use of either crawler tractors or wheeled skidders for moving forest products out of the brush.

Until the late 1800s all ground skidding was accomplished with animal power—horses, mules, and oxen. In some parts of the United States and Canada horses and mules are still used; however, this is the exception rather than the rule. Certain of the cable systems were already in use before mechanized ground skidding came of age. But cable systems utilized a stationary power source, and a mobile source was required. The first tractors were ponderous machines run by steam power and moving about on huge iron wheels. The iron wheels were replaced with tracks and over time the machine was streamlined— the crawler tractor came into its own.

The transition from animal power to mechanical power did not happen quickly. Loggers in the Pacific Northwest utilized the big tractors long before they became accepted in other parts of the country. In the eastern and southern regions the transition did not take place until the late 1950s and early 1960s. Even then horses and mules were not entirely replaced.

Crawlers are now commonly used all over the United States and Canada wherever ground conditions permit. They can be operated successfully on slopes up to 50 or 60 per cent. However, tractor operation on steep slopes is not considered practical from a production standpoint. It is also discouraged because of the adverse environmental impact resulting from soil damage and erosion. Even on gentle slopes with shallow soils tractors can cause severe soil compaction problems during wet weather. They can be used for skidding and also for road building, landing construction, and skid trail construction. They are versatile machines. On the other hand, crawlers are heavy and slow. In small timber or sparse timber, a crawler tractor is not the most economical power unit.

The wheeled skidder, which appeared on the scene during the 1960s, works well in small timber and can skid a given quantity of wood twice as fast as a crawler tractor. Because it can pick its own skid trail with a minimum of trail building the wheeled skidder offers significant savings in that area—up to 60 percent savings in skid road building. Its speed allows logging operators to skid greater distances and to concentrate more logs on a single landing rather than on several smaller landings. Finally, wheeled skidders cost less than crawlers. Not only is the initial cost less for comparable machines, but maintenance costs are also lower.

The wheeled skidder is a versatile vehicle, but it does not replace the crawler tractor. Each has its place. In northern California, among the big redwoods, the big tractors have no equal. There, the tractor's 3-mile-per-hour pace, with a real payload, is not a disadvantage. In the southern pine forests or elsewhere, where trees might range from 8 to 12 inches in diameter, the speedier skidder is obviously superior where tree sizes are small.

Not everyone wants or can use a wheeled skidder. A small company with limited resources and large timber to log would need and want a tractor even though it is more costly. Such a company requires a machine that can build roads, clean landings, remove stumps, and skid logs. This combination of jobs calls for a crawler, not a skidder. If a second skidding machine were required and could be afforded, the operator might purchase a wheeled skidder.

Just which machine should be used depends on a number of conditions and variables, such as stand density, volume per stem or per log, thickness of brush,

A wheeled skidder heading out for another turn.

and general terrain characteristics. Seldom do these variables act alone. In most cases two or more will interact in a logging operation. Under wet ground conditions skidder logging can cause the same soil damage as was described earlier with crawler tractors. The critical variables will be discussed in this chapter.

Stand density

Stand density is considered critical in both tractor and skidder operations. In low-density stands, travel time increases, production becomes lower than average, and unit cost increases. The positive relationship between density and cost reflects the time required to gather a full payload. Bunching, the gathering of stems for a load, takes a greater proportion of available productive time. The greater the distance between the trees, the longer the time required for bunching a load. Bunching time accounts for between 33 and 70 percent of the total skidding cycle time. Stand density has the greatest effect on bunching time as a percentage of total cycle time.

One author tends to minimize the effect of stand density. He contends there is little difference between productivity in high-density stands and low-density stands. With low volumes per acre more ground must be covered to bunch a load. In very high-volume stands the many logs on the ground and the stumps become obstacles that also increase bunching time. Extremes in either condition can be a disadvantage (Pope 1954, p. 66).

Stand density, the number of stems per acre, is always important, but it is especially important in highly mechanized operations. Productivity in cunits* is related to log size and skidding distances. With either long or short skid distances, a sparsely populated stand will reduce average load size and lower, therefore, the overall productivity. The table which follows clearly illustrates the relationship between volume per acre and bunch size.

ccf/acre	Bunch size (ccf)
5	.20
10	.40
15	.60
20	.80
25	1.00

The relationship is established in the following way: bunch size equals 0.04 times volume per acre in ccf (Dibblee 1965).

Slope

Slope is another variable which has considerable effect on skidding productivity. At the extreme, steep slopes may preclude the use of either tractors or skidders. This is the case in much of the commercial forest land on the Pacific Coast. Even though crawler tractors may be operated on some of the steeper

* A cunit (ccf) equals 100 cubic feet.

slopes, environmental damage resulting from building skid trails on steep ground dictates that crawlers not be used. The damage results from erosion and from difficulties in regeneration.

Slopes can be adverse (uphill) or favorable (downhill). The rule is to skid downhill to the landing whenever possible. Skidding uphill should be avoided if possible—if it is necessary, the payload should be lightened somewhat. Just how light the load should be is established by trial and error, but in the long run time will be saved skidding lighter loads rather than maximum payloads on steep adverse slopes.

The reason for the lighter loads is that both tracked machines and wheeled skidders lose usable pounds of pull on adverse grades. Grade resistance will result in the loss of approximately 20 pounds of pull for each ton of weight (tractor plus load) per percent of adverse grade. Grade assistance applies in the same magnitude, adding 20 pounds of pull for each ton of weight for each 1 percent of favorable grade (Mayfield).

There are accounts of crawlers working on 80 percent slopes, but the generally accepted limit is between 50 and 60 percent. On extremely steep slopes skid trail or skid roads are built diagonally across the slope so the tractor can maneuver. However, the skid roads built on such steep slopes are costly, remove ground from tree production, and cause erosion. In addition, the practice is dangerous. For these reasons the practice is not acceptable.

Wheeled skidders can also operate on steep slopes and have been known to do so. However, a 30 percent slope is about the limit for safe, productive work. Beyond 30 percent it is a question of operator courage—or foolishness.

When ground exceeds 30 percent, tractors are used to build skid trails, while the skidders do the skidding. This is sometimes preferred to using cable systems in extremely small areas. Table 9.1 ranks slope in terms of operability for both tractors and skidders.

Undergrowth

Under normal conditions, if there is such a thing, undergrowth offers little difficulty. Ground-skidding equipment usually follows the same trail from the woods to the landing numerous times during the skidding cycle. Most of the

Table 9.1 *Operability and Slope*

Operability class	Tractor slope	Skidder slope
Good	up to 30%	up to 15%
Poor	up to 50%	up to 25%
Impractical	50% plus	30% plus

Source: Mayfield.

Note: Even though slopes up to and exceeding 50 percent can be negotiated with tractors, it is not desirable in terms of environmental damage.

heavy brush is trampled or removed during the first few cycles. The place where underbrush can have an effect is in bunching and choker setting.

More time is required to prepare a load for skidding under heavy brush conditions. Primarily, the time required increases because it is more difficult to find the felled trees. On long skids the effect is not felt as seriously since a larger portion of the total cycle time is spent skidding. With shorter skids the full effect of the underbrush is felt.

Soil

Soil conditions, if they affect operations at all, also affect long skids. In soft, wet ground ruts are gouged out of the skid roads. As the condition worsens the operator must develop another road or suffer delays while attempting to negotiate mud holes. Also, the skidding cycle time will increase. Soil damage under these conditions must also be considered.

Rolling resistance, caused by tires penetrating the ground surface, increases as skidders negotiate wet, muddy ground. This increase in rolling resistance robs the skidder of traction power and reduces usable pounds of pull at the rate of about 30 pounds for each inch of tire penetration times the total weight (in tons—machine minus transferred load) on the drivers. Ground penetration and rolling resistance are negligible with crawlers (Mayfield).

The no-spin differential built into the rear axle of some skidders helps eliminate spinning due to loss of traction (Table 9.2). Hydraulics and articulated (hinged) frame construction permit these skidders to duck walk through soft spots. Duck walking involves steering first left and then right in an attempt to gain advantage by forcing one wheel and then the other onto firm ground. This is a good method for maneuvering a machine out of a spot, but it is not very effective for pulling a load. It also churns the soil, an undesirable side effect.

Volume per tree/stem

Another critical variable is volume per tree. The rule is simple, and pertains to nearly any logging system—the smaller the tree, the higher the variable operating cost per unit of production. The larger the tree, to a point, the lower the variable operating cost. The reason for this is also simple—in the brush, small logs are no more difficult to handle than large ones, but more pieces are required to make up a payload. This holds true for tree lengths or logs although tree-length logging is one way of offsetting the small volumes. As tree size (and therefore log size) increases beyond a certain point, depending on the machine used, some of the advantage is lost. Extremely large logs cause the equipment to be less maneuverable and to require more power.

Tree size, in fact, is the variable that has the greatest impact on skidding cost. Volume of a stem is a function of the diameter squared. As previously cited, a 5-inch stem contains about half the volume of a 7-inch piece.

A major problem is always the power capability of available equipment. An operator with small- to medium-sized equipment, either skidder or tractor, will

Table 9.2 *Approximate Coefficients of Traction Factors for Rubber-Mounted and Tracked Vehicles*

Soil type	Rubber	Tracked
Clay loam, dry	.55	.90[1]
Clay loam, wet	.45	.70
Rutted clay loam	.40	.70
Dry sand	.20	.30
Wet sand	.40	.50
Gravel road (loose)	.35	.50
Packed snow	.20	.50
Ice	.10	.10
Firm earth	.60	.90[1]
Loose earth	.45	.60

Source: TEREX 1981, p. 21.

Note: A spinning wheel or track does not deliver power to the ground. Two factors that reduce spinning are loading and traction. The degree of traction between tires or track shoe and the ground is called the *coefficient of traction*. The result of multiplying the weight of the machine (on track type) or weight of the rear axle times the coefficient of traction represents the maximum force which can be transmitted before the track or tire spins out.

1. Assumes full grouser penetration.

have to push the capability of his machines in order to log large timber. An alternative is to use two machines—one pushing and one pulling—to move the logs. However, this type of operation is obviously expensive. A basic solution to the problem of gravity is to skid downhill wherever possible. Uphill skidding costs money. If, in skidding 10-cubic-foot pieces 600 feet, you have to drop one log per turn to maintain cycle time, your cost goes up 11 percent.

Within reason, however, and with the right equipment, volume per tree or volume per log has a positive effect on productivity. One study indicated an increase of 0.1 to 0.2 cunits per hour production will be achieved with each cubic foot increase in tree volume (Winer 1967, p. 68).

A second study, performed for the Canadian Forestry Service, clearly demonstrates the impact of load volumes on productivity (McGraw and Hallett 1970, pp. 5-6). All things being equal, the larger the volume per tree, the larger will be the logs cut from the tree, and the larger will be the load volume. Figure 9.1, from the work cited, shows the relationships between total cycle time, load volume, and productivity.

The scatter diagram shown in Figure 9.2 illustrates the results from observations in Douglas fir timber. The 54 observations were taken on ground varying from level to slopes of about 45 percent. While the observations are not directly applicable to other areas, they do offer some indication of the effect of average log size on total payload. Up to about 600 board feet Scribner the relationship is linear, or nearly so. Beyond 600 board feet, the trend appears to become curvilinear, but the variation is greater.

Skidding distance

Skidding distance is perhaps the single most important variable affecting skidding costs and productivity. The cost of skidding almost any log size is directly dependent on skidding distances. All other things being equal, the farther a machine has to travel from the logs to the landing, the lower will be the productivity and the higher the unit costs.

Skidding distances vary, depending on such variables as setting size, road location, terrain, and slope. In some cases, such as when reducing the road-building requirements, there is ample justification for long skidding distances. On rugged terrain, with steep slopes, an operator will have to travel farther to skid a given horizontal distance. One logger, operating on 80 percent slopes, was forced to work on 3,000-foot skidding distances simply because it would have been too costly to build roads on such steep terrain.

There are many instances where a tractor or skidder has pulled a load beyond half a mile. In extreme cases 1-mile skids have been necessary. Most operators, however, agree that between 400 and 600 feet is a good skid for a

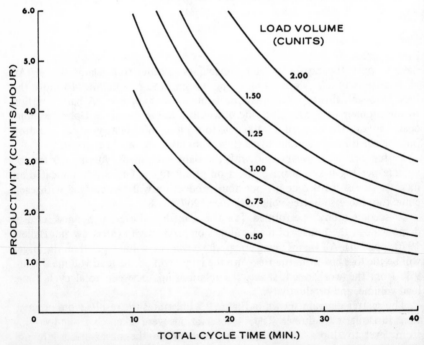

Figure 9.1. *Relationship between productivity, load volume, and total cycle time (McGraw and Hallett 1970, pp. 5-6).*

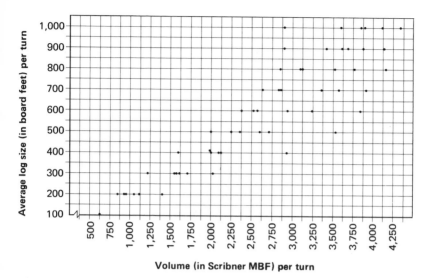

Figure 9.2. *Relationship between log size and load size.*

crawler tractor. Wheeled skidders, because of their speed, can skid up to 1,300 feet economically.

Optimum skidding distances vary with terrain and other physical conditions, as well as with the type of machine being considered. Skidders can be used on longer distances because they are faster than track-laying machines. Track machines, because of slower speeds, must have a full load on every drag.

The relationship between road-building costs and skidding distances is important. Total cost is minimized when variable skidding costs and spur road construction costs are equal. Longer skidding distances reduce the number of landings required and thus reduce landing construction costs. As skidding distances are reduced, more road building and landing construction is required.

Figure 9.3 shows the incremental cost of increasing ground skidding distance by 330-foot (5-chain) increments. Volumes of 15 cunits per acre and 21 cunits per acre are assumed for a 60-acre rectangular block in southern pine. Note the steep rise in cost for the final 330 feet. By holding slope and volume constant, the impact of skidding distance can be isolated. Changing the other variables through the same range of distances will push costs upward or downward depending on the movement of the variables.

None of the operating variables discussed thus far acts alone. That is, there is always an interaction among the variables, affecting both cost and productivity. For instance, a change in slope from 20 percent to 40 percent will increase the cost per cunit more when skidding 200 feet than when skidding 600 feet, given constant timber size.

The logging operator using ground-skidding methods must be aware of the effects of the operating variables. As the variables change from setting to setting, or even as they change on the same setting, the operator will have to

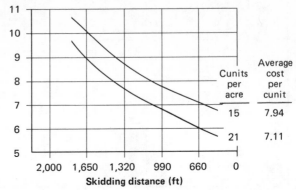

Figure 9.3. *Skidding distance versus unit cost in southern pine (from Conway 1977a, p. 35).*

devise ways of offsetting the effects. Longer skidding distances may be offset by an increase in load size. Scattered timber may indicate the need for an extra machine for bunching or forwarding. Extremely heavy and large log size may be offset by using larger equipment. Care must also be taken to ensure that the medicine is not worse than the illness. For example, increasing the load size in order to reduce skidding time can result in an increase in bunching and hooking time, thereby increasing, not reducing, costs.

Landings

The skidding operation has a very real effect on interfacing components—especially loading. The effect of short skidding distances on road- and landing-building costs has been discussed. In addition to the costs of construction there must be a balance in the loading component. Without proper balance there is likely to be an inordinate amount of truck waiting time at the landing.

Skidding can be a limiting factor in any logging system. Much of the limitation has to do with the size, layout, and location of the landing. Landings should be as carefully chosen and laid out as the skid roads. Layout varies with skid distance, haul road location, spur road construction, and in some cases, terrain.

Landings should be level, well drained, and large enough to accommodate, if necessary, skidder activity, truck loading, log storage, and sorting. In some areas, landings can be ideally located to balance skidding and loading activities. In other areas affected by terrain constraints, the landings may be small. Small landings cause problems in loading, as do wet landings. Finally, landings may be spaced too far apart, thereby resulting in skidding distances which are longer than necessary.

A landing need not be a carefully cleared area adjacent to or on the haul road. Sometimes landings are nothing more than wide spots along the road where logs are decked. When logging is completed a mobile loading machine moves in and works from deck to deck, loading logs on a haul truck.

Finally, landing activity must be smooth, since it relates to the skidding cycle. When a skidder arrives it must be moved back out with a minimum of delay. Several machines skidding to the same landing should be staggered to avoid congestion. Staggering means that the skidders operate on different skid distances so they do not all arrive at the landing at the same time. Conflicts between loading and skidding must also be resolved. If loading is going on simultaneously with skidding, the landing can become a bottleneck.

Work elements

The reader now has a basic understanding of the major operating variables, and the next step is to introduce him to what ground skidding is all about. Perhaps the best approach is to divide the skidding component into its basic elements. This division is the essence of the systems approach introduced in Chapter 4.

Conceptually, log skidding is not difficult to understand. A crawler tractor or wheeled skidder goes into a felled and bucked area, picks up the logs, transports them to a landing, and begins the process over again.

The full cycle begins at the landing or roadside, with the machine unloaded. The machine (let us assume it is a skidder) will travel to the site of the felled and bucked timber. There the skidder will be turned and backed into position at the selected tree or log. The object is to get as close as possible.

Mounted on the rear of the skidder is a winch drum holding between 75 and 150 feet of cable, called the *winchline* or *bull line*. The winchline may run from the drum directly to the logs or through an arch that is either attached directly to the machine (called an *integral arch*) or mounted on wheels or tracks and pulled by the machine. The arch will have a fairlead with rolling guides, which guides the winchline and reduces friction and wear on the cable.

Attached to the end of the winchline by a shackle are one or more chokers between 20 and 25 feet in length. The chokers are made into a sort of noose which is attached to the logs or trees for pulling.

The winchline on skidding machines is larger than the choker cable. Hence, if anything breaks it will generally be a choker rather than a winchline. This is not always true, however. With a full load it is entirely possible that a choker will still be within its breaking-strength while the winchline will have exceeded its breaking strength. Under these conditions, it is the winchline, not the choker, which may break.

There are generally between 3 and 10 chokers traveling with the machine. The actual number depends on machine size, timber size, and management policy. It is important to use a sufficient number of chokers to maximize skid volume. In some cases one man both operates the machine and sets the chokers. In other cases there may be a hooktender (straw boss) or a hooktender and one or more choker setters. When choker setters or hooktenders are used with crawlers or wheeled skidders, two sets of chokers are employed. The extra set is left in the brush for setting while the skidder pulls a load of logs (called a *turn* or *drag*) to the landing. When the machine operator sets his own chokers, only one set is used.

It is usually more efficient to preset chokers on long skids—over 300 feet. This reduces total cycle time while maximizing turn size. Under 300 feet, it is more efficient to have the operator set his own chokers.

Whether or not choker setters are used, the operator may have to move the machine several times before a full turn is accumulated. In other cases several chokers may be attached to the winchline by special sleeves which allow them to slide up and down the winchline. In this case the winchline can be stretched out among the logs and all the chokers hooked without moving the machine. The trees and the logs must be thick in order for this latter system to operate efficiently.

When he has accumulated a turn the operator will use the winch to pull the turn to the tractor and will then elevate the leading ends of the logs. Even a little elevation is helpful and reduces hang-ups. If a drawbar is being used rather than a winch drum then it is impossible to do anything but drag the turn along the ground with the full length of the logs in contact with the ground. A drawbar is a hook attached to the machine frame in the rear.

Grapple attachments for wheeled skidders and crawlers are widely employed in place of chokers. Grapple skidders are used to best effect on short skid distances, making the most of the mobility provided.

With a full turn the operator will start for the landing. The path he takes is called a *skid trail*. A skid trail may be nothing more than a way relatively clear of obstructions, which required little or no preparation, or it may be a rough road built by a crawler or the skidder itself.

Once at the landing the chokers are unhooked either by the operator or by a man called the *chaser*. One of the chaser's main functions is to unhook chokers. Once unhooked the chokers are either pulled through the fairlead until their ends are clear or nearly clear of the ground or they are placed in a rack welded to the side of the arch. The entire cycle is now ready to begin again. Unless a number of choker machines are skidding to one landing, it is usually more cost effective to have the operators unhook their own chokers.

The sequence of events described in the past few pages can easily be broken down into individual elements clearly distinguishable from each other. Once again, because of the similarities between tractor and skidder systems, both will be discussed simultaneously. Hereafter, the word *skidder* will mean either wheeled skidder or tractor.

There are five major work elements in the skidding cycle: return empty, bunching, skidding, landing, and delays. One additional element could be included and that is skid road construction. However, in this text skid road construction is included with the return element.

Return: This element begins the instant the skidder leaves the landing and continues as long as it is on the skid trail. The skid trail may have been previously built or the skidder or tractor may have to build as it goes, opening up a new skid road. Return ends when the machine leaves the skid trail to begin accumulating logs.

Bunching or Loading: Both these terms mean the same thing. The element begins when the machine leaves a recognizable skid trail. It includes all maneuvering necessary to accumulate a turn of logs as well as attaching the logs to the machine with chokers, chain, or grapples. (Grapple logging will be completely described in the next chapter.) The element ends when the machine and turn are back on a recognizable skid trail and moving to the landing.

Skidding: The skidding element begins when both the skidder and the turn are on a recognizable skid trail and ends when the turn arrives at the landing or beside a roadside log deck.

Landing: The landing element is a combination element which includes chasing or unhooking and decking. *Unhooking* includes the maneuvering required to get the logs into position for unhooking. The maneuvering may include several starts and stops to even-end the logs or to separate species. With grapple skidding, maneuvering simply means positioning the logs and opening the grapples to release them. Decking the turn once it has been positioned and unhooked is a matter of pushing the logs onto the deck.

Delays: Delays occur all through the skidding cycle and are described as *productive* or *nonproductive*. A productive delay might involve pushing debris out of the skid trail or off the landing. In other words, a productive delay is one that contributes to the operation. A nonproductive delay might be any one of several things—mechanical breakdown, personal time, unscheduled work breaks, and waiting on the choker setters when two sets of chokers are used.

These five elements can naturally be broken down even further. For instance, if an operator were using a large tractor with more than one set of chokers, then choker setting could be separated from bunching. Time spent while bunching operations were held up by choker setting would constitute a nonproductive delay. Another example might be prebunching, that is, the use of a second machine to accumulate the logs into bunches that will be picked up and skidded by a second machine.

Certainly, many other examples could be cited, but the main point is already clear. The only sure way to break a specific operation down into its elements is to examine it in detail. However, the skidding elements described in this chapter are sufficient for a general understanding.

Ground Skidding: Operations

Skidding is generally considered to be a major cost in logging, if not *the* major cost. Because of the cost relationships between functions, as noted in Chapter 6, a small incremental cost increase in cutting can be afforded if it will decrease the skidding cost. One study of seven pulpwood operations in South Carolina revealed that skidding accounted for between 32 and 52 percent of the total unit cost per cord (*Pulpwood Production* 1971*a*, p. 22). The cost of skidding and its effect on the total logging system are reason enough to pursue the subject further.

Since the descriptions of the working cycle and elements given in Chapter 9 were fairly general, this chapter will begin with further discussion of the work elements, particularly as they relate to improved efficiency. Specifics will be discussed in such a way that the reader will understand more about the what, how, and why of ground skidding. In addition, certain new functions and methods will be described. Grapple skidding will be discussed in detail, as will forwarding, which was briefly described in Chapter 4.

Although some elements may appear more important than others, they are all equally important in offering an opportunity for improving the operation in terms of both time and cost. A case in point is the return, or travel empty, element.

Return

The return element may appear unimportant but this is not the case. An alert operator can shave half a minute from his cycle time simply by using the correct gears in his machine. If the return distance is short he may elect not to shift at all since shifting always loses a little time initially. Also, during the return trip the operator can observe the skid road and, if necessary, make repairs. Or he can choose to take a different route and save the repairs for another time when the delay for road building will not slow up the entire operation.

The approach to the next turn of logs also is important. If with a little forethought the operator can avoid some maneuvering, time will be saved. Fractions of minutes subtracted from average cycle time can make a big difference in total productivity.

Bunching

A glance at Table 10.1 shows that the bunching element requires the largest block of time. One source, different from that used to develop Table 10.1, indicated the bunching element required an average of 40 percent of total cycle time (Perotto 1967, p. 62). Since bunching accounts for the largest block of time it probably represents the greatest potential for improvement.

Bunching really describes several different job elements, from minor skid trail construction after the skidder has left the main skid trail, to setting chokers. In one text bunching is defined as the preassembly of tree products for immediate or later skidding (Wackerman, Hagenstein, and Michell 1966, p. 129). This means bunching can apply either to accumulating a turn of logs and then skidding them with the same machine that did the bunching or to accumulating the turn with one machine and skidding them with another. Both approaches will be developed here.

Looking ahead

When the skidder operator heads back into the brush on the return element he should know where the next turn is. And while he is working in the brush

Table 10.1 *Time Distribution for Skidding Elements*

Element	Time in minutes
Return	1-5
Bunch	2-45
Skid	2-10
Landing	1-3
Delay	0 and up
Total cycle time	6-63

Source: Mayfield.

Note: The delay element has been left out because it is open-ended. Average delay time will depend on a great many factors that are unpredictable. Under favorable conditions the total cycle time could be as little as six minutes. Favorable conditions might include favorable slopes, moderate to short skid distances, and firm earth to skid on. Unfavorable conditions and inefficient operations could push the cycle time to more than an hour. Time distribution may vary with the type of machine. A crawler will have longer skidding and return times while a wheeled skidder will have less.

he should be looking for the next turn. When hooktenders and choker setters are used it is their job to locate and hook the turns before the skidder or tractor returns. The ideal situation is to develop a turn within mainline reach of the skidder from a single location. If this is not possible the objective is to minimize the starts and stops.

When working in heavy brush, locating a turn is sometimes difficult, especially if the timber is small. This is especially true if the skidder operator is setting his own chokers or if skidding grapples are being used. In a selective cut the problem is further complicated by the standing timber. Whatever the case, the operator must strive for a full turn on every trip.

Load size

The size of the turn or load is essential in both skidder and tractor operations. It depends on the slopes being traversed, skidder weight and horsepower, soil conditions, and log size.

Slope, of course, limits the load size that can be skidded. On adverse grades the load must be lightened somewhat. On favorable grades an additional log or two can sometimes be skidded without affecting the entire cycle. The objective is always to maximize load size, but good judgment must prevail. A slight load reduction may reduce the total cycle time enough to make up for the loss of volume.

Timber products are usually measured in cords, cunits, or board feet. However, it is best to think of load size in terms of weight. In fact, weight is the only unit of measure that makes sense when considering skidder capacity. When weight is applied to the question of load size, many differences become obvious. Table 10.2 shows the weight of selected species.

One cunit of western white pine from Idaho weighs approximately 3,500

Table 10.2 *Weight per Cubic Foot of Selected Commercial Species*

Species	Lbs/ft³	Species	Lbs/ft³
Red alder	46	Western larch	48
Yellow birch	58	Red oak	63
Alaska cedar	36	White oak	62
Western red cedar	27	Longleaf, shortleaf,	
Douglas fir (coastal)	38	and slash pine	62
Noble fir	30	Loblolly pine	62
Red fir	48	Lodgepole pine	39
White fir	47	Ponderosa pine	45
Black gum	45	Western white pine	35
Red gum	50	Yellow poplar	38
Eastern hemlock	50	Black spruce	32
Western hemlock	41	Sitka spruce	33
Hickory	63	Sweetgum	50
		Black walnut	58

Source: Caterpillar Tractor Co. 1981, p. 476.

The winch and integral arch are clearly shown. The roller on the arch is called a **fairlead** and turns as the winchline is used. This reduces wear on the line. Note the bull hook to which the chokers are attached. The man is setting the chokers while the machine waits.

Grapple skidder working as a bunching machine. The logs are skidded into piles. A larger cable skidder will complete the trip to the landing.

Table 10.3 *Drawbar Pounds Pull Required to Overcome Skidding Resistance*

Grade	Load weight (lb)									
	1000	2000	3000	4000	5000	6000	7000	8000	9000	10000
	Drawbar pounds pull required									
	Ground skidding – log lengths									
30% upgrade	1205	2410	3615	4820	6025	7230	8435	9640	10845	12050
20% upgrade	1100	2200	3300	4400	5500	6600	7700	8800	9900	11000
10% upgrade	995	1990	2985	3980	4975	5970	6965	7960	8955	9950
Level	890	1780	2670	3560	4450	5340	6230	7120	8010	8900
10% downgrade	784	1568	2352	3136	3920	4704	5488	6272	7056	7840
20% downgrade	679	1358	2037	2716	3395	4074	4753	5432	6111	6790
30% downgrade	574	1148	1722	2296	2870	3444	4018	4592	5166	5740
	Arch skidding – log lengths									
30% upgrade	1365	1928	2491	3054	3617	4180	4753	5306	5869	6432
20% upgrade	1025	1499	1973	2447	2921	3395	3869	4343	4817	5291
10% upgrade	685	1070	1455	1840	2225	2610	2995	3380	3765	4150
Level	333	634	935	1236	1537	1838	2139	2440	2741	3042
10% downgrade	–	209	414	625	833	1041	1249	1457	1665	1873
20% downgrade	–	–	–	–	59	199	339	479	619	759

Source: Caterpillar Tractor Co. 1981, p. 455.

Note: Data taken on dry, smooth clay loam slopes. Resistances are less on wet soil.

pounds. Another cunit of pine, this time shortleaf pine from Arkansas, weighs closer to 6,200 pounds. This difference in weight can make a difference in the cubic feet of payload being skidded.

The pounds of pull required to move a turn of logs, or any load for that matter, depends on the weight of the load and the friction between the load and the ground. Slope affects the efficiency of the pulling machine, as was pointed out in Chapter 9. There is also a difference when part of the weight is transferred to the rear wheels of a skidder, as when an integral arch is used. Table 10.3 illustrates the combined effect of slope and weight on the amount of drawbar pull required for ground skidding and arch skidding.

Orientation of the load can also have a dramatic effect on skidding efficiency. Loads can be carried with butts leading or with tops leading. The impact of load orientation depends on the type of machine being used, tractor or skidder, and whether the machine is equipped with some sort of arch.

Ground skidding or *drawbar skidding* means pulling the logs flat on the ground without the benefit of either arch or fairlead. When ground skidding it is best to orient the tops forward, since this reduces the plowing experienced when the logs are skidded with butts leading. Ground skidding with tops leading requires less power, reduces choker wear, and allows larger turns to be skidded.

When an arch or fairlead is used, the rules change. If the load is suspended with butts leading there is a decrease in power requirements. One study showed a power reduction of nearly 32 percent when the load was suspended with butts forward (Garlicki and Calvert 1968, p. 62). The rule of thumb is to skid butts forward with an arch for increased production. With wheeled skidders the butt-first orientation places more weight on the rear axle, thus increasing efficiency. In the case of either tractors or skidders the butt-forward and suspended orientation reduces plowing and damage to the site.

It is important to utilize the full capacity of the machine while at the same time maintaining optimum cycle times. Thus the effect of load size and orientation must be considered along with the effect on total cycle time. For instance, assume a machine can operate at 80 percent of capacity and still maintain an efficient skidding cycle. With 80 percent of usable capacity the skidder skids a turn in 90 seconds. If, to utilize the remaining 20 percent of capacity, an additional 30 seconds is added to the cycle time, the end may not justify the means, since skidding time has increased by 33 percent.

Shuttle skidding

The relationship of load capacity to time holds whether the load is skidded to the landing by the same machine that does the bunching or whether one machine is bunching the load for another to skid. In either case the load will end at the landing. When one machine accumulates and bunches a load for another to skid, it is called *shuttle logging* or, in western parlance, *swinging*. In the case of ground skidding, shuttle logging consists of using one or more small skidders, either conventional or grapple-equipped, to accumulate a load and using a larger skidder to pick up the bunched logs or trees for the long run to the landing.

Shuttle logging is justified by a reduction in total cycle time, an increase in payloads, and a reduction in costs. It also allows better utilization of equipment, since the bigger machines are assured full-capacity loads. The skidding is generally done by large-sized wheeled skidders, which allow for greater speed. Some operators might use small skidders and larger skidders, while others might use tractors.

Winching

The *winch* is a powered drum located at the rear of either a skidder or a tractor. The power is applied to pull in logs, and the drum can "free-spool" so that the line can be pulled off the drum by hand.

Winching time, the time required to pull a log or group of logs a given distance, is influenced by slope, distance, undergrowth, and other variables. Maximum efficiency is achieved in a straight-line pull. As the vertical or horizontal angle between the line and the skidder increases, efficiency drops. If winching at an angle is necessary, extra care should be taken so the skidder will not overturn, especially when winching on a slope. The angle also causes the line to spool unevenly and results in line damage.

As already mentioned, the use of arches reduces plowing and site destruction and, in general, improves the efficiency of the machine. Skidders are usually equipped with integral arches welded or bolted to the skidder frame over the winch. Although tractors are also sometimes equipped with integral arches, the arches do not offer the same benefits. The arch on a skidder transfers weight to the rear axle and improves efficiency. The pulling efficiency of a tractor, however, depends on ground pressure being the same over the full length of the tracks. Thus, although an integral arch on a tractor will help elevate the load, it will not increase pulling efficiency—in fact, it may reduce it.

Most often if anything is used with a crawler it will be either an arch or a sulky. An arch is track-mounted and a sulky is wheel-mounted. Neither the arch nor the sulky transfers weight to the machine. Rather, the tongue, like a wagon tongue, is pinned directly to the crawler's drawbar. The arch and sulky, like the integral arch on a skidder, result in less skidding resistance, higher skidding speeds, cleaner logs, less site damage, and reduced cable wear. On the other hand, there are some disadvantages, such as a loss of maneuverability, which result in increased bunching time and unhooking time.

Some operators, using ordinary farm tractors, still pull directly off the drawbar. Drawbar skidding, in which the chokers are attached to a hook welded on the skidders frame at the rear, is inefficient. More time is required to build a turn because the tractor must be maneuvered close to each log. Also, the turns are smaller.

Still others skid with only the winch drum and no arch at all. This method offers the advantage of greater maneuverability than can be had when towing an arch. However, payload is reduced and skidding resistance is higher. But probably more important, or at least as important, is the damage to the ground that occurs when the log is dragged with both ends on the ground rather than with one end elevated.

Logging tractor with sulky. Note that the front ends of the logs are elevated to make skidding more efficient.

Choker setting

The choker setter or hooker, where one is used, is a key man. He chooses the logs that will make up a turn, directs skid road location, and selects the path the logs will travel as they are accumulated for a turn. Obstacles such as stumps, rocks, or root wads must be avoided. At all times the hooker should attempt to choose routes that will minimize road building and damage to the residual timber if logging is in a selective cut.

Choker setters are used most effectively on longer skids. For distances under 300 feet, it is more efficient for the operator to set his own chokers.

The bunching element, when the hooker and/or choker setter is used, is a function of his expertise. A slow hooker who does a poor job of selecting turns or directing the skidder will increase bunching time and create skidding delays. It takes experience to minimize delays and hang-ups in making loads.

The tools of the hooker are his experience with machines and logging in general. His knowledge will enable him to do a better job and will lead to a more efficient operation. One of his primary tools is the choker.

A conventional choker used in skidding is a single length of cable between 5/8 and 1 inch in diameter. Skidding chokers are smaller than chokers used in cable logging. They are generally no larger than 7/8 inch in diameter and are between 20 and 25 feet long. The choker passes around one end of the log or

Setting a choker is shown in these four photographs: (1) The choker bell. (2) Choker setter has the nubbin in one hand and the bell in the other. (3) The nubbin is placed in the bell and the cable is laid in the slot. (4) Choker is set and ready to go. The choker was placed over the log so that it will not come unhooked while the log is being skidded. Highlead chokers are the same as Cat chokers, only larger in most cases.

logs and is secured by the use of a *Bardon hook* or *bell*, which forms a noose that tightens as the choker is pulled. The Bardon hook is molded with an eye that slips up and down the choker cable. A ferrule called a *nubbin* is pressed on the end of the choker to prevent the hook from falling off. The ferrule is also the means by which the noose is made, as the noose fits into the ferrule eye. The other end of the cable, opposite the nubbin, is equipped with a spliced or pressed eye. The eye is slipped over the hook, called a *bull hook*, which is attached to the winchline.

Another type of choker assembly, which was mentioned earlier, involves five or more tractor hooks, each with a choker attached. The hooks and chokers slide up or down the winchline as required. With this choker system the winchline is extended behind the skidder to the farthest log to be picked up. Following this, one of the hooks is slid along the winchline until the choker is close enough to the log to be set. The bight of the winchline is pulled to each successive log, the hook positioned, and the choker set until there is a full turn.

When the bull-hook system is used, all the chokers are set and the turn is then accumulated by pulling the winchline to each successive log. The machine is positioned as close as possible to the first log or to the majority of the turn. The first log is hooked and pulled up to the machine or to the next log. Then the winchline is pulled out to another log. Each stem added requires not only more time and effort, but more cost as well. For this reason the hooker will attempt to locate a full turn of logs as close together as possible in order to minimize movement and time.

The winchline choker assembly is used with skidders, especially when the operator sets his own chokers. The last method described is used with or without choker setters, as a matter of preference. When choker setters are used it is best to use two sets of chokers. One set is hooked on the logs while the other is being used to skid a turn.

Choker setting appears, to the casual observer, to be a simple task. After all, all that needs to be done is place a steel noose about a log. It is not extremely difficult in terms of mental requirements, but it does take a good deal of

Notice the size of the turn that is being skidded. There are 10 chokers hooked. The operator sets his own chokers.

common sense. A good, experienced choker setter is worth his weight in gold.

There is a right and wrong way to set a choker. The choker should be hooked as close to the end of the log as possible without risking the loss of the log during skidding. Preferably, the leading end of the log should be hooked, to avoid having to swing the log into lead with the tractor. When two logs are end-to-end, as they were bucked, the chokers are set at the break, one on each log. When the logs are winched out of the lay, they will form a V, and will move parallel to each other as winching continues or the skidder moves ahead.

With a three-log tree, the skidder is run parallel to the tree to the first log to be picked up. The choker is hooked to the winchline and the tractor is pivoted to pull the log out of its lay and free it for forward movement. The two remaining logs are picked up in the V break technique just described.

The sequence in which logs are hooked depends on the orientation of the large end. Large ends should be leading when a fairlead or arch is used. Small ends should be leading when doing drawbar skidding. When attaching the chokers to the bull hook, the tightest choker should be attached last, even if it means rearranging the chokers on the hook. With the tightest choker on top, the other chokers on the hook will not drop off. The tightest choker is the choker with the least amount of slack.

When the final log is hooked to the winchline, making up a full turn, the

The D-8 Caterpillar logging tractor is using the V break technique to pull the turn out of its lay prior to skidding.

skidder should always be pointed in the direction of the skid trail, if possible. This will reduce the amount of maneuvering necessary and will result in less trouble getting the turn started.

Hang-ups

If an obstacle can be avoided by moving the skidder into a better position, this is desirable. It is always better to lead a log around an obstacle or to use a high-angle pull rather than fight a hang-up. However, there are times when hang-ups cannot be avoided. At these times the choker setter earns his keep.

Three basic maneuvers can be used to work around a hang-up—the jump, the kick, and the roll. These maneuvers are not difficult to apply and can be extremely helpful under the right conditions. Each of the three methods will now be described, using the same hang-up condition.

Picture the butt end of a log directly facing a stump with the butt close against the stump. In this position, if the log had to be pulled into the stump, a choker would probably be broken. The chokerman could rechoke the log on the far end and pull it around, but let us here assume that rechoking is not possible. Nothing but a right-angle pull can move this log.

A *jump* could be used to apply the right-angle pull up over the stump. This is achieved by setting the choker in the usual fashion and then slipping the choker bell under the log. The leading end of the choker is forced under the log as close to the center as possible and is laid over the top of the stump. When the skidder winch is engaged, the bight of the line over the stump will pull straight up on the end of the log. The stump is used to change the direction of pull upward. Figure 10.1 illustrates the jump.

A *kick* utilizes the same principle as the jump except that the force is made to apply from the side of the stump rather than up and over. The choker bell is placed opposite the direction of pull and the leading end of the choker is again led around the end of the log, but around the stump instead of over it. When the winch is engaged the log will be pulled sideways away from the stump. Figure 10.2 illustrates the kick. A variation of the kick is used when a second stump to the side of the hang-up is burned by throwing the bight of the winchline over it to change the direction of pull (see Figure 10.3).

The *roll* is also effective in helping to free a hang-up. It is simple to apply, providing there is a little daylight or space between the log and the ground. The choker is set in the normal fashion. Once set, the noose is slid around the log, counter to the desired direction of pull and in screw fashion. As the noose is screwed onto the log, the leading end of the choker is automatically wrapped around it. The same effect results from setting the choker bell low on the log against the ground and wrapping the leading end of the choker around the log.

Once the roll is set, the choker eye is slipped on the bull hook and tension is applied rolling the log in the direction of pull—hopefully, away from the stump. The roll is often used with a kick when the kick alone will not free the hang-up. The roll is illustrated in Figure 10.4.

It is always better to think about a hang-up before it happens. If it looks as if a log will hang up on some construction, use one of the methods described

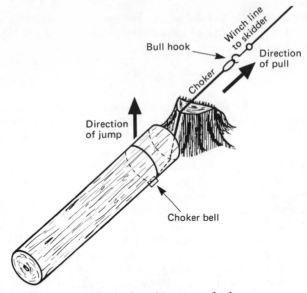

Figure 10.1. *The jump method of working around a hang-up.*

before the skidder tries to pull on the log. This will save time in the long run since some of the logs will not hang up at all. Also, it may be possible to use the blade on the skidder to push the log clear of a potential hang-up.

A second type of problem encountered by the choker setter is oversized logs—logs too large for the chokers. The winchline could be used in these cases, but the method limits the turn to one log and is hard on winchlines. An alternative is the bridle.

The *bridle* requires the use of two chokers. One is laid over the log with the nubbin on one side and the bell on the other. The second choker is pushed

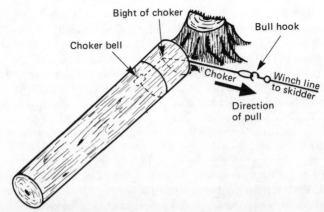

Figure 10.2. *The kick method of working around a hang-up.*

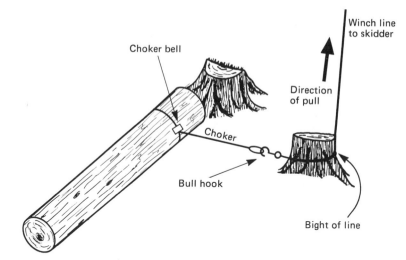

Figure 10.3. *A variation of the kick.*

beneath the log so that the nubbin is positioned directly below the bell on the first choker. The nubbin of the top choker is hooked in the bell of the bottom choker and vice versa. The eyes of both chokers are attached to the bull hook and the log is then ready for skidding.

A *swede* can also be used on large logs; however, not simply because one choker will not fit around the log. Perhaps one choker could be used but the choker setter may think the choker is too light (small in diameter) and will break when the log is pulled. In this case, two chokers can be used to provide greater breaking strength.

With a swede both nubbins are pushed under the log and leading ends and bells are laid over the logs, just as a choker would normally be set. However,

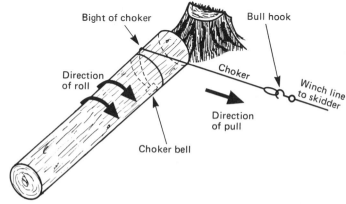

Figure 10.4. *The roll method of working around a hang-up.*

the nubbin from the first choker is hooked with the bell from the second choker and vice versa. A swede nearly doubles the strength of one choker.

Though there are a number of other helpful choker-setting techniques, we will not go into them here. The techniques described thus far apply equally well to cable yarding, and the reader should keep them in mind, as they will be referred to in the cable yarding chapters.

Just as it is desirable to minimize the movement of the skidder it is also desirable to minimize, as much as possible, the work of the choker setter. One way his work can at least be made easier is to avoid making him drag the winchline uphill to a log or turn of logs. The skidder should always be positioned so that the winchline can be pulled downhill to the logs. The time lost in winching logs uphill will be made up in lower choker-setting time.

One last point regarding winching. It is sometimes difficult to move a log out of its lay once the choker is set. Some operators try to move a stubborn log by taking a run at it with the skidder. Another technique used by skidder operators is *impact loading*, which involves tightening up the winchline until there is no slack left. The skidder is then moved ahead until the front end of the machine is lifted from the ground. The winch is then released, letting the front end drop. But, before it reaches the ground, the winch is reengaged, catching the skidder in midair. The sudden loading on the winchline can jar a stubborn log from its lay. However, it can also break chokers and winchline, and puts a terrible strain on the machine's frame.

The correct way to winch a load is to ease into it. Ten percent more force is required to start a load than to sustain movement at skidding speed. Before starting to winch, the machine should be as close as possible to the turn. Finally, winching should not be done at an angle. It is always better to pull straight into the load since in this position less force is dissipated.

Skidding the turn

The skidding element accounts for between 16 and 30 percent of the total cycle time. It is in the skidding element that the difference between tractors and skidders becomes obvious. Crawlers are slower than skidders but can negotiate steeper slopes efficiently. Skidders, while they are the faster of the two, lose efficiency with large loads. A load that is too small will increase cycle time but at the expense of unit costs. The skidding element is another area where improvements can be made with a little effort and some forethought.

Skid road layout is probably among the most important considerations. The rule is simple. The skid trails should be kept as straight as possible and should follow a favorable grade to the landing, if possible. For one thing, curves require the operator to slow his machine. In addition, large loads, whether or not arches are being used, tend to bind on curves, further increasing cycle time.

Primary skid roads must be laid out with environmental considerations in mind, to minimize soil disturbance and erosion.

Skid roads with many curves are dangerous, especially with skidders that have a higher center of gravity than a tractor. It is easy to overturn a skidder on sloped curves. If curves are unavoidable, the operator should always reduce

FMC skidder moving a turn to the landing. Equipped with a blade, it can perform cleanup and decking work. The arch swings back over the logs for hooking and then picks the turn off the ground, elevating leading ends.

speed when approaching them.

Skid trails may be laid out in several different patterns. Some, of course, are quite random—the least efficient practice. Others allow for a systematic clearing of felled and bucked timber. A systematic layout of skid roads increases the productivity of skidding equipment significantly, and this is the preferred practice. An efficient logger will plan his skid roads before cutting begins so the timber can be led to the trails, marking the layout with ribbons. When mechanical cutting equipment is used the skidders simply follow the same pattern. The T and L pattern have already been described in Chapter 8.

If the block has been clearcut before skidding begins, the skidders must develop patterns consistent with the lead of the timber and in relation to main skid roads, the haul road, and landings. Three such patterns are illustrated in Figures 10.5, 10.6, and 10.7.

Whichever pattern is used, when several skidders are involved the skid distances should be varied. The skidders should never all make long skids together. Some of the machines should be on long skids, while others are on short skids. This way, production is balanced, and there will always be logs on the landing for loading on the haul trucks. Failure to balance the skidding operation will produce a "feast or famine" situation at the landing.

Load height

Load height is a factor that affects skidding efficiency. Tests have shown that it requires 17 percent less power to skid a turn while holding the butt ends 5 1/2 feet off the ground than it does when skidding top first with the logs

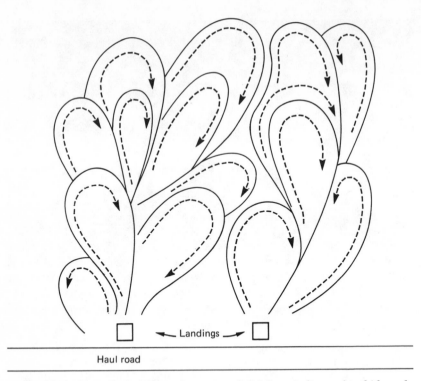

Landings

Haul road

Figure 10.5. *Cloverleaf skid road pattern. Solid lines indicate the skid roads, and arrows indicate the skid direction.*

elevated 2 1/2 feet (Garlicki and Calvert 1970, p. 23). The reason for the reduced power requirements in high loads is that the load adds traction to the skidder's rear axle. With a four-wheeled vehicle care must be taken not to transfer too much load to the rear axle. The more weight transferred to the rear, the more traction will be lost by the front axle.

When skidding downhill the load should be carried a little lower than normally. Once inertia is overcome, the logs are apt to take off on their own. On favorable grades, and at prudent speeds, the logs can easily swing into the machine's fenders. Also, at lower heights the danger of overturning is reduced. On a downhill skid, if braking is necessary, the load can be dropped to the ground quickly and will act as a brake for the machine.

Paying attention to load height pays dividends. Kept low on grades, the logs serve as brakes and make skidding less dangerous. Kept high (when possible), power requirements are reduced. In addition, plowing and hang-ups are reduced, while cycle times are increased.

When skidding logs over the same trail again and again, deep ruts are often dug in the ground. This is especially true in wet country. The ruts, after a time, turn into mud holes that can scar the land for a long time and create a problem in skidding. The best way to solve the problem is to avoid the mud holes.

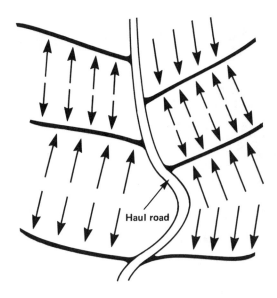

Figure 10.6. *Parallel skid road pattern. The skid roads are shown by solid lines, and arrows indicate the skid direction.*

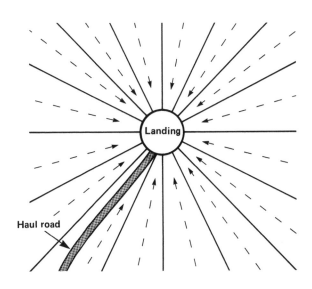

Figure 10.7. *Sunburst skid road pattern. The skid roads are indicated with solid lines, and arrows show the skid direction.*

Ground skidding: operations   179

The logging tractor is skidding uphill and couldn't pull the load. The operator ran the machine ahead and is winching the turn over the rough spot.

However, this is not always possible and the obstacles must sometimes be negotiated. Without a load, neither a skidder nor a tractor would have any trouble. With a load, however, it is a different story. There are three solutions, all of which also apply to pulling a load up a steep grade.

When forward motion is stopped because of mud or a steep grade, do not reef or jerk on the load. First, reefing will not solve the problem; second, it is hard on the machine. Instead of reefing, release the winch, dropping the load, and run the machine ahead while spooling out line. When the machine is on solid ground use the winch to free the load.

If winching does not work, the only alternative is to drop off part of the load—as many logs as may be required to free the balance of the turn. The abandoned logs must be picked up, of course. This should be done later on with another turn. Do not interrupt the current skidding element to attempt picking up the dropped logs. When the remaining logs are free of the obstacle, the machine should continue to the landing.

If a great deal of winching is required during the skidding cycle, the load is too large. In this case a large load is not economical, because it is increasing the total skidding cycle time. When the next turn is picked up one or two logs fewer should be tried. A little experimenting will result in an optimum load size for the conditions at hand.

Grapple skidding

No treatment of ground skidding would be complete without a description of grapple or chokerless skidding. Grapples are simply a hinged set of jaws which are opened and closed hydraulically. They are lowered over a log or logs, closed, and the turn is skidded, being held together by the grapples rather than by a choker.

Three major types of grapples are available. One type is mounted on a swinging boom that can swing in a 180-degree arc. The grapple can rotate 360 degrees from the end of the boom and can thus be lowered, raised, or otherwise positioned over the logs. The advantage, of course, is maneuverability and control. With this type of grapple it is possible to load a set of bunks simply by swinging the boom to one side and driving the logs into the bunk. Decking works in much the same way, but a certain amount of pushing must be done to build the deck in height.

A second type of grapple also has the advantage of rotation but the boom can move only vertically. A little maneuverability is lost with this system, but there is no trouble picking up a turn if the machine is carefully positioned.

The last type involves a set of grapples mounted on a rigid boom. If the logs are within reach, the wide-throated grapples can pick up the logs. Otherwise the machine must be repositioned.

The operation of grapple skidders does not differ much from conventional ground skidding with chokers. The grapple operations are subject to the same constraints and variables—skid distances, terrain, and log size. There are, of course, some differences within the elements themselves. The most notable difference is the absence of choker setters. Choker setting time is eliminated, a major savings attributed to the use of grapples.

Using either a swinging or a nonswinging grapple, the operator either will drive over the logs to be picked up or will back into them until the grapples are in position. The swinging boom eliminates some of this maneuvering since the grapples can be swung in an 180-degree arc. When rigid grapples are used positioning must be fairly exact and the log must be within reach of the opened grapples. When the grapples are closed the log is pulled into position for pickup.

Like cable skidding, grapple skidding requires a full load to the extent it can be gathered without losing efficiency in the total cycle. The number of loading points has a serious effect on total cycle time. With each additional loading point, roughly one minute is added to total cycle time (Bredberg 1970, p. 6). Because of the time consideration, loading points should be minimized. Grapples work best in prebunched logs.

In some operations grapple machines are used to bunch for a conventional skidder. This is a good application but the system does not always result in increased efficiencies. One report indicated that the combined production of a grapple skidder bunching for a conventional skidder was no greater than the production of the grapple skidder alone (Arthur 1967, p. 62). The efficiencies arise when the skid distance is long. With one machine doing the bunching, delays for choker setting and bunching are eliminated, and the larger, conventional skidder always has a maximum payload.

During the skidding element the logs should be oriented just as they are in conventional skidding—butt forward with the load raised. The hydraulic boom, when one is used, keeps the leading ends of the logs off the ground.

There is one disadvantage in skidding with grapples, which occurs during the skidding element. If the machine becomes stuck it cannot move ahead like a conventional machine, leaving the load for winching. Instead, the entire load must be dropped. Because of this problem special care must be taken to build a load of optimum size and to choose skid trails that avoid obstacles.

Care must be taken when skidding with a boom-type grapple machine. The boom, extended behind the machine, tends to transfer more weight to a point in back of the machine, making the machine relatively unstable. Also, with a swinging boom, skidding with the boom swung to the side has the same effect as winching to the side—it increases the danger of overturning the machine.

Additional efficiencies are gained when landing logs that have been grapple skidded. Chokers do not have to be unhooked. When approaching the log deck the machine pulls into position and simply opens the grapples to release the load. The use of swinging booms allows logs to be loaded directly into log trailers. In this case truck-loading efficiency is also improved since the trailers are already loaded when the truck arrives and a loading machine is not required.

The benefits derived from using grapples come from not setting chokers and from reducing the time required to bunch. Load size is set by the size of the grapples used; therefore reduction of the skidding distance is a major objective when making a grapple skidding layout. It is generally agreed that grapples work best combined with conventional skidders and mechanized harvesters. Harvesters allow for tree orientation at the least and for prebunching at the most. Bunched logs allow the grapples to secure a full load without delay. In addition, when rough, boggy ground is encountered the conventional skidders can take over. Of course, if the operator has a choice of equipment, he should use the grapples on high-production jobs where the machine's capabilities can be fully utilized. Such jobs involve high volume per acre, prebunched logs, and relatively short skidding distances.

Forwarding

During the skidding process at least part of the tree or log is dragging on the ground. Forwarding, also called *prehauling*, is an intermediate transportation function which involves transporting the stems, generally free of the ground. Forwarding is performed by several vehicles, including a skidder equipped with a trailer or integral bunk, and wheeled front-end loaders. Of course, there are other systems; one example is the Koehring system described in Chapter 8. In this system the bolts are carried in bunks to the landing and loaded on a set-out trailer.

A Timberjack forwarder with loading boom.
Photo courtesy Eaton-Yale Ltd., Forestry and Construction Equipment Div.

When a skidding device equipped with a trailer or bunks is used, the stems are loaded into the carrier for further movement. Loading of bolts may be done by hand; however, many forwarders are equipped with mechanical loading equipment. Forwarding of this type is a logical extension of the days of logging with horses, when logs or bolts were loaded on a dray for the haul to mill or landing. As the degree of sophistication increases, so does productivity. Of course, the required investment is also higher.

For instance, in one case involving short-wood operations a crawler and dray achieved man-day production of 4.25 cords. Rubber-tired forwarders achieved 6.00 cords per man-day. The crawler-dray required one or two men while the skidder operation required four or five men. Capital cost for the skidder operation was also about 60 percent higher than the crawler-dray (Jarck 1971, p. 98).

Another type of forwarding system utilizes pallets and either a tractor or a

Front-end loader being used as a forwarder in Ontario, Canada.

skidder to transport the loaded pallets. A pallet is a metal frame into which bolts are stacked—generally by hand. The pallets have a one- to two-cord capacity. Some machines have pallet loaders, such as a knuckle-boom loaders, to load the bolts in the pallet.

When mechanized loading is used the bolts may be prebunched. Otherwise the machine moves from stump to stump loading the cut bolts. Once loaded the pallet forwarder transports the load to a landing, where the load is either transferred to the ground or loaded directly on a set-out trailer. At times the pallets may be loaded directly on a waiting truck.

One of the advantages of the pallet system is that the skidding and hauling component are divorced. With pallets, especially if set-out trailers are used, the haul trucks need never be delayed. Also, in comparing short-wood systems, pallets offer the further advantage of loading trucks or trailers in units rather than in pieces. Another advantage is a uniform load size both in primary and secondary transportation.

The principal difference between the first forwarder system described and the pallet system is the means of retaining load integrity. With a conventional forwarder the pieces are loaded individually and then piled down or cold-decked at the landing. The exception is when the load is transferred to a waiting trailer or truck. At any rate the stems are handled a piece at a time, or perhaps two or three pieces at a time if mechanized loading is used. The pallet system deals in units, as was explained, and allows for more efficiencies in forwarding and in secondary transportation all the way to the final concentration point.

In general there are many advantages to the use of forwarding systems:

1. The systems assure full payload and thus reduce the delay potential in skidding while also allowing maximum use of machine capability.

2. Skidding or hauling distance may be increased. Forwarders are often used on skid distances up to a half mile.
3. Less damage is done to the residual stand than in either grapple or conventional skidding operations.
4. Because of the ability to self-load, move wood longer distances, and reduce damage to residual stand, forwarding systems are well suited to thinning operations.
5. Forwarders can often transport stems or bolts over ground that would be difficult for a conventional skidder to negotiate.
6. Production and costs are not as affected by tree size when forwarders are used. This is because of the consistent load size.
7. When equipped with a mechanized loader the forwarder can be used for both primary transportation and loading.

There are undoubtedly other advantages; however, these are the most important. Forwarding systems are not without their disadvantages, however. Some of the main ones are:

1. The systems require high capital cost relative to other systems.
2. For the optimum application wood should be prebunched for the forwarder operations.
3. Forwarders are limited, generally, to 20-foot log lengths. Some, of course, can carry only bolts.
4. Because of the size limitation the machines are not well suited to integrated harvest operations where sawlogs, plywood logs, and pulpwood bolts are all produced.
5. Cost per unit appears to be higher than in other short-wood operations.
6. Manpower requirements and skill requirements are higher for forwarders equipped with loading devices.
7. Forwarders are not as versatile as conventional skidders and are limited as to use.

The bobtail system

The bobtail system, which might be described as a variation of the forwarding system, is another system widely used by small pulpwood producers. The system involves a short-bed truck, a power saw, and one to three men. In a two-man system, one man cuts the timber while the other loads and drives the truck. In terms of unit costs the bobtail system is probably the least expensive. However, it is also the least productive. Further, cutting trees into short bolts in the woods does a disservice to the raw material value in cases where sawlog lengths could otherwise be extracted from the stems.

The truck, generally a single-axle vehicle, moves from stump to stump loading the bolts. Loading is sometimes done by hand. However, many trucks are equipped with big-stick loaders, which are composed of a gin pole or swinging boom and a powered winch. A cable runs from the winch up through the boom and out to the bolts. Bolts are picked up either by simply stringing the loading line around one or more of them or by using tongs that are shackled to the loading line. The short gin pole can also be used to pull bolts from an area surrounding the truck, thus reducing truck movement. Seldom will the bolts be cable skidded over 200 feet.

Stump-to-stump logging is among the crudest of methods. It is extremely weather-dependent and is limited to relatively dry ground and easy slopes. Cut per acre and tree size affect this operation as they would any other.

The one major difference between a bobtail system and a forwarding system is that once the bolts are loaded on the truck they remain there until delivered to a concentration yard. All of the other systems—forwarding, conventional skidding, and grapple skidding—require that the logs, tree-length stems, or bolts be delivered to a landing area for either transfer or decking.

The landing element

The landing element involves dropping the load at a preselected spot, unhooking the turn, decking the logs, and returning to the brush. With forwarders the load may be transferred directly to a waiting set-out trailer or truck.

The position of landings and the use of multiple landings should always be considered in relation to skidding distance. As skidding costs are strictly a function of distance, care should be taken to determine tradeoffs. Moving a landing may be preferable to longer skidding distances involved otherwise.

The approach to the landing should be without curves and should require minimum turning of machine and load. The skidder should always approach the landing at reduced speeds. Landings are dusty during dry periods, limiting visibility, and sometimes other operations are also being performed there. More than one skidder may be working into the landing; machines may be sorting logs; or loading may be going on. In any case, these conditions dictate extra caution on the part of the skidder operator.

When operating in mixed-species stands it may be necessary to sort for species at the landing. Some of the sorting may be performed in the brush, especially if a mechanized cutting system is being used. The remainder is done by the loader.

Landing a turn is accomplished by releasing the winch and driving ahead when the load is positioned correctly. Next the skidder is backed up to throw some slack into the winchline. The entire load can be released or some logs can be left hooked and then dragged to a second location. This is sometimes done to help sort the logs. With grapples, of course, the entire load is released when the grapples are opened.

When chokers are used, either the operator or a chaser unhooks the chokers. If there are a number of skidders hauling to one landing, using a chaser may be unavoidable to maintain efficiency. On the other hand, with only one or two skidders, it might be best for the operator to unhook his own chokers. Each operator has to measure the value of a chaser in terms of his own operation.

If there are several skidders, each skidding to the roadside for decking, the situation is the equivalent of several one-skidder operations, providing each is decking in a different location. One Canadian operation visited by the author used three or four two-man gangs in the brush. One man felled and bucked the timber while the other set chokers, skidded, and unhooked the turns. Each gang was decking in a different location. Nothing would have been gained by providing a chaser under such conditions. Indeed, the chaser would have cost more than he was worth.

Ordinarily forwarding operations do not require chasers, since loads are already unitized and are transferred from one vehicle to another at the landing. The exception is when bolts are hand loaded in the brush. They they must be removed from the forwarder and either decked or stacked on a waiting trailer.

Decking

Once the logs are unhooked and released in the proper position at the landing they have to be piled or decked. Decking of some sort always takes place even when logs are to be loaded out the same day. A log deck is the equivalent of temporary log storage. With either crawlers or skidders, both of which are generally equipped with blades, decking is accomplished by pushing the logs up onto the decks.

The logs are positioned in front of the proper deck or decking area and the choker is unhooked. (Grapples are opened to release the turn when a grapple skidder is used.) The machine is turned for a front approach to the logs. The blade is lowered into position and the logs pushed up the side of the deck. In pushing, the logs have a tendency to roll. The pushing and rolling of the logs upward is accomplished by raising the blade as the machine is moved forward. Each machine, as a rule, does its own decking.

Wheeled skidder waiting to land turn. Skidding on wet ground caused ruts.

When a front-end loader is used for forwarding, the load is picked up in the brush and carried to the landing. Upon arrival the loader need only lift the load to deck level, tip the forks slightly, and release the logs while backing slowly away from the deck. When decking in high decks, the loader approaches the deck with the forks relatively low and tipped forward. The forks are opened as they are raised. This action allows the logs to slip out of the forks, each one being released at little higher than the next. About a third of the load is saved for topping off. A small front-end loader is thus able to deck logs 6 or 7 feet high with little difficulty.

In some skidder operations the logs are decked perpendicular to the road rather than parallel, which is the method used in the descriptions thus far. When logs are decked perpendicular to the road, the skidder drives over the deck and drops the load when it is in position. This practice, while used extensively in some areas, can be extremely hard on machines, especially with high decks.

Landing a turn and making a deck by using a tractor are shown in this series of three numbered photographs: (1) A grapple tractor with an integral arch approaches the log deck to land a turn. (2) The tractor swings around in order to put the logs into a position in front of the deck. Once the logs are in that position, the tractor will drop the turn. (3) The tractor uses the blade to push and roll the logs up on to the deck.

Landing and loading

The landing element interfaces with truck loading. Because of this interrelationship, loading should always be considered when building log decks. One such consideration is log orientation.

When trucks are level loaded and the logs are large or fairly large, orientation is not so important. This is because the large and small ends are alternated to build a level load. In tree-length logging the butts are often oriented to the front of the truck or layered on the truck. Layering means a tier of stems is oriented butts forward and a second tier is loaded butts to the rear. This allows larger loads but can cause excessive breakage if stems are not decked properly.

With front-end loaders there are few problems. However, when a boom loader is used the butts must be oriented toward the road in perpendicular decking and one-way in parallel decking.

If possible, roadside decks should be built on both sides of the road. This, of course, depends on the terrain and the road width. Such a decking procedure allows the loader to load from either side of the truck.

Another important consideration is timing. Wood should be decked ahead of loading if possible. At least one day's run is necessary to avoid delays caused by skidder breakdowns, moving, inclement weather, and so on. When working on dirt, wet conditions may keep trucks away from a landing temporarily. This way, the loader can be moved to an accessible deck and it will continue to produce. Some operators schedule their operations so as to have one million board feet of stems decked before the loading machine is moved in. This sort of system means the logging takes place in stages.

In the first stage, the cutting begins. Enough timber is felled and bucked ahead to support the skidders. Any skidding that can be accomplished along the rights-of-way and skid trails is done during this time. With sufficient wood on the ground the skidder fleet moves in and moves the wood to the landing. After a large-enough surge deck is built, loading can commence. This sort of operation avoids delays and represents the ideal situation. Unfortunately, the system described will not always work, because of problems in machine scheduling, availability of areas for building surge decks, and lack of volume.

Another reason for not building surge decks may be the stain that occurs in some timber, especially in southern pine operations. Within a two-week period, and less under some conditions, blue stain attacks the logs. In dry summer months, leaving timber down a week will produce stain. Logs are therefore loaded out within two or three days after cutting. This calls for hot loading— the logs are loaded nearly as fast as they arrive at the landing.

Avoid delays

Throughout the chapter, regarding all skidding elements, awareness of delays has been emphasized. It is important to minimize delays throughout the cycle. The typical skidding operation seldom achieves more than 75 percent efficiency. Substantial savings can be accomplished by improving the efficiency of the operations. One way to do this is to minimize delay time so that more time is available for skidding.

One Canadian study (McGraw and Hallett 1970, pp. 13ff.) identifies two types of delays—productive and nonproductive. Nonproductive delays involve waiting for logs in the brush (logs not bunched), skid road building and waiting on the landing while another skidder decks a load. These are supporting activities, which in some cases can be improved. A second type of nonproductive delay involves mechanical downtime, personal time, and delays caused by, for instance, visiting on the job. The first type is called a *nonproductive* component, while the latter is called a *delay* component. Table 10.4 shows the distribution of productive delays, nonproductive delays, and delay time.

Table 10.4 *Analysis of Delays*

Component	Percent
Productive	82.0
Nonproductive	5.2
Delay Time	12.8

Source: McGraw and Hallett 1970, p. 13 ff.

Delay time is the largest component and therefore the area where the largest savings can be realized. The total time involved in mechanical, personal, and unnecessary delays averages 2.47 minutes per turn. The unnecessary delays alone account for 1.16 minutes per turn (McGraw and Hallett 1970, pp. 13ff.). Another study showed that delay time averaged 22 percent of total cycle time and averaged 4.5 minutes per delay (McGraw 1966, p. 53).

Delays, both nonproductive and delay time, obviously make up a good portion of the total time available during a working day. This time, if converted into productive time, can be translated into increased production. Here are some suggestions for reducing both nonproductive time and delay time.

Nonproductive time

1. Do not work a skidder in slash cleanup and landing cleanup during productive hours. Perhaps a Saturday shift will be necessary; the feasibility of this solution must be considered on its own merits at each operation. Otherwise such cleanup should be performed during break times if possible, or while other delays shut the operation down.
2. Try to keep landings built ahead so as not to shut the skidders down while a landing is being constructed.
3. If the machines either are waiting long periods of time for logs in the brush or are waiting for turns to be unhooked at the landing there may be a need for additional help. The additional wages might well be a good investment and pay off in additional production.
4. If the bunching element is taking too much time, perhaps an extra set of

chokers will remedy the problem. Chokers should be set, if choker setters are being used, before the skidder returns for the next turn.
5. Skid trails should be built ahead whenever possible. Skid trail cleanup and repair should be done when it will have the least effect on production.
6. Keep the landing clear so the skidders do not queue up. This can be accomplished by building the landings as large as possible; by staggering skidding distances so all the machines do not arrive at the landing at once; or by colddecking logs so that loading and skidding do not interfere with each other.
7. Never sacrifice load size for speed. This is especially true in tractor logging. A tractor will go just as fast with a full load as without, so load it down.

Delay time

1. A preventive maintenance program for all equipment will help reduce downtime that occurs for mechanical reasons. Never run a machine that has serious mechanical problems even though it can still operate.
2. Personal delays are tied to fatigue factors. The longer the hours of work the less efficient the workmen will be. Ten-hour days, six days a week, for instance, might increase production but will lower productivity. Look for another alternative—an extra machine to fill in during a production crisis might be rented or leased, or an extra choker setter might help. Better maintenance might give more machine availability and thus result in more production.
3. Close supervision will help keep visiting time to some acceptable level. If skidding time increases beyond an acceptable level the supervisor should be looking for the reasons. Perhaps an inexperienced operator is the cause. Poor workmanship can also result in increased cycle time. Innocent 'bull sessions' play havoc with productivity if allowed to go on unrestrained. The remedy may be to rearrange the crew, or perhaps a heart-to-heart talk will do the job.
4. Good supervision is essential to any operation. Any company that thinks it can reduce costs by increasing the span of control of their foremen is making a mistake. High production is almost directly related to constant and good supervision. Just the presence of the foreman will make a difference in production.
5. Good planning is essential and will help reduce delay time. Know where the machines are going and which men should be in the crew. Being aware of potential problems helps avoid them in the majority of cases. Otherwise, problem solving becomes nothing more than a knee-jerk reaction.

There are undoubtedly many other solutions to the problems that have been discussed and many problems that have not been touched on. Each operation must be examined on its own merits and an analysis made. There are few operations that cannot be improved, in terms of production, if one looks carefully at what is happening. In the following chapters the reader will recognize situations where suggestions made here will be applicable. In general, it does not make any difference whether the subject is cable yarding, loading, or hauling—the descriptions of nonproductive and delay time will not change, nor will the approaches to reducing these delays.

Section Four

Cable Yarding: Introduction

In our discussion of ground skidding, the systems described were characterized by machines that traveled on the ground to the logs and then transported the logs to some concentration point. With cable systems, the logs, in general, are moved to the machine. The machine, or yarder, is stationary and transfers power to the logs by means of one or more flexible steel cables called *lines*. The process of moving logs to the machine or a landing while the machine is stationary is called *yarding*. The term *yarding* is generally applied to cable systems that, in theory, are capable of vertical lift. This means the logs can be at least partially suspended part of the time during the yarding cycle.

There are two general types of cable yarding systems—*highlead* and *skyline* systems. Several variations of the skyline system will be discussed in this section. Though there is a third type of cable system that is often mentioned in logging texts, this system, called *ground lead*, might be more appropriately called cable skidding rather than yarding. In fact, in the Intermountain region, including Idaho, Montana, and Wyoming, the term used to describe ground lead logging is *long-line skidding*. Ground skidding has many of the characteristics of cable yarding systems but lacks the vertical lift capability.

Common elements

All the cable systems have certain elements in common, although some of the elements may have different names when different systems are used. In the next few paragraphs the common elements will be briefly introduced and described. Any difference in nomenclature will be explained later so as not to confuse the reader.

One element all cable yarding systems have in common is the *yarder*—the power source of the system. Yarders are generally diesel-powered, with engine ratings ranging from about 90 to over 700 horsepower. A yarder will have at least one drum and as many as four, depending on how the machine is used.

The weight of the yarder, fully equipped, might range from a few thousand pounds on a single-drum machine to more than 100,000 pounds on a four-drum machine. The old railroad skidders, which have long since ceased to exist, weighed as much as 200,000 pounds. Modern yarders can be mounted on steel or wooden sleds, a tracked undercarriage, or wheels. Those on wheels are either trailer-mounted and must be towed by truck or tractor, or are self-propelled.

The pulling power of a yarder engine is transmitted to the lines, which are wound on the drums. The *drums* serve to store the lines and to transfer power. Most yarders have three drums, although those used in slackline logging, a type of skyline system, have four drums. The drums on a yarder are named for the lines stored on them. A three-drum yarder has a mainline drum, a haulback drum and a haywire or strawline drum.

The skyline will always be the largest line in the system, with a diameter of up to 1 7/8 inches. With a highlead system the *mainline* is the largest line used and will generally have a diameter between 1 and 1 1/4 inches. The mainline pulls the logs to the landing. Highlead and many other cable systems also utilize a *haulback*, which pulls the mainline out to the logs.

A mobile yarder. Note the side of the drums next to the cab. The drums are for the mainline, haulback, and haywire. The smaller drums near the tower are self-tightening guyline drums. The yarder must be towed for moving.

In highlead logging, where just the mainline and haulback are used, a device called a *butt rigging* is used to connect them. The butt rigging, which is constructed with a series of heavy steel links, swivels, and shackles, not only connects the mainline and haulback, but also carries the chokers.

When a skyline is being used the mainline and haulback are connected by a *carriage*. A carriage is a trolleylike device that rides up and down the skyline on two grooved steel wheels that fit over the skyline. The wheels are called *sheaves*. The carriage, carrying the chokers, and the mainline are pulled up the skyline by the haulback. The mainline pulls the haulback, carriage, and logs back to the landing.

When a skyline is being used in uphill logging (in which the ground slopes downhill from the landing), a haulback is not needed. The carriage moves down the skyline by the force of gravity. The mainline, under these conditions, serves as a *snubbing line* to control the speed and position of the carriage along the

The butt rigging. This one is carrying three chokers. The rigging normally carries two or three chokers but occasionally, depending on the conditions, more than three chokers may be used.

Simple type of shotgun, or gravity feed, carriage. Chokers are fastened directly to the carriage, which rides on skyline.   Photo courtesy Forest Industries

skyline. It also serves as a pulling line. This entire system is called a *gravity feed* or *shotgun* system.

The *spar* is perhaps the most important element in cable yarding. Spars, along with the natural terrain and some other physical factors, allow the vertical lift so important to cable yarding. The first spars to be used were wooden and were called *spar trees*. In the early 1960s yarder manufacturers began to fabricate steel spars of either tubular or box construction. Today, steel spars or *towers* have nearly replaced wooden spars. Steel spars are, of course, much more expensive than wooden ones and are more vulnerable to damage. On the other hand they are mobile and can be placed anywhere a landing can be built. In addition, steel spars can be used over and over again and they can be prepared or *rigged up* for operation much faster than wooden spars.

Spars, whether wooden or steel, are between 90 and 120 feet high and are supported in an upright position by guylines. *Guylines* are cables, like skylines or mainlines, that range between 7/8 and 1 1/4 inches in diameter, depending on the system being used. The specifics of rigging up will be covered in detail later.

The *running lines*, the mainline and haulback, run from the drums on the yarder, up through blocks (large pulleys) located at or near the top of the spar, and out to the brush where the logs are located. If a skyline is being used the spar closest to the yarder is called the *head spar* and the skyline will be suspended from the head spar to another spar some distance away (from 1,200 feet to 5,000 feet) called the *tail spar* or *tail tree.* A skyline may be anchored at one or both ends depending on the skyline system being used.

The elevation of the running lines and of the skyline allows the vertical lift that defines cable yarding. In a highlead system, when the running lines are tightened while pulling in a turn of logs, two forces are applied to the logs. One force is horizontal, which moves the log forward, and the other force is vertical, which lifts the log at least partially off the ground.

Origins of cable yarding

There is nothing new about cable yarding systems. They have been used in the Northwest, in one form or another, since the late 1800s. Today they are used almost exclusively in the same region.

Despite its identification with the Northwest, true cable yarding with vertical lift capability did not originate in the West. The first cable system was invented by Horace Butters in Michigan. In 1886 he patented a skidding and loading machine that consisted of two spar trees, a carriage of sorts, and a steam yarder. With his invention Butters could lift logs out of the woods and transport them to waiting railroad cars (Holbrook 1949, p. 58).

John Dolbeer is generally credited with the invention of cable logging in the Northwest. Dolbeer invented the first steam donkey, a small yarder with a vertical boiler, one cylinder, and a single drum. The cable was pulled out to the logs by a horse, appropriately called the *line horse,* and then skidded the log on the ground back to the landing.

Dolbeer's brand of logging was called *ground lead.* The logs had to be guided around every stump on the setting. It did not take long for the loggers to start cutting the stump higher and hanging a block in the top of the stump. This allowed a slight lift on the skidding line pulling the log in to the landing.

Short stumps led to higher stumps, which led to full trees with blocks and guylines. The single drum became two drums, the second containing the haulback, which replaced the line horse in pulling the mainline out to the logs. The *lead,* the direction in which the logs are pulled, now got off the ground. In fact, it got so far off the ground that it was called *highlead.*

Highlead was a great improvement over ground lead. The next improvement came in the form of the skyline. There is, of course, a story about how some loggers decided that two trees with a cable strung between them would be better than one tree. However it happened, the skyline first invented by Butters made its way to the West Coast. By the early 1900s skyline logging was being used extensively, especially in conjunction with railroad logging.

The original mobile spar skidders were mounted on railroad undercarriages and used in skyline logging. They were first used on the West Coast in 1915—forty-five years before the invention of the modern steel spar mentioned earlier (*The Timberman* 1949, p. 54).

The original skyline operations required a crew of 26 men to operate at full efficiency, were equipped to reach out 2,000 feet, and required nearly seven miles of cable. The high manpower requirements of the system caused it to fall from favor in the 1940s because of World War II. It has regained its popularity only in the past few years (Peterson 1969, p. 7).

The genesis of cable logging is not unlike the genesis of ground skidding discussed briefly in Chapter 9. One difference is that cable logging cannot be operated by animal power–line horse excepted The mode of power changed from steam to gasoline to diesel. Horsepower ratings increased, equipment became lighter and more mobile, and steel cable became stronger. Today, thanks to the ingenuity of a great many loggers, the woods industry has a viable group of cable systems that fill a real need.

In the Pacific Northwest and parts of the Intermountain region, cable systems make it possible to harvest timber on ground that would otherwise be considered inoperable. In the South, some companies have begun to use cable systems in extremely wet and swampy areas where ground skidding is difficult or impossible for all or part of the year. In addition to operability, environmental considerations are also critical. During the past several years great emphasis has been put on aesthetics, prevention of soil damage, and water purity. Cable systems, with proper planning, are uniquely suited to harvest operations that leave the natural environment in good condition.

Comparison of cable yarding and ground skidding

Cable yarding is not the only way to preserve the environment. Nor can it replace ground skidding on an operational basis. In fact, the two types of systems, cable and ground, are not even comparable. The conditions, both external and internal, under which they work are entirely different.

For one thing, slope is not considered a deterrent to cable systems. While ground skidding can be used on slopes up to a maximum of 50 percent, much efficiency is lost and environmental damage is likely as that maximum is approached. Cable systems enjoy freedom from ground conditions, and slope is not a determining factor. Level ground, however, is not considered the best of conditions for cable yarding. It is on steep ground that cable systems have a distinct advantage. For skid road construction is unnecessary, and when the correct cable system is used, the logs can be entirely suspended over the ground, resulting in little if any soil damage.

Cable logging can be used on wet ground where tractors and wheeled skidders would have a difficult time. This is one important reason for the use of cable yarding in the South.

The big yarders used in cable systems have significantly higher horsepower ratings than even the largest ground skidding machines. But the additional power is provided at the expense of mobility. All modern yarding equipment is portable, but not as portable as ground skidding equipment.

The size of the equipment, in a way, dictates the application. Yarding equipment operates from a stationary base. Since it does not have to move to the logs, it can be used in essentially all-weather operations.

Under the conditions for which yarding systems are designed to operate, such systems are faster than ground skidding. In fact, yarding might be faster

Jammer logging can be done with various types of equipment. Here, a cable loader working in Idaho has been modified for the job. Other jammers are simple A-frame or pole-beam arrangements mounted, along with a yarding winch, on a truck or other carrier. Photo courtesy Forest Industries than ground skidding under any conditions. The cycle time for ground skidding ranges between 6 and 63 minutes per turn. The average cycle time on a highlead system is seldom over 5 or 6 minutes per turn.

Of course, the two systems are not directly comparable as far as cycle times are concerned since they do not operate on the same types of ground. Even if they were operationally comparable, there is a very large difference in both capital and operating costs.

The capital investment in a yarder-spar combination, along with the necessary lines and rigging, can run as high as $800,000. A highlead machine will cost around a half million dollars. A slackline machine, fully equipped, is significantly higher.

The high capital cost and the large size of the equipment suggest high fixed costs of moving and setup. Landings must be built, spur roads must be constructed from the main haul road to the landing, and it takes from one to several days to move and rig up if the distance moved is very far.

In addition to fixed costs, the operating costs of a cable logging side are substantial. While a skidding side utilizing three skidders and a loading machine can get by with as few as 5 men, a cable side always requires between 5 and 10 men including the loading function. In addition, the cost of supplemental equipment such as lines, blocks, and other miscellaneous equipment not required in ground skidding is substantial.

There are obviously differences between the two types of log-moving systems. These differences do not suggest, however, that one system is better than the other. Each has a set of applications to which it is uniquely suited. The problem, if any exists, is in attempting to apply one system, whether it be cable or ground skidding, to all situations and conditions. Many loggers have been heard to say, "If it can be logged with a tractor, it can be logged better with highlead." Of course, the proponents of tractor skidding have been heard to maintain the opposite point of view. Depending on the specific conditions, one of the systems may be more effective than the other. But if, for instance, a logging operator has only one system available to him, and that is highlead, then highlead is the best system for the job—because it's the only system the operator has.

Types of cable yarding systems

If all the different makes and models of equipment available—yarders, portable spars, carriages, grapples, and support equipment—were combined with an infinity of external variables, the number of identifiable yarding systems would be nearly endless. Since the objective of this book is not simply to identify systems but to describe how they work and why, the endless combinations will be avoided in favor of describing the two primary cable systems and a few variations of those two systems.

The two primary systems to be discussed are highlead and skylines. Highlead will be discussed in general terms along with some of the types of equipment used in that system. Of the skyline systems, three basic variations will be described—the tight skyline, the live skyline, and the running skyline.

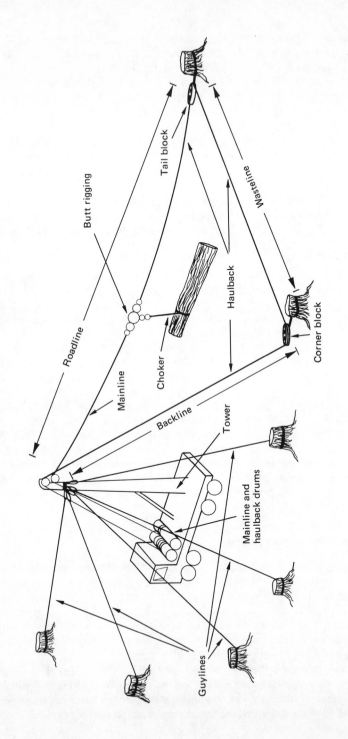

Figure 11.1. *Illustration of highlead layout.*

Highlead

Figure 11.1 illustrates the layout of a highlead system. The layout is triangular, with the mainline and haulback forming the two sides of the triangle. The base of the triangle is formed by the haulback running between the corner block and the tail block and the apex is the tower or spar tree. In operation, the yarder operator (called the *yarder engineer*, or simply, engineer) releases the brake on the mainline drum and applies power on the haulback drum. This moves the mainline and chokers back to the logs. The turn is set by the choker setters. They attach the chokers to the logs in the same manner as described in ground skidding. When the turn is set and the choker setters are in the clear, they signal the engineer, who pulls the turn in. The inhaul is basically the reverse of the outhaul. This time the haulback drum is slacked and power is applied to the mainline drum.

The area over which the mainline runs and from which the logs are removed is called the *roadline*. The other side of the triangle, over which the haulback runs, is called the *backline*. The line between the two corner blocks is called the *wasteline*. As one roadline is completed, with all the merchantable logs being removed, the lines are moved in, as shown in Figure 11.2. Since the spar remains stationary and the roadlines radiate from it in roughly a circular direction, the pattern formed is fan-shaped.

Highlead yarding is generally associated with yarders and wooden spar trees or portable spar-yarder combinations. However, the industry has adapted other types of equipment to highlead or near highlead operations. The mobile yarder-loader is one such piece of equipment.

A *mobile yarder-loader* is basically a wheel-mounted loading machine that has been adapted to yarding as well. The machine has two drums—one for the mainline which also doubles as the loading line, and one for the haulback. In

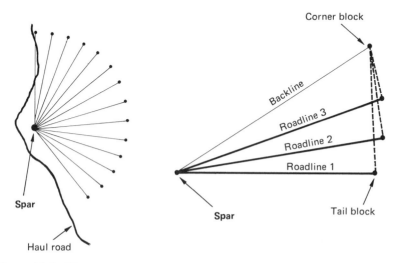

Figure 11.2. *Illustration of fan-shaped yarding pattern.*

Cable yarding: introduction   203

place of a spar, the loading boom is used. However, the loading boom has to be beefed up so that it can stand the stress of yarding and is supported during the yarding cycle by two guylines.

The yarder-loader-type machines are used for highleading small logs in clear-cuts, for thinning operations, and for salvaging down timber. In both the latter cases yarding takes place through standing timber.

The line layout is the same as that of conventional high lead. However, the *yarding distance* (the distance the logs must be pulled along the road line to transport them to the landing) is slightly shorter than conventional high lead because the line capacity of the drums is limited. Also, since loading booms are much shorter than towers, the lift that can be generated is limited. In fact, the

This yarding crane, a Washington Skylok, is being used with a running skyline for grapple logging. It can also be used with chokers. The butt hook hanging behind the grapples is for a choker.

A snorkel is a cable loader with an extended boom. Used for yarding and loading, it is an inexpensive way to log a strip along a road.

Photo courtesy Forest Industries

machines can generate little or no lift beyond about 100 feet. With good deflection and the right operating conditions, these machines are very efficient.

About 60 to 65 percent of the total time available in the day is spent yarding. Yarding continues until an area has been completed and a move is necessary, or until a truck arrives. When a truck does arrive the butt rigging is removed and replaced with loading tongs. The machine loads its own logs. The reverse change to yarding involves removing the tongs and hooking the butt rigging back up. The same crew is used for yarding and loading.

The advantages of yarding-loading equipment are lower capital costs, high speed, and high mobility. Since the machines are self-propelled they can be moved relatively short distances under their own power. Long moves require a lowboy trailer. The time for moving and rigging up for yarding is a little over an hour, depending on the distance involved.

Yarding cranes represent another type of equipment used for highlead logging. These machines are more specialized than the yarder-loaders in that they perform only one function—yarding. The crane's mast is longer than that of the yarder-loader and utilizes two guylines positioned opposite the direction of pull. These machines, with their taller booms, can yard efficiently at distances up to 700 or 800 feet. Yarding cranes are also used for grapple yarding, a type of skyline system that will be discussed in Chapter 12.

The *snorkel* is essentially a cable loader with a fabricated boom extension, often simply a pole with a block attached to the end (see photo). This type of machine can provide the cheapest means of yarding within 50 to 150 feet of a road.

No discussion of highlead logging would be complete without some mention of *jammers*. For years jammer logging has been a predominant cable system in the Intermountain region. The jammer features a double drum mounted on a truck, a short mast, and one or two poles forward of the mast, which are the equivalent of the spar. The haulback runs through a block hung in the mast, while the mainline runs through a block hung in the spreader bar between the poles.

In most cases the jammer line layout is the same as conventional high lead, but there are exceptions. Instead of a butt rigging a small four-inch block called a *monkey block* is used. The yoke of the block is attached to the haulback, and the mainline runs through the block. Instead of chokers, tongs are used, and the hooker drags the mainline (which is about 9/16 inch in diameter) and tongs to the logs 25 or 30 feet on either side of the roadline. Maximum yarding distance is about 400 feet for jammers, which means a lot of road building and moving is required. Landings are situated about every 50 or 60 feet along the haul road and the moving from one landing to another takes no more than 15 or 20 minutes.

The monkey block replaces the butt rigging in some areas. The mainline is threaded through the block and the chokers keep it from pulling back through. The haulback is shackled to the block.

Jammers offer a serious disadvantage in that road requirements are very high. The type of high-density roading necessary for jammer logging is not acceptable in terms of environmental impact.

It is obvious that with all the equipment available many different systems could be developed. However, they all utilize the same basic principles of highlead logging.

Highlead logging is the predominant system used in the Northwest. When the use of skylines declined in the 1940s it was highlead that replaced them—mainly because of decreased manning requirements. However, highlead is not a particularly efficient logging system. Yarding distance is limited by line length. The generally accepted maximum yarding distance is between 1,000 and 1,200 feet. The most efficient yarding distance is about 700 feet. The limitation of yarding distance leads to increased road requirements, which can only add to the total cost of harvesting and remove more acres from the production of trees.

While highlead has the capability of vertical lift it is generally limited. The system is capable of partially lifting logs from the ground in order to avoid hang-ups such as high stumps. The lift is accomplished by holding the haulback tight while pulling with the mainline—a condition called a *running tightline*. With a conventional yarder it is difficult to maintain a running tightline for very long without losing braking efficiency. As a result, lift is used only when required to avoid some obstruction. The rest of the time the logs are simply dragged full-length along the ground causing unwanted soil damage.

Soil damage resulting in serious erosion problems will under certain conditions be intensified by the yarding pattern shown in Figure 11.2A. The roadlines have a tendency to converge as they approach the landing. In any one quadrant this convergence results in considerable soil damage.

The severity of soil damage and the requirements for expensive roading caused by highlead systems has led in recent years to the resurgence of interest in skyline logging. With modern equipment, skyline logging is a viable alternative to highlead in some places. And, in areas where yarding distances are very long and the ground extremely rough and broken, skyline logging is the only system that can be used efficiently.

Skyline yarding

A skyline, for which the system is named, is a wire rope suspended between two or more points. Those points may be trees or trees and stumps. The yarding systems involve a carriage that moves logs laterally to and longitudinally along the suspended cable. During the lateral yarding stage the logs are dragged along the ground, at least on one end. As the turn gets closer and closer to the skyline the amount of dragging can be decreased. In the longitudinal yarding stage the logs can be either fully suspended or partially dragged, depending on the system and ground conditions.

The types of equipment used in skyline systems are, for the most part, similar to those used in highlead logging. Aside from the skyline itself, which is unique to skyline systems, there is the mainline and haulback. As suggested

Yarding with a radio-controlled carriage called a skyline crane. The turn is entirely suspended over the ground.

earlier in this chapter, the haulback is not used with a gravity feed system. The yarder is similar to a highlead yarder. The difference is that the horsepower rating of a yarder used on a skyline may be higher, although the size of the machine depends on the system. The carriage also is unique to skyline systems. All skyline systems use a carriage of some type. Chokers, of course, are common to both systems and are used the same way in both systems. The only difference is that skyline chokers, under certain conditions, may be longer on skyline systems than on highlead systems.

The *span* is defined as the distance between two supports that suspend the skyline. If only two supports are used there is only a single span and the setup is called a *single-span skyline*. If there is a head tree and tail tree plus intermediate supports the setup is called a *multiple-span skyline* (Figure 11.3). Of the two types of skylines the single-span is most commonly used.

In addition to the type of span, skylines are also classified by movement. If a single-span skyline is securely anchored at both ends so the line cannot move it is called a *standing or tight skyline*. If the skyline is *live* (that is, if it can be

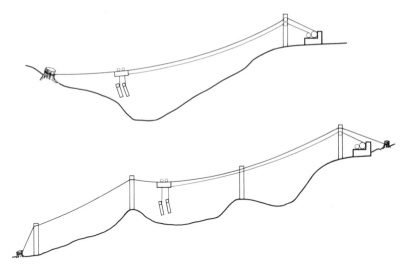

Figure 11.3. *Profiles of single-span (top) and multiple-span (bottom) skylines.*

raised or lowered at will), it is called a *slackline*. On a slackline operation there is generally a fourth drum on the yarder, which controls the movement of the line up or down. An exception is when the mainline is used as the slackline in a gravity feed system. Then the fourth drum is not required. A third classification of skyline is the *running skyline*. The running skyline does not utilize a skyline per se; instead, the carriage rides the haulback and both the haulback and mainline support the load.

There is one condition that is an absolute necessity in any skyline system and that is deflection. *Deflection*, or sag, in a skyline is the vertical distance between the chord and the skyline. The *chord* is an imaginary straight line between the tops of the supports at either end of the span (see Figure 11.4).

A skyline with no deflection has no load-carrying capability. In fact, a no-deflection condition is impossible with or without a load. The capacity of a skyline to support a load increases with an increase in deflection. A decrease in deflection will increase line tension and reduce its load-carrying capacity. The

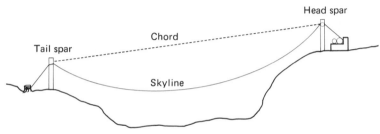

Figure 11.4. *Arrangement of elements in skyline system.*

A standing skyline. The maze of lines is guylines, skyline, and running lines.

minimum deflection acceptable in a skyline is said to be 5 percent, but that is a fairly high-tensioned line, and 10 to 15 percent deflection would be much more acceptable.

The amount of deflection in any line is limited by the length of the span, line size, slope of ground, and spar and support heights. If insufficient deflection is available the span must be shorted or the tailholds must be placed at a higher position. The method for calculating tension and deflection in skylines will be discussed in detail in Chapter 12.

Two types of ground profile are applicable to single-span systems. These are *concave* and *constant*. With the concave profile as pictured in Figure 11.5A, the landing is at the top of one ridge while the tailhold is at least partway up the opposite ridge. This profile generally allows for plenty of deflection. In addition, if there are stumps of sufficient size on the opposite slope a tail spar tree need not be rigged.

The constant slope profile (Figure 11.5C) is also applicable to single-span skylines although it is less desirable than the concave profile. With the constant profile a tail tree is required to develop deflection. A third profile is described as *convex*. This is a concave profile with a reverse slope, as shown in Figure 11.5B. When a convex profile is present the logging operator must either use highlead on the reverse slope and swing the logs to the primary landing or use a multiple-span skyline.

There are several specific skyline systems that apply to the descriptions offered thus far. They utilize different types of carriages and some different terminology to describe the rigging but they are nonetheless skyline logging systems. Furthermore, while the systems can be used under a variety of conditions they each have some specific applications to which they are best suited. In this text only three basic skyline systems will be described: the North Bend, the slackline, and the running skyline. In addition, various types of carriages used in the industry are introduced.

The North Bend system

The North Bend is the most commonly used tight skyline system. It utilizes a single-span skyline anchored at both ends. The skyline passes through a tree shoe hung in the head spar, is suspended over the ground and is anchored at the back end after passing through a second tree shoe hung in a tail spar. The skyline is stored and transported on a large single-drum machine that also can be used to tighten the skyline. The skyline carrier naturally must be anchored.

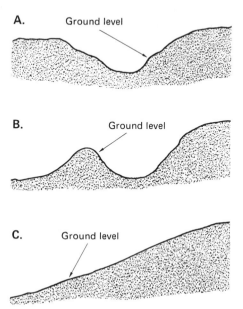

Figure 11.5. *Ground profiles: concave (A), convex (B), and constant (C).*

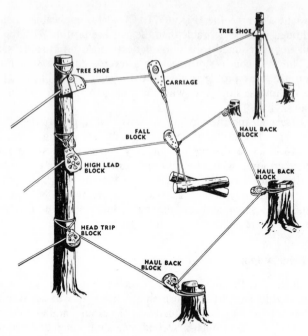

Figure 11.6. *The North Bend skyline system.*

The mainline runs up from the yarder, through a bull block, then through a fall block, and is finally shackled to the carriage that rides the skyline. The haulback runs through three or more brush blocks and is shackled to the fall block. The chokers are also attached to the fall block. The North Bend system is illustrated in Figure 11.6.

The position of the tail block determines the roadline and is positioned to pull the fall block and chokers down to a turn of logs. While the haulback is pulling, the mainline is being slacked allowing the fall block to be pulled to the side or at an angle away from the skyline.

When the turn starts into the landing it will drag along the ground until it hits an obstacle or reaches the skyline. The direction of yarding relative to the skyline is diagonal rather than perpendicular. If the turn encounters an obstacle the mainline will continue to pull, creating an upward pull that will lift the end of the log over the obstacle. Additional lift can be developed by braking the haulback while going ahead on the mainline. Once under the skyline, with tightline maintained, the turn can be completely suspended.

The North Bend system operates best in uphill logging and can also be used on moderate downslopes. If a tight skyline system is required over steep slopes (downhill) the system can be modified to generate more lift. The Modified North Bend system, or South Bend system as it is sometimes called, differs only in the way the mainline is rigged. The mainline, in the Modified system, runs through the fall block up through a block in or attached to the carriage

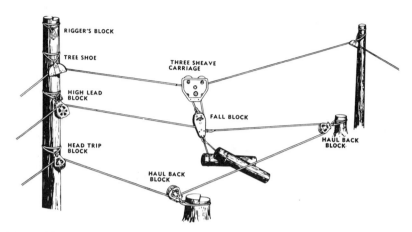

Figure 11.7. *The modified North Bend skyline system (sometimes called the South Bend system).*

and back down to be shackled to the fall block. The Modified North Bend system is illustrated in Figure 11.7.

The North Bend system is generally used over long yarding distances (longer than can be negotiated by highead) or for swinging timber away from a highead operation. Swinging might be required in a convex profile situation such as that shown in Figure 11.5B. The system offers better control of the turn than does the highead system, but it is more expensive to operate, mainly because of the large amount of time that is necessary for rigging up and for making cable changes.*

The slackline system

The slackline system (Figure 11.8) is used fairly often in the Pacific Northwest and is especially adaptable to downhill yarding and yarding terrain with a concave profile. However, it is also effective under almost any condition where there is sufficient deflection. It is definitely a better yarding system than any of the tight skyline systems.

The slackline requires a fourth drum on the yarder since it is raised and lowered in operation. The skyline runs from the drum up through a block that hangs just above the mainline block in the head tree. The skyline can then be strung out and either run through a shoe or block hung in a tail tree or anchored directly to a stump at the back end of the setting. Stump anchors will be discussed in a subsequent chapter. The mainline is shackled to the leading end of a carriage which rides the skyline. The haulback is attached to the trailing end of the carriage. The chokers are attached to butt hooks shackled to the bottom of the carriage.

---

* When the skyline road is changed it is called a *cable change*. When a roadline is changed in a highead system it is called a *road change*.

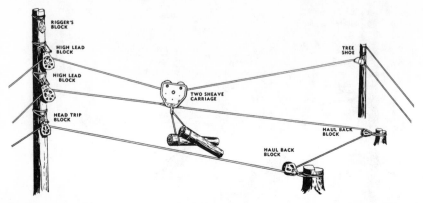

Figure 11.8. *Illustration of the slackline system.*

In operation, beginning at the landing, the skyline is raised to get the carriage in the air and to lift the chokers off the ground. Once the chokers are clear of the ground the haulback is engaged to pull the carriage back to the brush. Once there, the slackline is lowered and the chokers are set. When the turn is set the slackline is tightened and the mainline pulls the turn in. As the turn is being yarded the skyline can be raised or lowered as necessary to adjust to the ground profile.

One type of slackline system is the shotgun or gravity feed, which requires no haulback. When the shotgun system is used a highlead yarder can also be used, since the mainline is sufficient for use as a skyline. The haulback becomes the skidding line—the mainline—and no haulback function is required.

If a wooden spar tree is used there is no problem in converting a highlead system to a slackline system with gravity feed, since the bull block either is or can be placed over the haulback block. With a steel tower the haulback block is at the top of the tower, which results in the haulback and mainline sawing each other. If such a tower is to be used for a gravity system the blocks may need to be modified so that the mainline block is below the haulback block. One company manufactures a special carriage with a gooseneck that allows the haulback to run above the mainline. As another alternative a block can be hung on a guyline to put the mainline below the haulback.

Shotgun systems are used for uphill yarding only, since the carriage is moved by gravity out to the yarding area. Lateral skidding is accomplished by chokers that are up to 40 feet long. Cable changes for a slack skyline are discussed in a subsequent chapter.

The running skyline

The running skyline is another general type of logging system that needs to be discussed. However, this system is covered fully in Chapter 14, along with grapple logging. Since grapple logging is fairly common in the Northwest the

A set of grapples used in a running skyline with a yarding crane. A line opens and closes the grapples. Note the small block hung over the haulback. Both the mainline and haulback are supporting the load.

treatment of the subject will be a little more detailed than the descriptions offered thus far.

The carriage

Many carriage types can be used for running skylines and gravity systems. They range from simple mechanical devices to radio-controlled units containing their own drums and diesel engines. All are relatively heavy, ruggedly constructed, and simple to use. Either the chokers hang from the carriage itself or they are attached to a skidding line, which must be pulled by hand in most cases. A carriage of the non-motorized variety may weigh from 600 to 3,800 pounds, depending on its complexity.

Carriages that have a lateral yarding capability are called *skyline cranes* and are generally used with a standing skyline. The skidding line, which is used to

A skyline machine running with a radio-controlled carriage.

reach out on each side of the skyline for the log, is either pulled by hand or by some type of slack-pulling device. Depending on the type of carriage and on operating conditions, logs may be yarded laterally up to 250 feet to the skyline. Ordinarily, lateral yarding distances are no more than 75 to 100 feet on either side of the skyline. In some cases slack for the skidding line is pulled by the yarder from the landing by a slack-pulling line. In other cases the carriage is equipped with radio controls that operate the skidding line or tong line, whichever is used.

The Danebo M-110 *Spitter*, for example, is a two-drum slack pulling carriage with radio-controlled skyline clamp and skidding drum brake. The carriage is clamped to the skyline while the dropline—actually the haulback line spooled around a drum on the carriage—is paid out. When the chokers have been set, the yarder engineer reels in the haulback, raising the turn off the ground. The carriage drum is locked and the skyline clamp released by radio control, and the haulback pulls the carriage to the landing. A similar procedure is accomplished with other carriage designs, which use a mechanical ratcheting device, rather than radio control, to lock and release the haulback/dropline.

Typical of a larger and more complex type of radio-controlled carriage is the RCC-150, manufactured by the Skagit Corporation. It is equipped with a skidding drum in the body of the carriage, driven by a 150-horsepower diesel engine. The RCC-150 is rated at 40,000 pounds capacity, bare drum, at an inhaul speed of 24 feet per minute. Weight is 7,200 pounds with fuel, cable, and optional skyline clamp. The drum can carry 400 feet of 7/8-inch cable or 550 feet of 3/4-inch cable. The carriage unit is suspended from 24-inch sheaves grooved for a 1½-inch skyline cable.

Two previous Skagit models, the RCC-15 *Torpedo* (rated at 30,000 pounds) and the RCC-20 *Skycar* (rated at 40,000 pounds), were also equipped with skidding drums, and examples are still in service. Typically, the skidding drum contains a tong line that does lateral yarding. The mainline or snubbing line is attached to the carriage. Once the turn is yarded to the skyline the snubbing line pulls the carriage and the turn to the landing. The carriages are clamped to the skyline during slack pulling and lateral yarding.

Both carriages, the Skycar and the Torpedo, are equipped with 95-horsepower diesel engines to operate the winch. They also carry their own radio equipment. The Skycar weighs over 8,000 pounds and requires a 1 3/4 skyline. It is used with a multispan skyline and is designed to carry a turn downhill. Yarding distances may be as long as 5,000 feet.

The carriage is pulled up the skyline by a snubbing line and is stopped when the rigging slinger signals the snubbing operator to stop the carriage and set the brakes. The tong line is then lowered, also by radio signal, and carried to the turn. When the turn is set the rigging slinger signals for the carriage winch to pull in the turn and hold it suspended. Finally, the snubber operator is signaled to release the carriage brakes, and the carriage moves, by gravity, down the skyline toward the landing, sometimes at speeds of up to 70 miles per hour. The chaser also uses radio signals to lower the turn once it reaches the landing.

The Torpedo is somewhat lighter than the Skycar and is used to log uphill, generally with a single-span skyline. With the Torpedo the mainline is used as a skyline and haulback is used as the snubbing line. The carriage employs gravity to carry it out to the yarding area and is pulled uphill with a turn.

The two smaller Skagit radio-controlled carriages, the *Bullet* and the *Tracer,* are no longer in production, but units are still in operation. They are equipped with 24-horsepower butane engines and an air compressor and air storage tank to operate the slack pulling tramming sheave and cable clamp brake. The skidding line is threaded around the tramming sheave and out through the fairlead welded to the rear of the carriage. In this case the skidding line is the haulback, and the mainline is used as the skyline.

The carriages use a gravity feed to travel to the yarding area and are apt to travel fairly fast. If the brakes should fail, as they sometimes do, the carriage is lost. The author knows of at least two instances when the carriage was lost either because of brake failure or faulty radio signal. In one case the carriage was demolished. In the second case it was repaired but a large amount of downtime was suffered while the repairs took place. The operation of the Bullet and the Tracer is similar to that of the Torpedo except there is no tong line. The rigging slinger signals the yarder engineer and the load is pulled in by the yarder.

Electronically controlled carriages are also used with the running skyline in some instances. Experience indicates that when they are working properly, production can be very high. However, the radio controls take a substantial beating as the carriage moves up and down the skyline. And they tend to malfunction at the most inopportune times. The logging operator using radio-controlled carriages, either the type just described or the type used with a running skyline in grapple logging, should have access to experienced mainte-nance personnel.

Determining the right system

As the reader can see, there are many cable systems, each of which has its place. For example, certain blocks of land are difficult or impossible to yard using highlead. The conditions on these blocks exceed the physical capabilities of the system in terms of the line length and lift required to move the logs. Specifically, the limitation may be imposed by a need to protect the soil and water resources. Not only may the physical capabilities of the system be exceeded, but the economics of the system may be exceeded as well. In other words, road-building costs may be excessive or timber volumes may be located so as to make it impossible to yard the timber at a profit.

Here is an example of where the systems approach to logging can produce economic benefits, or indeed can make the difference between profit and loss on a job. It is essential to size up the problem and choose the right system for the right place at the right time.

When highlead logging is not applicable there are several alternatives:

1. More roads on steep ground.
2. Midslope roads located between the top and bottom of a ridge—roads that contribute to soil damage and erosion and are expensive to build.
3. An entirely revised transportation system.
4. Swinging the timber.
5. Some type of skyline logging.

Skyline logging is a proven logging method that can be applied where ground skidding or highlead is not a feasible alternative. Skyline logging, in general, requires fewer miles of road than highlead. On steep slopes using staggered settings, clearcutting, and conventional highlead, as much as 11 or 12 percent of the land is removed from timber production because of roads. For one thing, each mile of access road constructed takes about 5.5 acres of timberland out of production, and highlead requires significantly more roading than skyline systems.

The Pacific Northwest Forest and Range Experiment Station estimates that road requirements in the national forests alone can be reduced by 11,000 miles if skylines are used where possible. The savings would be in the hundreds of millions of dollars in road construction costs and about 60,000 acres of land that would otherwise be lost to timber production (Jorgensen 1969, p. 7).

The reductions in road-building costs are made possible by the long yarding distances allowed by skyline systems. Yarding distances range from 1,200 to 5,000 feet or more, depending on the system used. Skylines are also a feasible alternative in timbered areas where the soil is wet and extremely erodible, because the logs can be suspended over the ground, thus avoiding soil damage.

Skylines are also applicable where a residual stand is to be left. In mixed stands, the timber to be harvested can be lifted clear of the ground, thus accomplishing the harvest without inflicting excessive damage on the residual stand. Highlead, of course, can be and has been used in selective harvests. However, the system simply does not have the needed control, and damage is almost a certainty.

Under certain conditions, therefore, it appears that skyline logging is much preferable to highlead. However, lest the reader think skyline logging is the only way, we offer the following quote from Pete Rennie, a well-known figure in the industry: "The best advice a logger ever received about skyline logging was DON'T until absolutely forced . . . . Even if the economics looks excellent, the profit is gone when the tailhold turns upside down and wraps itself in the skyline" (Rennie 1969, p. 76).

Furthermore, skylining is more expensive than highlead. The equipment is generally bigger, moving costs greater, and crews larger. Skylines have many advantages to offer, but like other systems, the ends must justify the means. In this case, the end desired is a fair profit.

12.

Planning for Cable Systems

Nearly anyone newly introduced to cable logging is at first a little bewildered. From a conceptual point of view, cable system principles are not particularly difficult to understand. Anyone who has ever seen a pulley clothesline, ridden a ski lift, or watched an aerial tramway has seen the same principles at work. The difficulty lies in associating common experiences with a working system. Unless an observer is really thinking about what he is experiencing he does not relate his visual experience to such important questions as how large the lines are, how the load is supported, and how far can the load be transported. For that matter, how many skiers have given any thought to how the cable used in a ski lift is suspended? These very practical questions, and many more, will be answered in this chapter.

Operating variables

Like wheeled skidders, tractors, or any other operating system, cable systems are constrained by certain external variables over which the system, in the short term, has little control. In Chapter 9 the variables that affect ground skidding were discussed in detail. Many of the same variables will be discussed in this chapter. This is not a duplication of effort. While the general effects of these variables may well be the same in the various systems, there is a difference in their magnitude. Since capital costs and, in most cases, the direct operating costs of cable systems are significantly higher than those of ground skidding, the economic impact caused by operating variables is correspondingly larger. Also, since markedly different systems are being considered, the physical effects of the variables also differ in magnitude and, in some cases, in their very nature. Just as in skidding, piece size and distance are the two major variables here.

As was mentioned in Chapter 9, variables cannot be considered individually. More often than not two or more variables join in their effect on the system.

For instance, volume per log and yarding distance combine to constrain productivity. Yarding distance is constrained by such items as spar tree height, yarder size, slope, and volume per acre. In this chapter each variable will be described and discussed individually. Certain of the variables will be discussed jointly so the reader may understand the combined effect.

Total volume, volume per acre (stand density), and volume per piece are all related. Each of the three has an individual effect on the total system, which when added to the other variables means a plus or minus for the setting. The effects are felt in unit costs—both fixed and variable costs—and in productivity.

Total volume on a setting

The impact, in this case, is felt mainly in fixed costs per unit of output. The costs of spur road construction, landing construction, and moving and rigging are substantially the same whether an operator is dealing with a large total volume or a relatively small total volume. The following table shows the unit cost for moving and rigging when the total expense is $3,400, given a change in total volume.

Volume	Total cost	Unit cost
1.50 MMBF	$3,400	$2.27/MBF
1.00 MMBF	3,400	3.40/MBF
0.75 MMBF	3,400	4.53/MBF
0.50 MMBF	$3,400	$6.80/MBF

When total volume drops from 1.5 million board feet (1.5 MMBF) to 750 thousand board feet (.75 MMBF) the unit cost doubles. A decrease to .5 MMBF increases unit costs by 200 percent, from $2.27 to $6.80. There is no problem with unit costs unless the effect of total volume has not been considered. If the volume on a setting has been underestimated, the results can be disastrous, either reducing the profit margin or even resulting in a loss for the operation.

Volume per acre

In cable yarding volume per acre, or stand density, is critical for the same reason it is critical in ground skidding. There is an inverse relationship between unit costs and stand density. That is, when the density is relatively high, the unit costs are lower. When volume per acre is low, more cost must be distributed over smaller volumes, so unit costs are higher. There are more road changes and cable changes, therefore the average productivity will be lower.

Table 12.1 shows the effect of several variables on productivity, among them volume per acre. All other things being equal a decrease of 20 MBF per acre, from 85 MBF to 65 MBF, will result in a decrease in man-day productivity from 9 MBF per man to 8 MBF per man. The cost per day to operate a highlead logging side will be the same whatever the volume per acre. But 1 MBF per day can certainly make a difference in total cost of the job.

Table 12.1 *Productivity Differences in Cable Logging Systems Man/day Production*

Variables	Highlead, ≤400 hp (MBF)	Highlead, >400 hp (MBF)	Slack skyline, 500 hp (MBF)	Shotgun, 500 hp (MBF)
Deflection				
Good	6.0	6.6	6.0	7.2
Poor	4.0	4.5	4.5	5.0
Vol/acre (MBF)				
25	4.0	4.5	4.5	5.0
45	6.0*	6.6*	6.0*	7.2*
65	8.0	8.3	8.4	8.6
85	9.0	9.2	9.5	9.8
105	9.6	9.7	10.0	10.5
Vol/log (BF)				
30-140	4.0	4.5	4.5	5.0
141-415	6.0*	6.6*	6.0*	7.2*
416-550	6.6	7.5	7.6	7.8
551-700	7.8	9.0	9.3	9.4
Yarding distance (ft)				
800	6.0*	6.6*	6.0*	7.2*
1200	3.0	4.0	7.2	8.6
1600	–	2.6	5.8	9.0
2000	–	–	4.5	5.8

Notes: *Good Deflection* means that turn clears ground over 80 percent of the ground. *Poor Deflection* means that turn clears ground over 20 percent of the ground.

To determine total production for each variable multiply highlead factor times an eight-man crew; slack skyline and shotgun use appropriate factor times a ten-man crew.

*Indicates base case used with good deflection to estimate man/day production. For all other variables assume base case and good deflection.

For example, a logging operator has contracted to harvest approximately 2 MMBF of timber. The stand density on the setting is 85 MBF per acre. A highlead system will be used with a seven-man crew, a yarder-spar, and a loader. Cost per day of operations is $1,750. At an estimated productivity rate of 72 MBF per day, roughly 28 days will be required to finish the job, at a total cost of $48,650. If the volume per acre were 65 MBF rather than 85 MBF, the daily production would be 64 MBF per day rather than 72 MBF per day, and a total of 31 1/3 days would be required to complete the job. The additional time would cost the operator $6,125 more. Unless the operator has calculated his costs accurately and negotiated a favorable price for his work, the additional $6,125 will directly and negatively affect his profit margin.

Table 12.2 illustrates the relationship between productivity, total cost, and unit cost for the conditions described in this example.

Average volume per log or piece

If an operator has the correct piece of equipment, this variable is not a problem. A large firm that has several logging sides would be in a position to match each size of yarder with the conditions that would optimize productivity. If the firm has only one or two sets of equipment and they are being fully utilized, then there is a problem. Avoiding the effects of a problem by acquiring equipment equal to the task is generally a long-term process. In the short term the operator must make do with what is available.

In general, assuming the short term, costs will be higher if tree size is small, because more pieces must be handled per unit of output. If log volume is high and the equipment is not equal to the task, then the cost will also be high. The question of equipment size in the situation now being discussed is one reason why many logging operators, especially small- to medium-sized contractors with a limited equipment inventory, insist on buying equipment that is versatile and large enough (in terms of horsepower), to handle any set of conditions.

The most extreme case occurs in so-called YUM (yarding unmerchantable material) yarding, or "clean logging," where everything down to a small-end diameter of 6 or 7 inches is removed. A change from 25-cubic-foot pieces to 20-cubic-foot pieces increases the yarding cost 27 percent. Moving from the 10-inch class to the 5-inch class increases the cost 80 percent. As diameters fall below 5 inches, the cost increase is very steep, reaching over $1,000 per cunit for a 4-inch by 8-foot piece.

Log sizes affect productivity and, in turn, unit costs. Referring again to Table 12.1, if log sizes are small, between 30 and 140 board feet, man-day productivity will range approximately from 4 MBF to 32 MBF per day. An increase in log size significantly increases productivity.

Table 12.2 *For Yarding and Loading Effect of Volume Per Acre on Cost and Productivity*

Vol/acre (MBF)	Average daily production (MBF)	Days required	Total cost ($)	Unit cost ($/MBF)
25	32	62.5	109,375	54.69
45	48	41.7	72,975	36.49
65	64	31.3	54,775	27.39
85	72	27.8	48,650	24.33
105	77	26.0	45,500	22.75

Note: Most highlead cable systems load logs as they are yarded. Therefore, productivity and costs are combined.

Planning for cable systems   223

In small logs, it is advisable to fly a sufficient number of chokers to maximize turn size. For example, a skyline system reaching out 800 feet, and requiring a hook and chase time of 3 minutes, would produce 12 turns an hour. Assume its maximum capacity is 2.5 cunits per turn. If the system is under-utilized at 2 cunits per turn, production would be 24 cunits an hour. Boosting the load to 2.5 cunits per turn, with the same hook and chase time, would raise production to 30 cunits an hour, an increase of five loads per day.

Another example involves an operator, typical of those working in small logs, who adds a third choker to the pair normally used with larger logs. An increase in chokers will result in an increase of turn size. Production would be between 30 and 35 MBF per day. If only two chokers were used by the operator or the logs were on the lower end of the range, production might drop as low as 24 MBF per 8-hour day. Using a larger machine in this case would, on the average, increase production only slightly. However, because of small turn size, higher line speeds and more power do not buy a great deal of production in small logs.

Extremely large logs will not increase production in direct proportion to the increase in board foot volume per turn. Large, heavy logs require a little more choker-setting time, and, all other things being equal, an increase in inhaul time. Thus the number of turns per hour might be more like 9 or 10 rather than 12.

For instance, an increase of turn volume from about 950 to 1,250 board feet—approximately a 30 percent increase—results in an 18 percent productivity increase. As the volume difference per turn increases, the productivity will increase at a diminishing rate until the capacity of the yarder is exceeded. The major cause of the diminishing rate is an increase in turn time. One study showed that with a 600-foot yarding distance and 20 percent slopes, inhaul time increased by 43 percent when volume per turn increased from 500 to 3,000 board feet (Tennus, Ruth, and Berntsen 1955, p. 14). Of course, the study involved a triple-drum yarder in the low-horsepower range (less than 350 HP), but the information is still relevant.

When operating in large logs the machine should be loaded to capacity on each turn. In fact, the tendency might be to overload. Whether operating a cable system or any other system it does not pay to overload. In general, it is better to yard a lighter turn at maximum speed and control than it is to wood the machine down (overload it) and risk mechanical failure as well as decreased inhaul time.

Slope

Slope is a variable that becomes significant when combined with the effects of volume per turn and can result in an increase in inhaul time when both slope and turn size increase. However, most operators agree that with a machine that is large enough slope is not a critical factor as long as deflection is available.

While slope itself may not significantly affect inhaul time, it certainly does have an effect on the ability of the rigging crew to move around in the brush. On steep slopes it is simply more difficult to get around. On slopes up to about

50 percent, maximum production is possible in terms of manpower. Above 50 percent, 10 to 12 percent of normal production will be lost for each 10 percent increase in slope (up to the point where it is considered inoperable).

The direction of the slope relative to the landing is, perhaps, a more critical factor. If there is plenty of lift, a highlead system is very effective on moderate downhill slopes. However, as slope increases to a point where the turn overruns the mainline and is considered out of control there will be a loss of productivity resulting from hang-ups and lost logs. There is often severe soil damage and log breakage as well. Finally, equipment can easily be damaged when control is lost. To avoid this damage the yarder and loader should be shielded by the terrain whenever possible.

In general, highlead with conventional yarders is most efficient logging uphill, while skyline systems are best for downhill logging. One logging superintendent suggests that if the operator has a choice he should not log downhill with any kind of system on slopes exeeding 70 percent.

If the ground slope is at right angles to the road line it is called *side slope*. If sufficient lift is available, side slopes of up to 40 percent can be tolerated. Without lift anything over 20 percent is difficult. With highlead, side slopes should be avoided if possible. When logging on a side slope with poor lift, production is reduced because more time is spent fighting hang-ups. Also, since the logs are dragging, soil damage will be severe.

Skyline systems are often laid out to follow ground contours working on side slopes which may be extremely steep. However, when possible it is better to log across the contours and avoid the side slopes. It is much more difficult to work on steep side slopes and the area logged under any roadline on a side slope will be much smaller than it would be on flatter ground or constant uphill or downhill slopes. The reason is the difficulty in getting coverage on the upper side of the road line. Pulling chokers uphill is more difficult, naturally, than pulling them downhill. Since the roads on sidehill logging are relatively narrow, more time is spent in road and cable changes than in yarding.

Yarding distance

Yarding distance is a major variable whether speaking about skyline or highlead. It is also a deciding factor in choosing the system to be used. Highlead logging is limited to about 1,000-foot distances, although there are any number of operators who brag about using highlead systems for yarding distances over 1,200 feet. Such instances are obvious cases of using the only equipment available even though it is inefficient to do so. Actually, when selecting a system, keep in mind that 700 to 800 feet is a good maximum yarding distance for highlead operation.

Skyline systems require longer yarding distances than highlead. This is shown in Table 12.3 where the reader will note that at 800 feet the slack skyline has a lower productivity than highlead using a big yarder. The most efficient yarding distance for a slack skyline is around 1,200 feet. Beyond that point productivity decreases nearly 1.4 MBF per man-day for each 400-foot increment of yarding distance added.

Table 12.3 *Data for Fan-Shaped Settings*

Ideal conditions	Highlead	Slackline	Shotgun
Maximum recommended external yarding distance (ft)	1,000	1,600	1,800
Optimum external yarding distance (ft)	700	1,200	1,500
Average yarding distance (ft)[1]	467	800	1,000
Road width (ft)	50	300	300
Acres served each road[2]	0.4	4.13	5.17
Acres beyond average yarding distance (%)	55	56	56
Average acres logged per day assuming 50 MBF/acre[3]	1.0	1.3	1.5

1. Average yarding distance factor is 0.667.

2. Area computed using equation for a triangle with height of optimum yarding distance and base equal to road width.

3. Acres per day a fair estimate over long term only.

Note: A rectangular-shaped setting with the same height and width would have about 50 percent more acres on a road and would have shorter average road length.

The shotgun or gravity feed system begins to lose productivity beyond 1,600 feet. The difference between the gravity feed system and the slack skyline is mainly the speed of the outhaul. Gravity feed gets the chokers back to the logs faster than a slack skyline, which makes a difference when 44 percent of the volume is beyond 1,200 feet. As the yarding distance increases, so does the cycle time. Increasing the boundary on a shotgun system setting to 1,800 feet from 1,400 feet would typically increase cycle time to 6.2 minutes from 5.3 minutes. Table 12.3 gives the ideal maximum yarding distance, average yarding distance, approximate acres served by a single yarding road, and the percentage of the volume beyond the average yarding distance for the highlead, slackline, and shotgun systems.

Actual yarding distances may exceed both the optimum and the recommended maximum for any cable system. Unfortunately, necessity often dictates what must be done in any given situation. When the alternatives are weighed, a 1,200-foot yarding distance may be necessary in one quadrant of a highlead side. It may be less expensive than building 3,000 feet of spur road and landing for 10 or 11 acres of ground. The real difficulty comes when the operator does not weigh the alternatives that he has but rather goes blindly into a set of conditions that are suboptimum in both the operation and the economic areas.

Yarding distance is constrained by several variables, such as tower height, slope, direction of yarding, available lift, and equipment size. Also, there is a cross-correlation between yarding distance and variables such as log size. Table 12.4 focuses on machine size, yarding distance, and average log size. Using a smaller machine than is indicated will result in inefficient operations and loss of productivity.

Table 12.4 *Machine Size, Yarding Distance, and Log Size –
Highlead Yarding*

	Yarding distance		
Average BF per piece	*500 ft*	*800 ft*	*1,100 ft*
200	small	medium	large
400	small	medium	large
600	medium	large	large
800	large	large	large
1,000	large	large	large

Note: The table is based on estimated horsepower requirements made
by a knowledgeable operator. The numbers provide an indication of
requirements only.

Machine sizes:

small – up to but not including 350 hp.
medium – from 350 hp to 550 hp.
large – over 550 hp.

The relationship between yarding distance, log size, and machine size is
dominated by line speed. The increase in machine size relative to increases in
log size and yarding distance has very little to do with the turn size per se. The
smaller machines can easily yard the turns. Everything in logging productivity
relates to the time and the speed of the inhaul cycle. While a small machine is
quite capable of yarding turns of 3 MBF, it cannot do so as fast as a
medium-sized or large machine. In order to maintain line speed and minimize
the inhaul time, a larger machine is used. The relationship between yarding
distance and type of cable system* can be further understood by studying
Table 12.3.

Landings

Adequate landings are extremely important on all cable yarding operations,
especially on steep slopes. There must be sufficient room for a yarder-tower,
the loading machine, a truck for hauling, and for sorting logs to some degree.
And, finally, there must be room to store a certain volume of logs even if
yarding and loading operations are being performed simultaneously (called
hot loading).

Ideally, a landing should be built on a slight adverse slope from the tower.
The adverse condition will reduce the landing problems caused by logs slipping

---

* For highlead yarding the yarding distance on the upper side of a setting should be
limited to about 1/3 the optimum yarding distance. This, of course, is a rule of thumb and
should be tempered by slope and other considerations. In general, though, yarding
distances are the shortest on the upper side and the longest on the lower side.

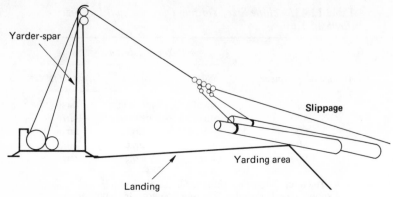

Figure 12.1. *Overhang on small landing causes slippage.*

back. This problem becomes serious when the landing is so small as to cause the logs to hang over on the downhill side (see Figure 12.1).

While a slight adverse slope is desirable, there should also be drainage away from the landing so that it does not become too muddy during the rainy season. Also, for winter yarding the setting for the yarder and the road to the landing close to the deck should be rocked.

Small landings are the bane of loggers. When space is limited there is not sufficient area for the loader to maneuver and to sort logs; logs cannot be landed without overhang; the log deck under the tree grows higher (which causes the chaser to take longer unhooking the turn), and the yarder-tower is often damaged by logs butting into it.

A landing should have a radius on the yarding side about equal to the longest log length cut on the setting plus 10 or 12 feet for a safety factor. If the longest log length is 40 feet, the radius should be 50 feet. As a general rule a 60-foot radius is the minimum landing radius acceptable (see Figure 12.2).

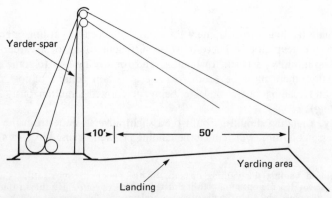

Figure 12.2. *Minimum landing radius acceptable.*

Skyline layout and planning

Thus far in this chapter the discussion has involved both highlead and skyline operating variables and their effects on the respective systems. Skylines present a special class of problems that are quite different from any found in highlead systems. A highlead cable system is not designed to fully support a load. Rather, the turn is tightlined (lifted) over obstacles as required. The balance of the time the logs are dragged full-length along the ground.

In Chapter 11 the reader learned that a skyline has the capability to fully suspend a load of logs. The reader also learned that the capacity of a skyline to support a load depends on deflection in the skyline. Deflection, in turn, is limited by length of span, line size, slope of ground, and the height of head spar and support spars or anchors. Before going on the reader should re-read the section in Chapter 11 dealing with skyline yarding.

Cable yarding is not like ground skidding. Because of the easy mobility of the equipment, errors in judgment or layout on ground skidding systems can be adjusted. The effort might be costly, but in most cases the operator can survive, providing he does not make the same mistakes frequently. With a cable operation, especially with skylines, a serious error in judgment could well mean financial disaster.

Let us suppose, for example, that a small operator has purchased the stumpage on a small woodlot. The operator planned to use wheeled skidders on the job but was forced to use tractors instead. He can still accomplish the job and the costs, while they may be a bit higher, will be acceptable. Now, in cable systems entire drainages are planned before any logging takes place. If the drainage was originally planned for skylines, and the roads laid out accordingly,

Three skyline strips that have been logged off. The strips are about 600 feet wide and 5,000 feet long. The crew logged about 300 feet on each road, 150 feet on either side of the skyline.

it might well be economically impossible to use highlead instead.

The point is that planning, while it is always important, becomes critical as the investment increases, for the simple reason that the cost of errors goes up proportionately with the size of the investment. This generalization is especially true of skyline systems.

In a great many cases, perhaps the majority of cases, the logging operator using skyline systems relies heavily on his experience to make decisions regarding layout of skyline roads, payload sizes, and line sizes. The amount of systematic planning that takes place is minimal. There is always a risk involved—even when planning is complete. Fortunately, in most cases, considerable error can be tolerated. And, when the ground is good, volume per acre high, and profit margin favorable the cost of error is simply a reduced profit margin. However, when a setting is marginal, good planning is critical. The man without knowledge and experience in planning is then at a loss.

The problem in any skyline planning exercise is to determine, skyline road by skyline road, the capacity of the skyline to transport logs—the payload—over the ground. Basically, three variables must be considered: the ground profile, allowable deflection at midspan, and allowable line tension, which dictates allowable payload size.

Calculating deflection

Deflection, as was discussed earlier, is the vertical distance between the chord and the skyline. The loaded deflection at midspan is critical in designing single-span skylines. At this point deflection must be such that the line can support the load and still leave adequate clearance for a turn of logs to be

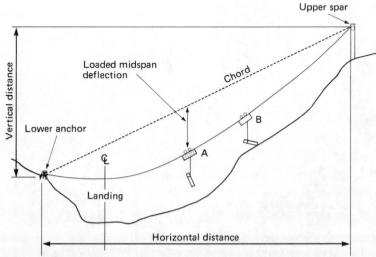

Figure 12.3. *Relationship of clearance to allowable loaded deflection.*

suspended over the ground if necessary. Midspan deflection, which allows for desired clearance, is called *allowable loaded deflection*.

Allowable loaded deflection is often dictated by clearance at some point other than at midspan. In Figure 12.3 there are two carriages shown for illustrative purposes. At position A the turn is suspended clear of the ground. At position B the turn is dragging. When the turn can be suspended at position B the allowable deflection can be measured at midspan. In the illustration, the skyline would have to be raised to allow for clearance if such clearance were desired.

The method used to determine allowable deflection is called the *graphic* or *chain-and-board* method. It involves using rectangular coordinate paper, a drawing board, a length of light chain and three heavy pins. The profile of the ground is drawn to scale, along with head spar and tail spar or tail anchor. The chain is draped between the head spar and anchor, and a light weight is hung from the chain with a paper clip or some such attachment. The length of the chain is adjusted until the suspended weight, which simulates a payload, clears the ground profile at all points on the graph. When this condition exists the deflection is measured to scale at midspan. This midspan deflection is the allowable midspan deflection. Since the graph is drawn to scale the planner can also determine both the horizontal distance, the vertical distance between supports, and the slope of the span. The measurements and resulting data are all found in Figure 12.4 (U.S. Bureau of Land Management pp. 37-40; see also Lysons and Mann 1967). The calculations were made as follows:

$$\text{Percent deflection} = \frac{\text{Allowable loaded deflection}}{\text{Horizontal distance}}$$

$$= \frac{280 \text{ feet}}{2,800 \text{ feet}} = 0.10 \text{ x } 100 = 10 \text{ percent}$$

$$\text{Slope of span} = \frac{\text{Elevation between spar and anchor}}{\text{Horizontal distance}}$$

$$= \frac{1,260 \text{ feet}}{2,800 \text{ feet}} = 0.45 \text{ x } 100 = 45 \text{ percent}$$

Breaking strength and weight

Armed with the data gleaned from the chain-and-board simulation, the planner now determines the breaking strength and weight of the skyline in pounds per foot. In most cases the operator will already have the skyline, so skyline diameter, which is needed to determine breaking strength and weight, is given. For the sake of the discussion assume the skyline is 1 5/8-inch-thick, extra-improved plow steel. The data needed can be found in any wire rope handbook made available by wire rope manufacturers. The specifications for the rope used in this example are shown in Table 12.5.

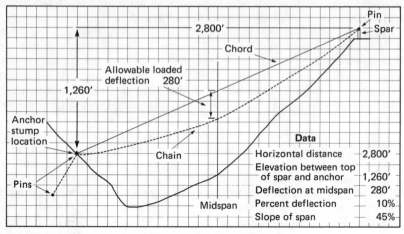

Figure 12.4. *Graphic method measurements and data.*

Type and weight of carriage

The type and weight of carriage is the next consideration. Some carriages are of the clamping type—they stop themselves and hold a position on the skyline. Others are controlled by the snubbing line (mainline or haulback) attached to the carriage. Because a clamping-type carriage increases tension on the upper end of the span it is considered separately from a nonclamping carriage. With a nonclamping carriage, part of the load is supported by the snubbing line. The carriage used in this example weighs 2,500 pounds (2.5 kips) and is nonclamping.

Calculating payload capability and unloaded deflection

Sufficient information has now been collected to calculate the payload capability and unloaded deflection of the skyline. The necessary tables and

Table 12.5 *Wire Rope Specifications Extra Improved Plow Steel*

Cable dimension (in)	Weight per foot (lb)	Safe working load (kips)	Breaking strength (kips)
1 1/8	2.34	43.4	130.0
1 1/4	2.89	53.3	159.8
1 3/8	3.50	64.0	192.0
1 1/2	4.16	76.0	228.0
1 5/8	4.88	88.0	264.0

Note: A minimum safety factor of 3 is recommended for skylines to take into account stress due to load acceleration and impact loading resulting from hang-ups.

graphs were extracted from the *Skyline Tension and Deflection Handbook* (Lysons and Mann 1967), which greatly simplifies the calculations and makes it possible to work them out without a computer.

The parameters of the skyline problem are:

Allowable loaded deflection	10 percent
Slope of span	45 percent
Horizontal length of span	2,800 feet or 28 stations
Cable diameter	1 5/8 inches
Cable weight	4.88 pounds per foot
Breaking strength of cable	264 kips
Safety factor	3
Safe working load	88 kips
Unclamped carriage weight	2.5 kips

The factors to be calculated are:

Tension due to cable weight
Tension for loading
Payload capability (less carriage)
Unloaded deflection

The first calculation necessary is tension due to cable weight. Given the slope of the span and the allowable loaded deflection the reader will find the factor for kips of tension per station per pound of weight in Table 12.6. Find deflection on the left side of the table and slope of span at the top of the table. Cable tension equals the factor times the cable weight per foot times the number of stations in the span.

Tension factor = 0.184 kips/station/pound
Cable tension = 0.184 × 28 × 4.88 = 25.14 kips

The tension left for loading is found by subtracting cable tension from the safe working load of the cable used.

Tension for loading = 88 kips − 25.14 kips = 62.86 kips

Before arriving at payload capability the loading of the carriage must be found. This is accomplished by dividing the tension for loading by the tension per kip of load attributable to the carriage (Table 12.7 shows the factor for carriage loading):

Tension for loading = 62.86
Tension/kip of load =   2.57

62.86/2.57 = 24.46 kips

24.46 − carriage weight (2.5 kips) = 21.96 kips

Table 12.6 *Upper-end Tension Due to Weight of Cable*

| | Deflection (%) | | | | | | |
Slope of span (%)	10.0	10.5	11.0	11.5	12.0	12.5	13.0
0	.137	.131	.127	.122	.118	.115	.111
5	.140	.134	.129	.125	.121	.117	.114
10	.143	.138	.133	.128	.124	.121	.117
15	.147	.141	.137	.132	.128	.124	.121
20	.152	.146	.141	.136	.132	.128	.125
25	.157	.151	.146	.141	.137	.133	.130
30	.163	.157	.151	.147	.142	.138	.135
35	.169	.163	.157	.152	.148	.144	.140
40	.176	.170	.164	.159	.154	.150	.146
45	.184	.177	.171	.166	.161	.157	.152
50	.192	.185	.179	.173	.168	.164	.159
55	.201	.194	.187	.181	.176	.171	.167
60	.211	.203	.196	.190	.184	.179	.175
65	.221	.213	.206	.199	.193	.188	.183
70	.231	.223	.216	.209	.203	.197	.192
75	.243	.234	.226	.219	.212	.206	.201
80	.255	.246	.237	.230	.223	.216	.211
85	.268	.258	.249	.241	.234	.227	.221
90	.281	.271	.261	.253	.245	.238	.232
95	.295	.284	.274	.265	.257	.250	.243
100	.309	.298	.288	.278	.270	.262	.255
105	.325	.313	.302	.292	.283	.274	.267
110	.340	.328	.316	.306	.296	.287	.279

Source: Lysons and Mann 1967, pp. 32-33.

Note: Tension in kips per station per pound of cable weight per foot.

The payload capability of the skyline is 21.96 kips. If the timber is very large the skyline will not be able to carry the load safely. By moving the anchor up the hill a little more than 200 feet, and increasing the horizontal distance by 200 feet, deflection increases to 13 percent and slope of span decreases to 35 percent. The payload capacity increases to 32.7 kips. A slightly larger skyline also increases payload capacity. But going to a larger line also increases the weight per foot of line, so the gain is not as great.

Assuming that a 32.7-kip-payload is large enough, the unloaded deflection must still be found. Otherwise, the skyline may have too much tension resulting in overload with the designed payload. The first step is to use Figure 12.5 to determine the decrease in deflection caused by removing the load. Next subtract the unloaded deflection from loaded deflection. In the example the load factor is 98.80 and is calculated using this formula:

$$\frac{\text{Remaining cable tension}}{\text{Tension/station due to cable weight x weight/foot}}$$

Table 12.7 *Tension Due to Load (Carriage Not Clamped to Skyline)*

Slope of span (%)	Loaded deflection (%)						
	10.0	10.5	11.0	11.5	12.0	12.5	13.0
0	2.54	2.43	2.32	2.23	2.14	2.06	1.98
5	2.52	2.41	2.30	2.20	2.11	2.03	1.96
10	2.51	2.39	2.28	2.19	2.10	2.02	1.94
15	2.50	2.38	2.27	2.17	2.08	2.00	1.93
20	2.49	2.37	2.27	2.17	2.08	1.99	1.92
25	2.50	2.37	2.27	2.17	2.07	1.99	1.91
30	2.50	2.38	2.27	2.17	2.08	1.99	1.91
35	2.52	2.40	2.28	2.18	2.09	2.00	1.92
40	2.54	2.41	2.30	2.19	2.10	2.01	1.93
45	2.57	2.44	2.32	2.22	2.12	2.03	1.95
50	2.60	2.47	2.35	2.24	2.14	2.05	1.97
55	2.64	2.51	2.38	2.27	2.17	2.08	1.99
60	2.68	2.55	2.42	2.31	2.20	2.11	2.02
65	2.73	2.59	2.46	2.35	2.24	2.14	2.05
70	2.79	2.64	2.51	2.39	2.28	2.18	2.09
75	2.84	2.70	2.56	2.44	2.33	2.23	2.13
80	2.91	2.76	2.62	2.49	2.38	2.27	2.17
85	2.97	2.82	2.68	2.55	2.43	2.32	2.22
90	3.04	2.88	2.74	2.61	2.49	2.38	2.27
95	3.12	2.95	2.81	2.67	2.55	2.43	2.33
100	3.19	3.03	2.87	2.74	2.61	2.49	2.38
105	3.27	3.10	2.95	2.80	2.67	2.55	2.44
110	3.36	3.18	3.02	2.87	2.74	2.62	2.50

Source: Lysons and Mann 1967, pp. 34-35.
Note: Tension in kips per kip of load.

$$\frac{67.5}{0.140 \times 4.88} = 98.80$$

The difference in deflection is 2.4 percent. When the skyline is hung the rigging crew will be looking for about 10.6 percent unloaded deflection or about 320 feet of belly in the line.

The calculations just made are often foregone in favor of quick estimates based on experience. Many loggers will simply hang the skyline as far up the hill opposite the head tree as possible and use a much larger line than is necessary. The result is increased payload capacity but a reduction in productivity. Deflection is increased at the expense of getting the skyline higher off the ground in a concave profile. This means it takes longer to lower the skidding line. Skidding-line capacity may be a controlling factor. Lower productivity is the price paid for poor planning.

Payload and tension calculations are especially important when a tight skyline is being used. If a slackline is being used, providing the skyline is long

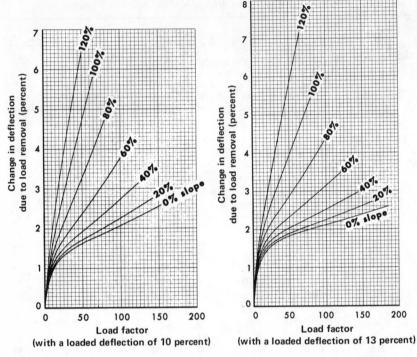

Figure 12.5. *Change in deflection due to load removal versus load factor for various slopes (Lysons and Mann 1967). Based on a modulus of elasticity of 14,000,000 psi. Load factor equals tension due to load (kips) divided by tension due to cable weight (kips/station).*

enough, a cable change can rectify an awkward situation. Also, if the skidding line is not long enough the line can be lowered.

Payload capabilities for a running skyline can be calculated in a manner similar to the calculations made for standing skylines by using the coefficients in the *Skyline Tension and Deflection Handbook* (Lysons and Mann 1967). The procedures will not be discussed here, but can be found in *Forest Service Research report PNW-75.*

Cable Yarding Operations

Differences between various cable yarding systems were described in Chapters 1 and 5. These differences were defined in terms of equipment size, system capabilities, operating variables, operating ranges, and productivity over a wide variety of operation conditions. Despite the differences, however, the basic operations of all cable systems are similar. Crew size varies depending on whether highlead or skyline is used. In one case, a highlead operation, crew size may be seven men; in another case, a running skyline with grapples, crew size may be only three men. As we shall see, however, job descriptions are essentially the same, as are work elements and operational sequences.

This chapter is devoted to the operational aspects of cable logging. It begins with a description of jobs, moves on to work elements and a description of operations, and ends with a brief section on grapple logging. Except for the final section, all descriptions are in terms of highlead logging. Where necessary, appropriate descriptions of skyline operations are also given.

Crewing

The rigging crew consists of those men required to operate a cable side. Actual crew size depends on the specific cable system being used. As a rule, a highlead side uses six or seven men. A slackline operation requires an eight-man crew, one or two of these men being responsible for cable changes, rig-up, and layout. A running skyline with grapples requires only a three-man crew. The jammer system used in the Intermountain region often operates with only two men. For a highlead operation, the following job descriptions are common.

The *hooktender* is the straw boss in charge of a single highlead side. He is responsible for all yarding and loading production, moving and rigging, and crew safety. The hooktender works under the supervision of a logging foreman who may direct operations for as many as three sides or more. The *rigging slinger* is the person responsible for the rigging crew, which consists of two or three choker setters and the chaser. The rigging slinger selects the logs to make up a turn and is responsible for getting them to the landing. He trains new choker setters and is responsible for the safety of his men in the brush. In the

Table 13.1 *Highlead Logging Whistle Signals*

Signals	Meaning
1 short	Stop all lines.
3 short – 3 short	Ahead slow on mainline.
3 short	Ahead on mainline.
2 short	Ahead on haulback.
2 short – 2 short	Ahead slow on haulback.
3 short – 1 short	Ahead on strawline.
3 short – 1 short – 3 short	Ahead slow on strawline.
4 short or more	Slack the mainline.
2 short – 4 short	Slack the haulback.
3 short – 1 short – 4 short	Slack the strawline.
3 short – 2 short	Standing tight line.
1 short – 1 short	Running tight line or break if running tight.
3 short	When rigging is in: strawline back on haulback.
3 short plus "X" number of shorts	When rigging is in: indicates number of strawline sections back on rigging.
3 short – 1 short – 2 short	Strawline back on rigging.
1 short	When rigging is in: Chaser inspect and repair rigging.
2 short	When rigging is in: no chokers back.
2 short – 1 short plus "X" number of shorts	Number of chokers back.
2 short – 4 short	When rigging is in: slack haulback – hold all lines until 2 short blown.
3 medium	Hooker.
3 medium and 4 short	Hooker and crew.
5 long	Climber.
4 long	Logging foreman.
1 long – 1 short	Start or stop work.
7 long – 2 short	Man injured, call transportation and stretcher.
1 long – 1 short repeated	Fire.

Source: State of Washington 1972, p. 59.

Note: Short, medium, and long whistles controlled by length of contact when whistle is engaged. The dashes indicate a slight pause between whistles.

absence of the hooktender, the rigging slinger is the logical candidate for the job. The *choker setters* perform the same task as choker setters on a ground skidding side—they wrap the chokers around the logs in preparation for yarding. Choker setters often work together in 'hooking' a turn. The rigging slinger sets chokers as well.

On the landing the *chaser* is responsible for unhooking the turns as they are yarded in and for building a safe log deck. The latter part of the chaser's job has diminished since the industry abandoned colddecking logs in favor of hot loading. In some operations the chaser also bucks logs on the landing (*landing bucker*), and fills in as a second loader in the loading operation (which will be discussed in the next chapter). He is also responsible for several miscellaneous jobs, such as coiling haywire for road changes and new layouts; exchanging new chokers for worn-out chokers in the butt rigging; and care of blocks, straps, and so on at the landing.

The *yarder engineer* is a key man on the landing—and in the whole operation. He is the man who operates and looks after the yarder. A good engineer can increase production by running smooth rigging and operating the yarder at peak efficiency.

The brush crew, including the rigging slinger and choker setters, is responsible for hooking the turns. Actually moving the logs through a maze of stumps, chunks, and other physical obstructions requires close cooperation and communication between the yarder engineer and the rigging slinger. Communication takes place through a system of long and short whistles generally transmitted by radio. Radio whistles replaced whistle punks several years ago. The whistle punk relayed signals from the rigging slinger to the engineer by means of a long electric cord strung from the operating area in the brush to the yarder. The signals used in highlead operations vary from camp to camp but are substantially the same. Those found in Table 13.1 are approved by the state of Washington.

Work elements

Cable yarding systems depend on serial operations. Certain steps or functions are performed in an ordered sequence and repeated over and over. The sequence is substantially the same for each system. The chokers are hooked to the logs, the turn is transported to the landing and unhooked, and, finally, the rigging is returned to the brush for another turn. Terrain, log size, and equipment may vary, but the sequence is always the same.

The sequence of operations is called the *yarding cycle*. The yarding cycle contains five regular elements and several irregular elements. The five regular elements are outhaul, choker setting, inhaul, chasing, and delays.

Outhaul: The outhaul element is the least complicated and requires the least time of all the yarding elements. It begins when the chokers are free at the landing and the rigging starts back toward the brush. And it ends when the engineer is signaled to stop the rigging at the yarding area. The time required for outhaul is simply a function of distance and line speed.

Choker Setting: This is the most time-consuming of all the elements, taking between 45 and 50 percent of total cycle time. The choker-setting element begins when the rigging is stopped at the yarding area and ends when the yarder engineer receives the signal to begin the inhaul element.

Choker setting is composed of several smaller elements. First, after the rigging is stopped, the rigging slinger signals to *position the rigging.* Next the *chokers are untangled*, either in the air or on the ground, depending on lift. Once the chokers are untangled they must be *pulled to the logs* that will be set. Actual *choker setting time* involves the time required to wrap the choker around the log. This element ends when the last choker is set. Finally, the brush crew must reach some safe place or *get in the clear* before the inhaul element begins.

Inhaul: The inhaul or yarding element begins when the engineer receives the signal to start the rigging ahead and ends when the leading ends of the logs are visible to the chaser at the landing. Distance is the key factor in this element, provided the yarder is large enough to maintain line speed. Turn size and slope are also determining factors.

Chasing: Chasing is also called *landing time.* It begins the instant the chaser can see the turn at the landing and ends when the turn is unhooked and the chokers are free. Chasing includes positioning logs and unhooking chokers.

Delays: Any number of delays may interrupt the various elements already defined. Delays are associated with equipment or the crew. For instance, upending the chokers* or replacing worn chokers is an equipment delay which takes place during the chasing element. Hang-ups are a major delay in the inhaul element. Another type of delay might involve picking up a log lost during the previous turn.

There are many delays that can occur at any time during an operation. They range from equipment breakdown to some of the physical delays mentioned in Chapter 12. Since a high incidence of delay time reduces the time during which productive work can be performed, it is important to find ways to avoid delays. More will be said about this later.

The time distribution for the regular elements will, of course, vary with conditions. For instance, when yarding distance increases, the percentage of time spent in setting chokers decreases. In addition, inhaul time and outhaul time increase, both in quantity and as a percentage of total cycle time. Adding an extra choker will increase choker-setting time slightly and will increase chasing time, but in most cases inhaul time will not be affected. However, with two chokers and 700- to 800-foot yarding distances, the following time distribution of work elements is representative for highlead operations.

---

* Chokers have a pressed nubbin on both ends, and the bell slides up and down the full length of the choker. When one end of the choker becomes worn or kinky from use the chaser puts the worn end in the butt hook so the brush crew can work with the unworn end. Changing ends is called *upending.*

Element	Time (%)
Outhaul	8
Choker setting	41
Inhaul	19
Chasing	22
Delays	10

The time distribution shown is for yarding time only, including delays attendant on the yarding cycle. The distribution does not account for 100 percent of the available time. Certain unavoidable operational delays, such as road changes (called *cable changes* in skyline logging) and moving into a new layout, are necessary but cannot be considered part of the total productive time. Total time available for yarding varies from 65 to 85 percent of total available machine time. The remainder of the time comprises downtime and unavoidable operational delays.

Moving and rigging up

Before any yarding can take place the equipment must be moved into the new setting and rigged for logging. These activities together represent a major time increment and a significant cost, which must be accounted for.

Moving begins at the old setting and involves pulling in the running lines and brush blocks, loading all guylines and equipment, and physically moving all men, machines, and support equipment from the old setting to the new. Rigging up includes unloading equipment, stringing guylines, notching stumps, tightening guylines, making a layout, raising the tower or, if a wooden spar tree is being used and it has not been pre-rigged, rigging up the tree.

A mobile spar-yarder gets an assist during a move. The machine is trailer-mounted and must be towed from one location to another.

The entire process of moving and rigging can take anywhere from a few hours to several days, depending on the length of the move, the preparation, and the experience of the crew. Experience is the most critical factor. An experienced crew can rig a standing tree in 8 to 12 hours. An inexperienced crew might take anywhere from 2 to 4 days. The author has seen a mobile tower moved and rigged in less than 3 hours with an experienced crew. He has also seen an inexperienced crew take 3 days to accomplish the same thing.

Moving is a straightforward process. A sled-mounted yarder must be loaded on a truck and trailer for moving. Self-propelled yarder-spars can move themselves over short distances. On a long move the self-propelled units must be broken down into sections in order to meet legal weight limits. Generally the yarder makes one load, and the tower, which can be removed, makes the second. Rigging up, however, is a bit more complicated than moving.

Procedure for rigging up

Rigging a wooden spar tree or a tower involves many similar work elements, which can be accomplished simultaneously or even in advance. One main difference is that rigging a tree requires a high climber. Another is that it requires the guylines to be spiked and tightened, usually by a rigging machine or crawler tractor. With a tower, the guylines are permanently attached and the machines generally have self-tightening drums on each guyline. In addition, the taglines, which must be strung for a tower, are attached to the guyline stumps with shackles, rather than spiked. Other differences will become obvious as the rigging process is described.

When the landing is located on a new setting, and if a wooden spar tree is to be used, the selection of the spar is a primary consideration. Prospective spar trees are marked with an X, which tells the timber fallers to leave the tree standing. Only one spar is used in highlead operations, and the final choice, if more than one tree is chosen, is made when the landing and guyline circle is being cut. Douglas fir makes the best spar tree. On the West Coast, a spar will be at least 110 feet high with a top diameter of at least 16 inches.

Rigging a wooden spar tree begins with a climber walking up the tree, removing limbs as he goes. He will cut the top out of the tree (called *topping*) at no less than a 16-inch diameter. With topping completed the climber will descend to a point approximately 16 feet from the top and remove the bark from a strip approximately 4 feet high all around the tree. The barked section is for tree plates, around which the top guylines are placed as shown in Figure 13.1.* The top guys are placed over the tree plates so that the tree will not be cut off when tension is applied to the guylines. Tree plates also have hooks welded to them to accept the bull block strap. The plates are nailed to the tree with railroad spikes.

Before hanging the tree plates the climber must hang the pass block. A pass block is a 6-inch block placed close to the top of the tree (see Figure 13.1). A 5/16- to 1/2-inch cable, called the *pass line*, is threaded through the pass block.

* In the woods, *guylines* is frequently shortened to *guys*.

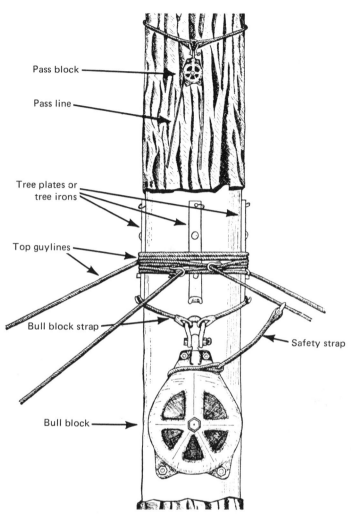

Pass block

Pass line

Tree plates or
tree irons

Top guylines

Bull block strap

Safety strap

Bull block

Figure 13.1. *Top of a rigged spar. The tree is barked to prevent loose bark from falling out of the tree and to prevent the irons from slipping down.*

Both the block and line are pulled up the tree by the climber, using a small-diameter hemp rope. One end of the pass line is attached to the haywire drum with at least three wraps before use. The other end, hanging at the base of the tree, has attached to it a special rigging chain which has two short chains attached to a steel ring. This chain, called the *pass chain*, is used both as a seat for the climber and as a means of attaching the pass line to the various pieces of rigging which must be carried to the top of the tree.

When the pass line is operational, the tree plates are hauled up and nailed into position. Then the top guylines are lifted to the top, one at a time. A highlead spar 110 feet or taller requires six top guys. A tree used for slackline logging requires eight guylines.

The rigging chain is attached to the guylines, allowing enough tail so the line can be wrapped around the tree plates. Each guyline has an eye spliced into both ends. A shackle is used to form a sort of noose around the tree by placing the U-shaped part of the shackle over the bight of the guyline and attaching it to the eye end of the line by placing a pin through the shackle and eye. Each successive guyline is wrapped around the tree in alternating directions. That is, if the first guy is wrapped clockwise, the second is wrapped counterclockwise. If all the guylines were wrapped in the same direction, the top of the tree would be twisted out in a pull.

The climber will hang all six top guys for a highlead tree and then drop down to hang the buckle guys. The top guys are located about 16 feet from the top of the tree. The buckles are about 24 feet under the top guys. Tree plates are not used for the buckle guys.

Guylines range from 1 1/8 to 1 3/8 inches in diameter. Since their primary function is to resist the force of the operating lines, they should be equal to the breaking strength of any component part of the rigging. The should be sufficient in number and spaced to provide lateral stability to the tree or tower, as well as providing the resistance required. If there is any doubt, hang on an extra guyline. This alternative is better than pulling over the tower, breaking the top out of the tree, or buckling the tree.

Self-tightening guyline drums on a mobile spar-yarder.

The guylines on a tower are self-contained, as explained in the next section, so *taglines* are used to rig the guyline stumps. Taglines are generally 100 to 150 feet in length and, when rigging is complete, are shackled to the guylines strung out from the tower. Guylines for a wooden tree or taglines for a tower are strung out to stumps before the tree or tower is rigged. The guylines for a tree are shackled to the tree first and then anchored to the stump. Taglines are shackled to the stump first and guylines from the tower are then attached. Guys for a wooden tree are tensioned from the ground by a tractor or the yarder. The tower's guylines are tensioned from the tower.

The distance from the tree to the guyline stump should be between 1.5 and 2.0 times the height of the tree, measuring from its base. (If the tree or tower is 110 feet tall, the stumps should be between 165 and 220 feet from the base of the tree or tower.) Guyline stumps should be large enough and sufficiently well anchored to withstand the pressures applied during the yarding cycle. The stumps are barked and notched prior to stringing the lines.

The tail of each guyline or tagline is pulled out to the stump from the landing by the haywire. Guylines are burned around the stump. With taglines, the haywire is run through a rigging block positioned behind the guyline stump. There must be enough tail on the tagline to wrap the stump or additional slack will have to be pulled.

The curve formed by running the line around the stump is called the *bight* of the line. With the bight around the stump, the line is tightened and spiked with at least eight railroad spikes. Another wrap is placed on the stump and spiked with two or three spikes. The third and final wrap is spiked with no less than six spikes. Each wrap is tightened before being spiked (see Figure 13.2). Taglines are shackled to the stumps as can be seen in the photograph later in this chapter.

Guyline spacing on horizontal and vertical planes is critical in rigging up. A guyline under tension produces two force components. A horizontal component resists the tension of the operating lines and a vertical component puts the tree or tower under compression. The amount of tension the guyline can resist and the amount of compression it will generate depend on the vertical angle,

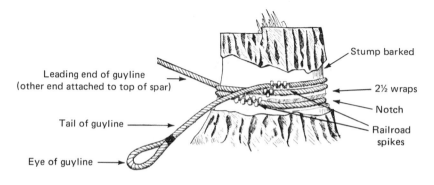

Figure 13.2. *Example of rigged guyline stump.*

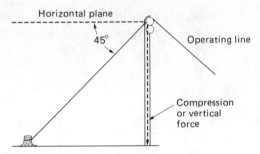

Figure 13.3. *Guyline at 45 degrees from horizontal equalizes operating line resistance and compression force.*

measured from the horizontal at the top of the tree or tower as illustrated in Figure 13.3. From 0 degrees to 45 degrees, the guyline provides greater resistance to the operating lines and less compression. From 45 degrees to 90 degrees, compression is greater than the resistance against the operating lines.

As the angle is increased from zero, the horizontal component or tension decreases and the vertical component increases. At 45 degrees, the two components are equal—the force resisting the operating lines and the force creating compression are the same. As the steepness of a guyline increases beyond 45 degrees, the force resisting the pull of the operating lines decreases and the guyline is less effective in providing stability to the tree or tower.

The horizontal angle of the lines as they radiate from the tree or tower is just as important as the vertical angle. The horizontal angle is measured from the lead of the operating line.

Once again there are two components, this time measured on the horizontal plane. The first component provides resistance against the pull of the operating

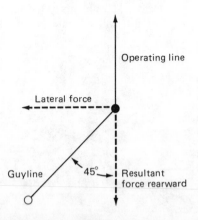

Figure 13.4. *Guyline at 45 degrees from rearward extension of operating line equalizes operating line resistance and lateral force.*

Table 13.2 *Guyline Effectiveness at Angles from 0 to 90 Degrees*

Guyline angle	Resistance to operating line	Lateral or vertical resistance
0°	100%	0%
15	97	26
30	87	50
45	71	71
60	50	87
75	26	97
90°	0%	100%

line. The second provides a force that gives lateral stability to the tree or tower. As the horizontal angle between the operating line and the guyline increases, the resistance against the operating line decreases and the lateral force component increases. At 45 degrees the two force components are equal (Figure 13.4). Table 13.2 describes guyline effectiveness at given angles measured from either the horizontal or the vertical plane.

Now that some appreciation of the function of guylines is accepted, guyline placement can be considered. For a spar tree, guylines radiate at a spacing of approximately 60 degrees if a full tree is to be logged. The 60-degree intervals can be measured roughly by using an ordinary pocket watch. A guyline is placed at 2, 4, 6, 8, 10, and 12 o'clock. However, more often than not such spacing is neither wanted nor possible. There is frequently one area of the setting–perhaps a steep face where the timber is exceptionally heavy–where the pull will be hard. Actual spacing will depend on the logging system and where the heavy pulls are located. Figure 13.5 illustrates guyline spacing.

In Figure 13.5B the guylines are spaced so the major pull will fall either opposite a guyline or between the two guylines. Locating the guylines this way provides more effectiveness than will be found in any single line. When six guys are evenly spaced, the line is more effective when the pull comes directly against either one guyline or halfway between two guylines. The same is true in any arrangement except when eight guys are used, in which case the maximum effectiveness comes between two guylines. In any case, adjacent guylines should not be more than 90 degrees apart.

There should always be at least six guylines in a tower and six top guys and three buckles in a wooden tree. This is standard for highlead logging. Additional guylines are necessary if a skyline system is to be used. The maximum tension on the mainline, and thus the guylines, in a highlead system is limited by the breaking strength of the mainline and the power of the yarder. With skyline systems the tension varies as the load moves through the span, as noted in Chapter 12. Even when the span has been planned carefully, an overload condition relative to the guylines can be created when the lines become impact loaded, as with a hangup. It is not at all difficult to create tensions in a skyline system that far exceed highlead tensions.

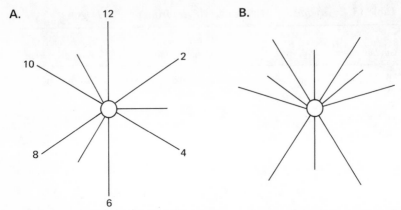

Figure 13.5. *Guyline spacing. The longer lines are top guys and the short lines are buckles. A shows the guyline spacing for a full tree with no area of particular difficulty. In B, four of the top guy lines and three buckles are on the back of the tree indicating the hard pull will come from the opposite direction. The extra buckle was added for stability.*

Guylines in the back quarter (Figure 13.6) are the most efficient in resisting operating loads. Guylines placed in the side quarter, but in the back hemisphere, resist the operating loads somewhat but are placed primarily for lateral stability. Guys in the front hemisphere act only to provide stability to the tower or tree during sudden load changes. As a general rule, there should always be three guylines against the pull, with at least two of these in the back quarter. There should be at least two guylines in the front hemisphere with at least one in the front quarter. Again, whenever there is a doubt, add another guyline.

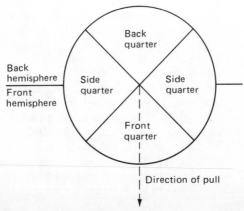

Figure 13.6. *Relationship of quarters and hemispheres to direction of pull.*

After guylines for a spar tree are all positioned correctly and properly tensioned, the climber can complete his job. The mainline, or bull block, is hung first. The block and strap are sent up the tree at the same time and the strap is threaded across the hooks welded to the tree plates. The ends of the strap are equipped with steel eyes that are attached to the block with a shackle. The bull block generally weighs between 400 and 1,600 pounds. As a safety precaution, a second strap is hung between the block and one of the top guys. If the main strap breaks, the safety strap will prevent the block from falling.

The haulback block, which is quite a bit smaller than the bull block, is hung several feet below the bull block. After both blocks are secured, they are each threaded with a length of haywire, which will later be used to pull the mainline and haulback through.

Mobile spars

Rigging a tower for logging is very similar to rigging a wooden spar tree, but there are some important differences. First, the tower and yarder are one unit. Where the tower goes, so goes the yarder. This means the unit can be located in the best possible position on a landing with no concern about spar tree location. The tower, which is lowered to a horizontal position across the top of the yarder while the unit is being moved, can be raised mechanically with the use of hydraulics, cable lift, or a combination of both.

Second, the towers may be in one piece or they may be telescoping. Telescoping towers are raised in two stages. A hydraulic ram pushes the first stage to a nearly vertical position and then the second stage is telescoped up hydraulically or with a cable hoist arrangement.

Third, there is no need to hang blocks in a steel tower. All of them have self-contained fairleads at the top of the tower for the mainline and haulback. The haulback fairlead is on top and the mainline fairlead is below it. Some towers have a third fairlead for skylines.

A fourth major difference is the self-contained guylines and drums. Most portable towers have at least six top guylines, and more if necessary. Towers shorter than 50 feet have fewer than six guylines. For each guyline there is a hydraulically operated guyline drum for tightening and slackening the guylines. Tightening guylines on a tower is much less time consuming than on a wooden spar tree. Each guyline runs from the drum located at the base of the tower, up through special blocks at the top of the tower, and back down to the bottom of the tower. When the tower is ready to move to a new setting, the ends of the guylines, which have eye splices, are shackled to the base of the tower.

Because the guylines are of limited length, an additional length of line, called a *tagline*, is used to reach the guyline stumps. Taglines are generally 100 to 150 feet in length and, when in use, are shackled to the main guylines.

Guyline stumps are prepared for taglines in the same manner as for guylines on a wooden spar tree. However, taglines are generally shackled to the stumps rather than spiked. Like ordinary guylines, taglines can be strung out and shackled to the stump either while the yarder is present or before it arrives. The difference is that the taglines do not have to be tightened, since they are simply shackled to the stump. A line truck, which is used to transport the taglines, or even a pickup truck, can be used to string them.

Pre-rigging is one way to speed up the rigging-up process with either a wooden spar tree or with a tower, although with a tower pre-rigging is much simpler. The stumps can be picked out, barked, and notched by a couple of

Tower raising. (1) The tower-yarder is located on its setting and leveled. The hydraulic rams above and to the left of the man on the machine raise the tower. (2) Still going up. There is a clear view of the guyline drum and raising cylinder controls in front of the man on the machine. The hooktender is responsible for raising and rigging the tower. (3) The tower is as high as the cylinders can set it. It will now be raised the remaining few degrees with the guylines. Photos courtesy Loggers World men in a few hours. If an extra set of taglines is available, the lines can be transported to the new setting, strung to the stumps, and shackled. All that is required for stringing is a small rigging block, several sections of haywire, a pickup truck, and a few other pieces of miscellaneous equipment, such as a rigging chain.

On short moves prerigging means the difference between a two-hour rig-up and an all-day rig-up. At a cost of $900 to $1,000 dollars per day, this difference can be crucial.

Top: Close-up of a guyline sleeve (shackle) which is used to attach tower guyline to stumps. Bottom: The guyline is on the stump and is tight. Note the notch in the stump so that the line cannot jump the stump.

Making a layout

A layout is defined by the position of the running lines in a cable yarding show. Generally the layout is triangular in shape, with the roadline and backline making two legs of the triangle and the wasteline making the base. The corners between the base and the sides are maintained by the corner block and tail block. The blocks, also called *brush blocks*, are 16 inches in diameter measured on the sheave. Figure 13.7 indicates that the backline is formed by the haulback, as is the wasteline. The roadline is formed by both the mainline and haulback. The roadline is defined by an area extending, between 70 and 150 feet, on either side of the mainline and haulback depending on the cable system being used.

The size of running line was given in Chapter 11. A haulback 7/8 inch in diameter weighs about 1.4 pounds per foot while a 1 1/4-inch mainline weighs

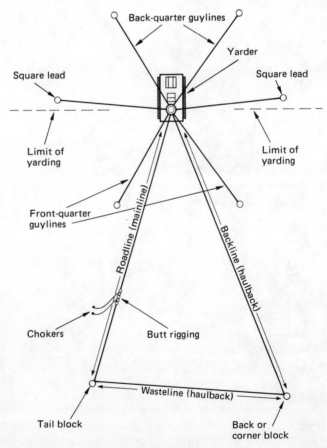

Figure 13.7. *Highlead yarding layout and line positions.*

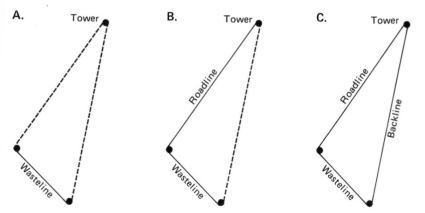

Figure 13.8. *Making a layout. A shows brush block hung and wasteline strung. B shows roadline strung. C shows haywire layout made.*

nearly 3 pounds per foot. The rigging crew, in making the initial layout, must arrange for the running lines to be moved into position.

Because of the weight of the running lines, the initial layout is made by stringing haywire all around the first area to be logged. Usually yarding will work downhill—that is, from high ground to low ground.* The haywire is pulled around the layout (which will be from three to four roads wide, if possible) by members of the rigging crew.

Depending on the situation, one or two choker setters will carry the corner blocks and straps to the back end of the proposed layout. In addition, one coil of haywire will also be carried back to make the wasteline.† The balance of the haywire is pulled, one leg of the triangle at a time, directly off the haywire drum from the landing. One man takes the leading end and the others pull slack for him. When both legs are formed the ends of the haywire are connected, the bight of the line is placed in the corner blocks, and the entire layout is surrounded by the haywire. Figure 13.8 illustrates the necessary steps.

The end of the haywire that makes the backline is hooked into the eye of the haulback. The other end is attached to the hayline drum. The haulback is pulled around the layout, first down the backline, around the wasteline, and up the roadline. Once the haulback makes the entire circuit and arrives at the landing it is shackled to the butt rigging and the system is operational—ready for yarding.

It is important, when stringing the haywire, to avoid tangling it by passing it under logs or around saplings and brush. (When a line is tangled thus it is

---

* Yarding moves downhill, as a rule, so that those logs that are dislodged from their lays or lost in yarding will roll downhill into an area which is unlogged. Those logs can be picked up as yarding progresses down the hill. A second important reason is safety. When logging downhill, the rigging crew always has the option of walking uphill to safety, out of the felled and bucked timber.

† A coil of haywire is also referred to as a *section*.

called, in logging parlance, a *siwash*.) Siwashes are not created intentionally, but they do happen. To avoid accidents that might result when the line is suddenly cleared of a siwash, the butt rigging is generally tightlined at the landing and then the rigging is run to the back end of the roadline. In this way any existing siwashes will probably be freed. It is important, however, for anyone working around the running line to stand safely out of the way, since clearing the lines causes the lines to jump, sometimes wildly.

Skyline layouts

A single-span skyline layout is made the same way as a highlead layout. The haulback is, in the case of a skyline, first used to pull out the skyline and is then attached to the carriage, which replaces the butt rigging used in a highlead operation. In cases where the ground is very brushy or for some reason there is not a clear line of sight to the tail hold, a compass may be used to run a true line and avoid siwashing the skyline. If a gravity feed system is being used, hanging the skyline is all that is necessary, because no line is needed to pull the carriage out to the logging area. In downhill logging, where the carriage must be pulled uphill, a haulback is needed.

If the skyline is used in terrain that results in sidehill logging—that is, logging across the contour—the haulback may be used as a dutchman. The haulback is run out above the skyline, through a brush block and downhill to the skyline. A large block through which the skyline can travel is attached to the haulback, and the skyline is placed in it. This extra line is called a *dutchman*. By pulling with the haulback the skyline can be pulled uphill over the logs. It allows for wider skyline roads and allows the rigging crew to avoid pulling chokers uphill. Figure 13.9 illustrates the dutchman.

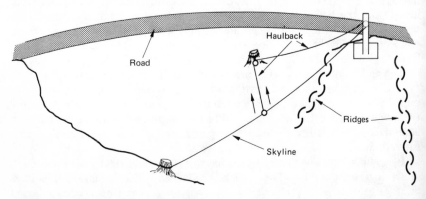

Figure 13.9. *The dutchman. Pulling the haulback will result in the skyline being pulled up the hill. This allows positioning of the skyline over a new road without making a cable change. Also, it will not be necessary for the choker setter to pull chokers or skidding line uphill to a turn since the skyline can be positioned above the logs by the dutchman.*

Landing a turn on a slackline show (gravity system). The top block is used for a dutchman. The second block is the skyline block and the lowest line is the haulback block—used here for a skidding line. Note the spacing of the guylines on the back side of the tree opposite the most severe pull.

Anchor stumps

Hanging a skyline is a bit more complicated than simply making a highlead layout. A skyline is anchored at least at one end. A standing skyline or tight skyline is anchored at both ends. Stumps are by far the most common type of anchor used. The skyline may be anchored directly to the stump or it may first lead through a head or tail spar. When a spar is used a tree shoe or a block is used to support the line in the tree. A *tree shoe* is simply a steel device which is grooved to accept the skyline and is hung in the tree by a strap. When a slack skyline is used a block is required, at least in the head tree, because the line is raised and lowered as required for yarding.

Anchor stumps are easier to rig at the bottom of a unit (downslope) than at the top. There is less tension on the stump, soils are generally deeper, and the stumps are larger. Unfortunately, the bottom of the unit is not always where the anchor is needed. In terms of cost, stumps on ridge tops are cheaper to rig when there is vehicle access. With vehicle access it is not necessary to hand carry all the tools and rigging to the stump location. Ridge-top stumps often do not have the holding power of stumps at the bottom of a canyon, because the soil on a ridge is shallower and the stumps are generally smaller. The following general rules should be considered when selecting and rigging anchor stumps.

1. A stump in deep soil has more holding power than a stump in shallow soil.
2. Holding power increases approximately with the square of the difference in stump diameters. A 48-inch stump is twice as large as a 24-inch stump and has roughly four times the holding power.
3. Stumps have a greater holding capacity and are more stable with an uphill pull than with a downhill pull, because of the heavier root structure on the uphill side (see Figure 13.10).
4. In wet weather or in areas with relatively shallow soil use a tieback stump as well as the main anchor stump, even if the main stump is large. In wet weather especially, stump pull is common. The tieback stump will relieve some of the tension on the main stump and will provide a safety margin in case of stump pull.
5. All stumps, both anchor stumps and tiebacks, should be notched to prevent the lines from slipping off.
6. Care should be taken to align the anchor stump as closely as possible with the skyline.

When available stumps are small, several may be rigged together as an anchor. Some operators get by with fairly frail anchors but the practice is risky, especially when radio-controlled carriages are being used. Stump failure will result in loosing the skyline and the carriage.

Two or more stumps are often necessary to anchor a skyline (see Figure 13.11). When this is the case the tieback angle formed between the skyline and the tieback lines should be kept as small as possible. The greater the tieback angle, in either a horizontal or vertical direction, the greater the load on the stumps. With a very wide angle it is possible to generate more tension on both the anchor and the tieback stumps than there was on the skyline to begin with.

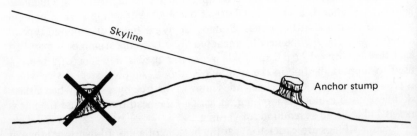

Figure 13.10. *Choice of stump. Use the stump that allows an uphill pull.*

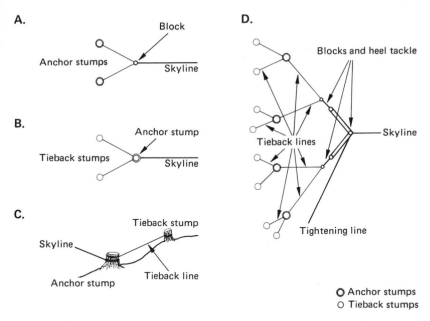

Figure 13.11. *Multiple anchors. A and B illustrate two types of two-stump anchors. C and D illustrate three-stump anchors.*

Anchor stumps can be pre-rigged just like guyline stumps. This is accomplished by rigging the anchor ahead of time and then using a strap 20 or 30 feet long and 1 1/8-inches in diameter for a skyline tagline. At times it may be convenient to use a tagline up to 200 or 300 feet long. When it is time for a cable change the skyline is simply shackled to the tagline rather than to the anchor stump.

Road changes

When the initial logging road is yarded clean the road must be changed to get the rigging over a new logging area. The system for making road changes on a highlead system is illustrated in Figure 13.12. Cable changes for single-span skylines are illustrated in Figure 13.13.

For a highlead road change the rigging slinger or hooktender will ask for a section or two (or however many are needed) before making the change. The sections, which are coiled in lengths of up to 200 feet, will be carried back to the new tail hold. Since there is usually a spare block and strap at the back of the setting it will probably already be hung where the new tail block is required to make a new logging road. The new tail hold should be chosen carefully so as not to make the road too wide or too narrow. The bight of the haywire will be placed in the block, and the yoke will be pinned shut. Then the haywire will be strung out as the dashed line indicates in Figure 13.12A. If desired, the haywire

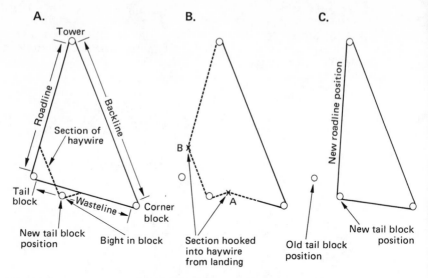

Figure 13.12. *Road change for highlead operation.*

may be sent back strung out, in which case the hooktender will have to pull the line through the brush rather than carry the coil. Either way is common and depends on the conditions.

With the haywire section or sections strung out, the present road is completed and the rigging slinger will call for the haywire. The chaser at the landing will hook the haywire from the drum into the eye of the haulback and then unshackle the butt rigging. The haulback will then be skinned back just as if the butt rigging were being returned to the brush. The difference is that the

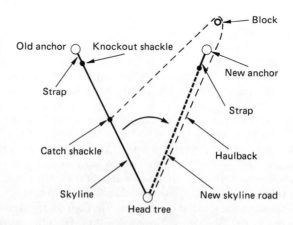

Figure 13.13. *Cable change for single-span skylines.*

258   Logging practices

haulback is now pulling out the haywire. If two men are working together to make the change one will be stationed at each X marked in Figure 13.12B. Generally at least two sections will be allowed to run by and the lines will then be stopped. The section that was strung out originally will be hooked in at either point A or B (A if there is a chance the haywire will run). The haywire will then be run ahead and when it reaches the other end of the original section it will be stopped again and attached.

When the original section is hooked in to the haywire from the landing, the haywire signal is given again and the haulback is run into the landing. The haulback is then shackled to the butt rigging, and all that is left to do is to clear the line down the new road. This is accomplished by tightlining the rigging at the landing and then running the rigging back until it is obvious that all the running lines are clear. In most cases the rigging crew will be able to see lines move over. At any rate they will make certain the lines are in their new position and there are no siwashes before they commence logging. If a siwash is left in the line and comes loose while the crew is near the rigging someone could be seriously injured or even killed.

If the new tail hold is very close to the wasteline, a haulback change can be made rather than a haywire change. Haulback changes are much faster than haywire changes but require either that the new tail hold is right next to the wasteline or that there is slack in the haulback. A haulback change is made simply by placing the bight of the haulback in the new tail block and throwing the bight out of the old tail block. The same precautions are required to be sure the line is clear before logging.

Cable change

A cable change in a skyline operation may require what amounts to making a new layout. The haywire is strung to the new anchor and is used to pull out the haulback, which in turn is used to pull out the skyline. The haywire is seldom used to pull out the skyline since it is very light line, not intended to pull such a load. When conditions are right the road change might be made as is illustrated in Figure 13.13.

A block is hung on a tree or stump behind the new anchor, and the haulback is strung out from the head tree through the block and back to the skyline. The eye of the haulback is shackled to the skyline with a catch shackle. The catch shackle has a bar welded across the U-shaped part of the shackle, which allows the skyline to move through until the eye reaches the shackle. The skyline eye will not fit through the catch shackle. The skyline will run a bit until it takes up all the slack in the haulback; then it will stop. The skyline is released by knocking the pin out of the knockout shackle used to attach the skyline to the anchor strap.

When the skyline stops running after being freed, the haulback is used to pull it up or down to its new anchor. It is shackled to the new strap, lifted to clear the line, and logging can commence. Siwashes on a skyline should be avoided, as it is not always possible to clear the line by tightlining. With a tight skyline it is imperative that the skyline be strung clear of siwashes.

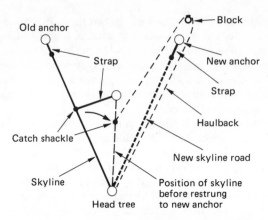

Figure 13.14. *Alternative cable change for single-span skyline. Strap catches the skyline and keeps it from running.*

The catch shackle is used for changing skyline roads. The bar is welded across the throat of the shackle to keep the skyline eye from running out of it when the skyline is turned loose from the anchor stump.

Figure 13.14 is an alternative to the cable change already described. A strap is hung on a stump partway between the old skyline road and the new position. The strap, which should be about 7/8 inch in diameter, catches the skyline, and the stump stops the skyline from running. When the skyline comes to rest the haulback is attached to the skyline and it is pulled to the new anchor as shown in Figure 13.13.

There is an infinite variety of conditions under which both road changes and cables changes must be made. While the general method of making these changes has been described in this chapter, the particular way these changes are actually accomplished depends on the specific situation the crew is working in.

14.

Cable Yarding Production Cycle

Careful planning, knowledge of operating variables, right choice of equipment, recognition of system limitations, and understanding basic operating elements are all critical to the efficient operation of a cable logging side. It must be remembered, however, that these are means, not ends. There can be several objectives for a harvesting system, but chief among them will always be production. Measured in cubic feet, thousands of board feet, pieces, or loads, production will spell out success or failure for any operation.

Manpower is responsible for turning the wheels of production. It is the yarder engineer who runs the yarder and skillfully utilizes its power capabilities. The hooktender's experience and knowledge keeps the entire yarding operation running efficiently and safely; the rigging slinger is the key man in the brush, planning the turns for maximum production. The choker setters, along with the rigging slinger, are the men who do the bull work—they move the rigging, dig the choker holes, set the chokers, fight the hang-ups—in short, they keep the rigging full and the logs moving toward the landing.

Whether or not the reader has worked or will work on the rigging as a choker setter or in some other capacity, it is important that he understand what is happening during the production cycle. That is the major purpose of this chapter—to explain what occurs in the production cycle and why.

Outhaul

A yarding cycle begins with the outhaul element—after the chokers are free of the logs at the landing. Sometimes the engineer will simply slack the mainline and engage the haulback drum, which will start the rigging back to the logging area. In most cases, however, the engineer will lift the rigging clear of the ground before starting back. This clears the chokers from the ground at the start. As the rigging is hauled back it is kept fairly tight so that the chokers are not dragged. Dragging the chokers causes them to tangle and increases the work of the brush crew, who must untangle them before they can be set.

To keep the chokers in the air during the outhaul element the main drum is braked lightly while the haulback is pulling, thus causing the lines to run tight. With an interlocking yarder the brakes are not necessary, because the main and haulback drum are synchronized—that is, the mainline is fed out at the same speed the haulback is pulled in.

In a shotgun system, picking up the rigging before the outhaul begins is especially important. The carriage must be off the ground for the gravity system to operate most efficiently.

A good engineer knows where the brush crew is by the number of wraps on the mainline drum. When the rigging is about to enter the yarding area the engineer will slow the rigging somewhat so as to avoid overrunning the crew. This will save time in spotting the rigging. Rigging is always returned as fast as possible to minimize the time required to perform the element.

When there is little or no lift on the rigging (as is the case on flat or relatively flat ground or in the back end when the tail and corner blocks are higher than the spar tree), the rigging will drag on the ground and the chokers will be tangled or lost. This is called *ground lead* and means more work for the brush crew.

Spotting the rigging

When the rigging arrives in the logging area a single whistle is used to signal it to stop. If the chokers are tangled and tightline is possible, the rigging slinger will signal for tightline. This will raise the chokers off the ground and make it easier for the choker setters to untangle them.

It is important to note that while the rigging is moving during the outhaul element and the initial phases of spotting the entire rigging crew should be in the clear. This is especially true when the chokers are either fully or partially suspended. When there is plenty of deflection the chokers will come back to the brush suspended, at least partially, off the ground. This is the best situation, since there will be little untangling. The rigging comes back fast, however, and when it suddenly stops, as signaled, the chokers 'dance'—swinging back and forth and in an arc. Anyone standing near them risks being struck and seriously injured.

When the rigging is fully suspended the lines will be stopped by the rigging slinger. When the lines have quieted enough so the choker setters can safely approach them, the chokers will be unwound, if necessary, and moved to the side toward the logs which are to be hooked and away from under the butt rigging. With everyone in the clear the lines are then slacked. Slacking the mainline will allow the butt rigging to drift toward the tail block. Slacking the haulback will allow the butt rigging and chokers to drift toward the landing (see Figure 14.1).

If the rigging is suspended far off the ground, simply slacking either the mainline or the haulback will cause the rigging to drift beyond the desired point. Such a condition exists when the crew is working in a canyon and the tail block is located up the back slope as in a concave profile. When this occurs

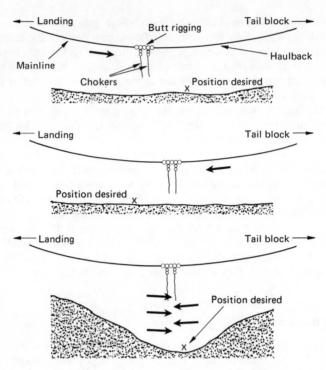

Figure 14.1. *Spotting the rigging by slacking the mainline (A), slacking the haulback (B), or alternately slacking both (C).*

the mainline and haulback are slacked alternately until the chokers are in the position that is desired.

Selecting the turn

The logs that will make up a turn should be selected by the rigging slinger before the rigging is returned to the brush. The number of logs in a turn depends on the number of chokers being used, the position of the logs, the opportunity for a bonus (more than one log in a single choker), and the safety of the crew. Always fly a sufficient number of chokers to maximize turn size relative to piece size.

Picking turns requires a bit of planning. The rigging slinger is responsible for planning. He will do some of his planning during the choker setting cycle and some during the inhaul and outhaul cycle. He must determine which logs can be reached, which have to be reached, where the choker holes are, and which logs can be included in the turn to maximize the turn size.

A rigging crew may contain anywhere from one choker setter and the rigging slinger to three choker setters and the rigging slinger, depending upon piece size and the number of chokers being flown. The normal configuration is

Rigging slinger using radio whistle (right hand) to signal "slack rigging."

two choker setters and the rigging slinger, working with three chokers. It is not necessary to add a choker setter with each additional choker; in fact, the cost per thousand of adding that man should be calculated against the increased production that will result. That cost will vary with the turn size. For example, the cost per thousand of an extra choker setter where the turn size is 2,000 board feet would be 75 percent of what it would be with a 1,500-board-foot turn. The rule is not to add people in the brush unless they will increase productivity to a significant degree. Otherwise, you are adding more to the cost than you are gaining in increased production.

Choker setting

Setting a choker is basically a simple task. The cable is wrapped around the log and the nubbin is placed in the bell. If possible, the choker should always be thrown over the log and pushed under from the opposite side. When hooked in this manner the bell opening is always down. When tension is applied the choker closes like a noose (see Figure 14.2).

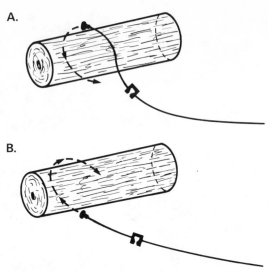

Figure 14.2. *Setting the choker—place it over the log (A) not under it (B).*

As the chokers are being slacked to the ground the choker setters carry the free end toward the logs the rigging slinger has chosen. Sometimes, when the logs are thick and well placed, both chokers can be set at once. In other cases only one choker can be set. Then the rigging is run ahead so there is enough slack to set the second choker.

Additional slack is developed either by running the rigging ahead very slowly or by calling for more slack in the mainline or haulback. Where there is lateral yarding the rigging is sprung to reach a log just barely within reach.

Springing the rigging can be accomplished in several ways. The most diffi-cult way involves slacking off the lines completely and picking up the butt rigging by hand and moving it sideways. This may be possible if only a few inches of slack are required, but if several feet of slack are required the task ranges from very difficult to impossible. In such instances the rigging slinger must use the machine to move the rigging, which in almost all cases is the best approach anyway.

Another method of springing is to hook the rear choker on a log, slack the haulback, and run the rigging ahead slowly. This will swing the rigging to the side—perhaps enough to get sufficient slack to set the front choker (see Figure 14.3A). A third method involves choking a stump as far to the side as possible, slacking the haulback, and running the rigging ahead slowly. The process can be repeated a second time, if necessary, to spring the rigging to the side (see Figure 14.3B).

If a problem requiring a spring is recognized in advance it might be easier to string out the chokers—two or three—to reach the problem log. This often happens at the back end where the yarding road widens because of the fan-shaped road layout. When stringing out is necessary it may save time to ask

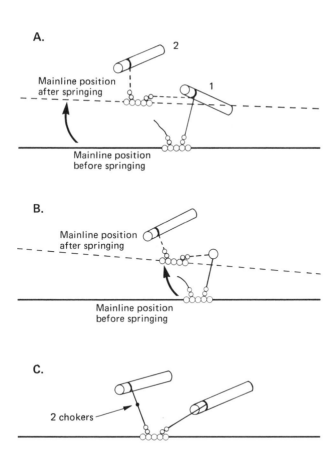

Figure 14.3. *Top views of springing the rigging and stringing out the chokers. In A, the rear choker is hooked to log 1 and the rigging is sprung to reach log 2. In B, a stump is used for springing the rigging. In C, two chokers are strung out to reach the problem log.*

for an extra choker when the rigging comes back. Then it might be possible, if the problem log is outside the layout, to set the choker while the turn is going in. When the rigging returns it is just a matter of hooking up the chokers. After stringing out, it is generally a good idea to shorten up again before sending the turn in, especially if there is a chance of hanging up on the way in to the landing.

Hooking the log

In hooking a log there are several problems. First, the chokers should not be set so close to the leading end that they can be easily pulled off in yarding. Logs lost in yarding must be picked up later—at a cost of additional time. If two chokers are being used the first choker should be at least 3 to 4 feet from the leading end. The rear choker should be set back about 10 feet. In this manner the two leading ends will be about even for landing. Remember, the two chokers hang in the butt rigging 4 to 6 feet apart.

Choker holes can sometimes be a problem. Either they already exist or they must be dug. In the past choker holes were often made using dynamite; however, this practice was eliminated, mainly for safety purposes. When a log is tight, if the rigging slinger is experienced, the digging part of the process may be avoided. If the log is lying right it can be humped around, thus making a hole. The humping is done either during the turn or during the previous turn (see Figure 14.4). The chokers are strapped (hooked to each other) and laid close to the ground at the end of the log. The haulback is slacked and the rigging is run ahead slowly. In many cases the log can be pulled from its lay in this manner. The other method is to use a technique similar to the kick described in Chapter 10. The choker is simply laid across the end of the log to be moved, the haulback is slacked, and the rigging is moved. Instead of pulling the choked log sideways, the tight log will be moved—hopefully, making a choker hole.

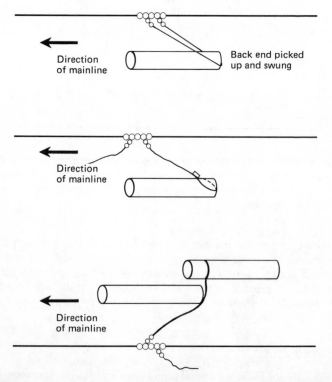

Figure 14.4. *Making choker holes with the hump or kick. Strapped (A) or single (B) chokers can be used for the hump. The kick (C) uses one choker.*

Sometimes there are no holes under logs. This choker setter is busy digging a hole. The choker is being pushed in from the other side of the log.

Moving into the clear

After the chokers are set and before the rigging begins to move toward the landing, the crew must have time to walk to a position of safety. Generally it is the rigging slinger's responsibility to choose a safe place. However, each member of the rigging crew is responsible for the safety of the other members.

The crew will move to a position out of reach of the running lines and on the uphill side of the logging. Sometimes the crew has to find safety inside the layout. However, this is rare. In any case it should never occur near the back end of the layout—near the wasteline or corner blocks. If either of the brush blocks should break loose with the crew inside the layout, the crew would be standing in the bight—a very dangerous position.

In any case, the crew should always remain standing while the turn is moving to the landing and should keep alert for logs, root wads, or rocks which could be dislodged from above. Only after the entire crew is in the clear should the signal be given to start the rigging.

Yarding the turn

Moving logs to the landing can be compared to running an obstacle course. If there is plenty of lift, or if some sort of skyline is being used the obstacles may be avoided. However, in any system problems can occur.

When the signal is given to go ahead the engineer will put his machine into gear and start the rigging moving. He will always start slowly, no matter what system is being used. With a skyline—either tight or slack—a carriage is used. When the signal is given the turn is first yarded laterally to the roadline and then up the road to the landing. With highlead, the logs are simply pulled out of their lays and started on their way.

If much slack has been used to get the chokers set—especially haulback slack—the engineer will be signaled to take the slack from the haulback before he is signaled to go ahead on the mainline.

Unfortunately, with highlead at least, the turn does not always enjoy an uneventful trip to the landing. The obstacles, and there are many, often cause hang-ups. When a turn gets hung up, on a stump for instance, the first course of action is to try to use the machine to fight the hang-up. Fighting a hang-up involves using a running tightline and/or skinning the rigging back with the log in an attempt to clear the obstacle. If this doesn't work, and it often doesn't, the rigging slinger, or possibly the hooktender, fights the hang-up by hand.

In Chapter 10 three basic maneuvers for fighting hang-ups were described—the jump, the kick, and the roll. The same methods are used in cable yarding. There is, however, a basic difference between ground skidding and cable yarding. In ground skidding the machine moves to the turn. This being the case, it is possible to avoid falling into the same hang-up trap on each successive turn. Cable yarding is different to the extent that the roadline remains relatively constant, with the logs running over the same ground and in the same pattern again and again. This means that the brush crew may wind up fighting the same hang-up on every turn.

If the hang-up is caused, as is often the case, by a stump located in a very inconvenient place the stump may be removed simply by cutting it off at ground level. If a great deal of timber is to travel over the same road, as in a skyline, and if there is no deflection at a critical point, a stubborn stump may be blown out of the ground.

A more commonly used method for avoiding hang-ups is the shear. A shear is a log strategically located to cause other logs in a turn to drag around the hang-up. Generally, when a hang-up occurs in a location where a shear will work, one of the logs is dropped off the turn and left in place. It is hoped that successive turns will hit this log and be sheared around the troublesome stump. Though the shear sometimes works, it is just one of several methods used. Often the brush crew must go on fighting the hang-up by hand.

Fighting hang-ups represents a large portion of the total delay time on a cable operation. For instance, on a highlead side, fighting hang-ups may account for 50 percent or more of the total working delays, which also include picking up lost logs. In most cases hang-ups cannot be avoided, so the problem becomes one of minimizing the possible delay. Most rigging men will try to free the hang-up by manipulating the rigging as just described. Frequently, however, they have a tendency to continue manipulating the rigging long after it is obvious the method will not work.

In the long run it is best to use the rigging only two or three times in attempting to free a hang-up. If no progress is being made an experienced choker setter should be dispatched to free the hang-up by hand. In cases where the same hang-up causes a delay turn after turn it may be necessary to leave one man at the hang-up location, thus saving travel time from the back end on each turn.

In general, there is a higher frequency of hang-up delays in salvage and thinning operations than in clearcut operations. The reason, of course, is that the yarded material must be pulled through and around standing timber.

Chasing

When the logs reach the landing they enter the domain of the chaser, who is responsible for landing the turn, decking the logs in a safe pile, and unhooking the turn so the chokers can be returned to the brush. The chaser may also saw knots off logs (called *bumping knots*), and perform second loading functions, which will be explained in following chapters.

The chaser and the yarder engineer must work together to land a turn of logs. Both men must be within view of each other while the work is being done, since they communicate with hand signals. There are two main signals: holding the arm straight up means the engineer should keep pulling the turn; dropping the arm, or holding it straight out, with the palm downward means to drop the turn.

The chaser will allow the turn to be pulled in until the leading logs are even with the rest of the logs in the deck. Then he will signal for the turn to be dropped. The speed with which he unhooks the chokers will depend on the number of chokers in the turn, whether the logs are fairly well even-ended, the slack he can get in the chokers, and the size of the deck. For example, if it takes 1.1 minutes to land and unhook two chokers it must take longer if more than two are flying. If the logs are hooked so they land even-ended, the unhook time is less. Otherwise, the chaser must unhook one or two of the chokers and then pull the remaining logs into the deck to be unhooked.

High log decks are not so much a problem today as they were in the days when entire settings were colddecked. However, the higher the deck the longer it takes to unhook the turn.

Perhaps the most frustrating experience for a chaser is to have no slack for unhooking the chokers. This occurs on very steep ground where the weight of the haulback strung out down the hill is sufficient to pull all the slack from the landing. In some cases the only way to create enough slack for unhooking is to use the haywire to hold the butt rigging on the landing. In other cases the loading crane may be used to hold on to the turn. With a slackline system on

Landing a turn on a slackline operation. Note the position of the choker on the carriage. This is a gravity system. The mainline is the skyline and the haulback is the skidding line.

steep ground the problem is diminished somewhat by lowering the skidding line block in the tree so the carriage can be sucked closer in to the tree.

One way to reduce the time required in unhooking is not to unhook at all. Instead, the butt rigging or carriage is slacked and the chokers are removed

A chaser unhooking the turn at the landing.

from the butt hook and replaced with fresh chokers. This system allows the chokers to be unhooked after the rigging has gone back to the brush.

Increasing production

An entire turn cycle may take anywhere from three to seven or more minutes depending on such variables as brush, log size, and yarding distance. Chasing, yarding and haulback times are constant. The choker setting time is the only element that can be manipulated. Speeding up the process will reduce cycle time and increase system productivity. A slight increase in the element which results in a larger turn size will also increase productivity.

In many cable systems chokers are added to increase production. The additional 0.5 minute it takes to set another choker can be worthwhile if the added volume brings the unit cost down. One study indicated that cost was 27 percent less per unit of volume when a third choker was used and carried the same amount of volume as each of the two others (Adams 1965, p. 11).

Another way to increase production is to reduce choker-setting time by presetting chokers. When the chokers are returned to the brush another set is already hooked to the logs. The empty set is removed from the hook and the filled set is hooked up. In this system a tagline is used with a carriage of some sort, perhaps a slack-pulling carriage. This is beneficial, of course, only if the chokers are set just as the rigging is returning. Close to the landing, where the crew may not be able to work safely and where the rigging is returned quickly because the yarding distance is short, presetting is not beneficial. Farther out, there is still some benefit, but not in direct proportion to the choker-setting time saved. However, in overall production there is a definite advantage.

There are some simple rules of thumb to remember for increasing production. To begin with, always arrange for the rigging to be pulled downhill to the turn, not uphill. This is a matter of laying out the logging roads in the right way. In most cases, the operator is better off, given normal yarding distances, to take a narrower road rather than to pull the rigging uphill. If the rigging has to be pulled uphill, let the machine do it. Some loggers, using a slackline system, insist they would rather use a dutchman to side-block the rigging uphill than use a slack-pulling carriage. With a slack-pulling carriage the brush crew still has to pull rigging, whereas the dutchman pulls the rigging uphill for you.

If there is any way to achieve tightline it should be done. Not only will the

Logging tractor spiking a highlead landing. This is often done where there is good tractor ground close to the highlead operation. Not only is production increased, but also the highlead loader can be used more effectively.

inhaul go faster, but the rigging crew can also get the chokers off the ground to untangle them. Ground-lead logging can nearly double choker-setting time because the chokers are badly tangled when they return to the brush.

Good equipment, including chokers, is essential for high production. When a choker gets kinked and jagger-laden it is difficult to work with and causes choker-setting time to increase. (A jagger is the result of a wire rope strand breaking. The individual strands of the rope stick out and can cause painful stab wounds on the hands.)

Reducing delay time is another way to increase production, simply because there is more time in which to produce. The application of technological advances and mechanization is also an important means of improving productivity. Grapple yarding is a good example of one such application.

Grapple yarding

The use of grapples in cable yarding is not unlike their use in ground skidding. The difference, of course, is that the grapples are transported to and from the brush by cables.

Grapples are used with a running skyline, which was mentioned in Chapter 11. The running skyline is not new—the application is new. Loggers have been using a type of running skyline, called a Grabinski, for a long time.

The Grabinski is used with highlead logging when extra lift is needed to clear some obstruction. A block is attached to the butt rigging and is hung on the haulback. When a turn is being pulled in, the haulback is run tight, which has a tendency to lift the turn or at least the leading ends of the logs. With a block on the haulback the lifting capability is increased, and both the mainline and haulback are supporting the load.

When a Grabinski is used the corner blocks are spaced about 50 feet apart. If the blocks are spaced more closely the running lines have a tendency to wrap, which causes excessive line wear. Also, to further reduce line wear, a 4- or 5-foot strap is used between the butt rigging and the block attached to the haulback. The use of the strap also helps to keep the lines from wrapping.

The running skyline is rigged in much the same way as a Grabinski. A carriage takes the place of the butt rigging and rides the haulback rather than a single block. The mainline is attached to the front or leading end of the carriage and the haulback is attached to the rear end. In addition, a third line, called a *slack-pulling line*, is used to open and close the grapples as is shown in Figure 14.5. In many cases only one brush block is used.

During both the inhaul and outhaul parts of the yarding cycle the running lines must be kept under tension. The yarding cranes mentioned in Chapter 11 are well suited for grapple yarding with a running skyline because the mainline and haulback drums are interlocking, as are many yarders. That is, both drums are synchronized so that as line is being pulled off one drum, it is pulled off at the same rate as line is being wound on the other drum. Other machines use special air brakes to accomplish the same thing. The yarding cranes, because of their swing capability, also allow the logs to be landed to the side of the machine rather than in front of it.

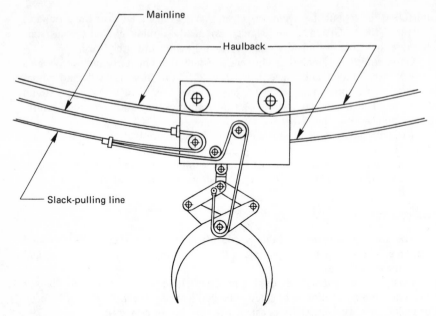

Figure 14.5. *Grapple attached to skidding carriage (Mann 1969, p. 3).*

There are several types of carriages used with the grapple system. Some are sophisticated, radio-controlled devices and some operate hydraulically. The simplest and most dependable seems to be a mechanical carriage similar to that pictured in Figure 14.5.

Like other logging systems, grapples work well in some conditions and not so well in others. Planning is essential with this system. The main factors to be considered are road location, deflection, yarding distance, ground profile, and training.

Deflection is as important in running skylines as it is in other skylines. Running skylines work best in a concave profile. The operator has better visibility, which helps when the grapples are being placed on a log. Also, yarding is faster because gravity is doing some of the work. The deflection is calculated in a manner similar to that described in Chapter 5. Deflection is necessary, since the grapples must be free of the ground during operations.

Mobility is likewise important with a grapple system. Again, the yarding crane is well suited to this requirement because it has the ability to move freely up or down a road. This is one reason for using the crane rather than adapting a conventional yarding machine to grapples. Using the cranes along the road offers certain advantages. Since the logs can be decked right in the road there is no need for landing construction, and moving and rigging costs are minimized.

Some operators use grapples in operations where yarding distances range up to 1,200 feet. In certain cases this practice is necessary and acceptable. However, grapples work best when the operator has a good line of sight to the turn. As the yarding distances increase the hooktender has to spot the turn, which tends to increase cycle time. The ideal yarding distance is between 500 and 600 feet.

In general, the shorter the distance the greater the production. However, as yarding distances decrease, road-building requirements increase. In most cases, the additional road requirements are not an advantage since they increase cost, take productive land out of production, and contribute to erosion.

The yarder, as already mentioned, works best from the road, although it can be used effectively from a landing when terrain conditions are appropriate. Several layouts can be imagined, but basically they fall into two types—the circular or fan-shaped layout and the rectangular layout.

The circular layout, illustrated on the left in Figure 14.6, has the yarder working around a pivot point. The tailhold is stationary and the yarding crane moves around it. The tailhold may be a tree, a stump, or a tailhold machine. This layout minimizes the yarding distance and is found in the curve of a road, as the illustration indicates. Note that by moving the yarder around the tailhold the maximum yarding distance is half the distance between the end points of the curve. An alternative layout would be to let the machine remain stationary and move the tailholds around the curve, but the average yarding distances would be significantly greater.

A second layout which is not uncommon is the rectangular layout. This involves moving the tailhold and/or the crane as each yarding road is completed. Once again, any kind of a tailhold can be used. If it is possible to build a rough Cat trail around the perimeter of the setting, a mobile tailhold is best. A mobile tailhold is nothing more than a tractor, perhaps with some sort of a mast built on it, that carries the brush block. It is not unusual to have as many as 10 road changes in one day with a grapple system. If the mobile tailhold is not used, the hooktender must be working ahead continuously prerigging

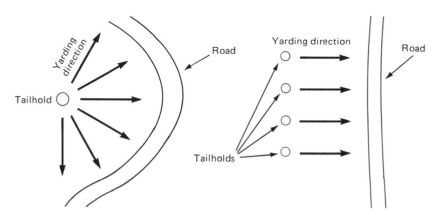

Figure 14.6. *Road layouts for a grapple system. The circular or fan-shaped layout on the left is most efficient because the machine pivots around the tailhold. The rectangular layout is shown on the right.*

The mobile tailhold used with a running skyline is simply an old logging tractor with a mast mounted on it. The line running through the right-hand block at the top of the mast is the haulback. The second line is a guyline.

tailhold stumps and making new layouts. Since the hooktender is kept busy prerigging for road changes, it may be necessary to have a second man in the brush spotting the rigging. Crew size with a mobile tailhold can be as small as two men—the yarder engineer and the hooktender. Most often the system works with a three-man crew.

The logs are landed in the road when a mobile yarding crane is used. With a stationary yarder converted to grapples the logs are landed at the foot of the tower just as they are in any other cable system. When the crane is used, the logs are often windrowed in a continuous deck either right on the road or on the shoulder. When sufficient logs have been decked, the loader moves in and begins to load out the logs.

Grapples, used in a proper location, are extremely productive and about half as costly as a highlead system used in the same location. With good conditions

Grapple yarder pulling in a turn. Grapples are spotted by a man in the brush. The system seems to work better when the operator can see the logs.

and an experienced operator, a cycle can be completed in the time required to set the chokers—that is, walk to the turn, set the chokers, and get in the clear—on a conventional highlead operation. In essence, the cycle time is reduced to that normally required for inhaul and outhaul. With short yarding distances, production of up to 600 logs per shift has been reported. However, the average seems to be between 125 and 150 logs per shift.

Grapple yarding has several advantages that conventional highlead operations do not have. For one thing, manpower requirements are reduced from a seven-man crew to a three-man crew or perhaps even a two-man crew. There are no choker costs, of course, since chokers are not used. The system is generally considered safer since fewer men are in the brush and those who do work in the brush are usually clear of the rigging. Because this system is safer it can be used for night logging, which allows for greater production per 24-hour day and faster amortization of the equipment.

Grapple yarding also has some disadvantages, of which high capital cost is one. Yarding cranes are expensive—as expensive as many conventional yarders.

This being the case, a small operator cannot afford the flexibility of having both. If such an operator does own a crane he must use it like a yarder and loses some of the flexibility and efficiency of a yarding crane with grapples. Fortunately, it is a simple operation to convert a crane into a highlead machine with chokers. In fact, chokers are used frequently with extreme yarding distances or when the machine loses deflection and it is not possible to use the grapple.

Grapple yarding can be very efficient or very inefficient, depending on the conditions. In this it is no different from the other systems described in the preceding four chapters. Each system must be used in the right place and under the right conditions to achieve maximum efficiency and lowest costs.

15.

Aerial Logging

In recent years aerial logging systems (helicopter and balloon systems) have become the hope of sale administrators, industry, and some independent logging contractors. Each system, in its own way, provides possible alternative solutions to various logging problems, including damage to the environment, escalating road-building costs, and the utilization of timber resources considered inaccessible with conventional cable systems. In some cases aerial systems provide economic alternatives to conventional systems. In other cases, though their application is not a feasible economic alternative, they do provide favorable advantages in terms of environmental impact. Like the use of other logging systems, the use of aerial logging systems must be based on their physical capabilities and the economics of their application to a specific set of conditions.

Both balloon and helicopter systems are characterized by their ability to fly forest products from a yarding area to a landing. That is, the logs are or can be suspended clear of the ground. Both are extended reach systems. Since they are not tied to the ground, these systems can negotiate significantly longer yarding distances than conventional systems can. Where helicopter and balloon systems can be used their capabilities result in less soil damage and in reduced road-building requirements.

Under the right conditions the productive capabilities of aerial systems far exceed those of conventional systems. For instance, the Sikorsky S-64 Sky-crane can easily produce 35 to 45 cunits per hour, and the balloon systems can be expected to produce in excess of 10 to 15 cunits per hour depending on which system is used. However, in spite of the high productivity potential, the systems are not widely used. Only a handful of firms offer helicopter logging services, and only four firms regularly use balloon systems.

Note: This chapter on the subject of aerial logging systems is the work of Dr. Jens E. Jorgensen, Associate Professor at the University of Washington, and is based on two unpublished reports he prepared in 1973-1974. Data have since been revised.

The lack of utilization within the industry can be attributed mainly to high capital and operating costs. Balloon systems require a total investment of about $1.5 million, a cost two or three times that of a highlead system or half again that of a slackline system. The capital cost of a helicopter suitable for heavy lift operations (not including support equipment) can range as high as $7.5 million, depending on payload capacity. Average hourly costs for logging helicopters, including support manpower, based on 1,200 hours per year, range between $1,248 per hour (4,500-pound payload) and $2,948 per hour (13,000-pound payload), as indicated in Table 15.1. Such systems typically achieve between 50 and 70 percent availability, and this should be taken into consideration when analyzing costs.

Economics notwithstanding, the aerial systems—balloons and helicopters—offer feasible solutions to serious problems. In the balance of the chapter each of the systems will be described in some detail.

Helicopter logging

Helicopters, as yarding tools, have achieved an almost dreamlike status over the years, for they have appeared to be unencumbered by many of the physical obstacles that limit the use of ground skidding and cable yarding equipment. In the very recent past, dreams have become reality as helicopter logging has taken its place as an effective system for harvesting timber crops. Helicopters indeed can lift trees out of the forest with relative ease—providing the trees or tree segments do not exceed the payload capacity of the machine. With operating ranges far greater than those of conventional equipment and an ease of maneuverability not enjoyed by tractors or skidders, highlead or skyline, the helicopter is often viewed with something less than scientific objectivity.

Closer scrutiny, however, reveals there is nothing magical or dreamlike

The Sikorsky S-64E Skycrane. Payload 18,800 pounds; fuel consumption 525 gallons per hour. Used in heavy timber logging and heavy construction.

Table 15.1 *Helicopter Logging Cost Data (costs based on 1980 prices and before-tax operating costs)*

	Bell 214B	Sikorsky S-61[1]	Vertol V-107[1]	Sikorsky S-64[1]
Acquisition cost (dollars)				
Helicopter[2]	1.80MM	5.00MM	6.00MM	7.00MM
Auxiliary equipment[3]	.12	.12	.12	.20
Total	1.92MM	5.12MM	6.12MM	7.20MM
Operating costs, 1,200 hrs (dollars)				
Hourly fixed	470	1216	1450	1725
Hourly direct	600	800	700	1000
Hourly labor support[4]	178	189	189	223
Total	1248	2205	2339	2948
Productivity/cost				
Average load/turn, lbs	4500	5500	6500	13000
Lbs/bd ft	10	9	9	8
Bd ft/turn	450	611	722	1625
Turns/hr, ½ mile	22	20	20	20
Gross M bd ft/hr	9.9	12.22	14.44	32.5
Net M bd ft/hr[5]	8.4	10.4	12.3	27.6
Cost/net M bd ft/hr[6] (dollars)	148	212	190	107
Turns/hr, 1 mile	20	18	18	18
Gross M bd ft/hr	9.0	11.0	13.0	29.2
Net M bd ft/hr[5]	7.65	9.35	11.0	24.8
Cost/net M bd ft/hr[6] (dollars)	163	236	213	119
Turns/hr, 2 miles	18	16	16	16
Gross M bd ft/hr	8.10	9.78	11.55	26.00
Net M bd ft/hr[5]	6.89	8.31	9.82	22.10
Cost/net M bd ft/hr[6] (dollars)	181	265	238	133
Turns/hr, 3 miles	16	14	14	14
Gross M bd ft/hr	7.20	8.55	10.11	22.75
Net M bd ft/hr[5]	6.12	7.27	8.59	19.34
Cost/net M bd ft/hr[6] (dollars)	204	303	272	152

Source: Bell Helicopter Textron.

1. Out of production. Kawasaki produces KV-107, an uprated version of V-107.
2. Assumed operator cost, although replacement may not be available.
3. Cost of fuel trucks, maintenance equipment.
4. Hourly labor support as follows:

	214B	S-61/V-107	S-64
Hooker @ $8.40/hr	$ 8.40 (1)	$ 8.40 (1)	$ 8.40 (1)
Choker setters @ $8.40/hr	16.80 (2)	25.20 (3)	33.60 (4)
Chaser @ $8.40/hr	16.80 (2)	16.80 (2)	25.20 (3)
Pilots @ $19.68/hr	78.72 (4)	78.72 (4)	78.72 (4)
A/c mechanics @ $10.93/hr	21.86 (2)	21.86 (2)	32.79 (3)
Total	142.58 (11)	150.98 (12)	178.71 (15)
Plus 25% due to non-flight delays	$178.00	$189.00	$223.00

5. Assuming 15% scaling loss.
6. Does not include Forest Service items such as taxes, interest, and forest management expenses.

The Sikorsky S-61L. Payload 7,100 pounds; fuel consumption 159 gallons per hour. Used in medium-to-heavy timber logging, fire fighting and control, fertilizing, general construction, and for passenger transport.

about helicopter logging. Like any other system, it is subject to limitations posed by certain operating variables and must interrelate with necessary support equipment.

When skylines were discussed it was pointed out that line tension and line speed are two key variables. The analogous variables in a helicopter system are net payload capacity and speed of the aircraft. Lift capacity ranges from 1,000 to 28,000 pounds, depending on the machine. The size or lift capacity requirement of the aircraft depends on the job. In the heavy timber of northwestern forests the lift requirement is from 8,000 to 20,000 pounds. In the Intermountain region or the eastern and southeastern United States, the requirements are much lower, ranging from 3,000 to 5,000 pounds. Table 15.2 lists the specifications of some helicopters suitable for logging.

In studying Table 15.2 the reader will note that price and operating costs increase as aircraft size increases relative to lift capacity. However, the price per pound of lift capacity or operating cost per pound of lift does not vary on the same scale among the aircraft. Attempting to select an aircraft on the basis of least capital cost or operating cost per unit of lift can be a mistake. The selection must be made on the basis of the operating variables likely to be encountered. Many of the variables are related to topography, just as they are in other logging systems.

Altitude and temperature

Helicopter payload—hence productivity—is dependent on operating altitudes and temperatures. More specifically, it is the variations in these parameters

Table 15.2 *Helicopter Basic Data*

Characteristic	Aircraft make and type			
	Bell 214B	Sikorsky S-61L	Sikorsky S-64E	Boeing Chinook 234
Power rating (shp)	2,250	3,000	9,000	8,150
Transmission limit (shp)	1,850	2,500	6,600	6,750
Gross weight (lbs)	13,800	22,000	42,000	51,000
Maximum payload (lbs)	8,285	8,500[1]	19,200[1]	28,000
Effective payload (lbs)	8,000	7,100[2]	18,800[2]	26,900
Rated cruise speed (mph)	115	139	109	170
Rated rate of climb (fpm)	1,740	1,300	1,300	1,585
Fuel consumption (gal/hr)	na	159	525	373
Purchase price ($)	1.8MM[3]	5.0MM[3]	7.0MM[3]	7.4MM[4]
Purchase price/effective payload ($/lbs-payload)	225	704	372	275
Operating cost ($/hr)	1,170[5]	2,117[5]	2,836[5]	886[6]
Operating cost/effective payload ($/lbs-hr)	.146	.298	.150	.033

Sources: Sikorsky Aircraft, Bell Helicopter Textron, Boeing Vertol Co. Some data estimated.

1. Maximum payload at sea level on a standard day (15°C, 59°F).

2. Effective payload at 2,000 ft altitude, 70°F with pilots and cargo sling. Calculations based on 40 min of total fuel (including reserve) and pilot weights of 200 lbs each for two pilots.

3. Prices are for basic aircraft with standard equipment and are in 1980 dollars. S-64E and S-61L are no longer produced, and prices shown are estimates of current market value.

4. 1979 dollars.

5. Aircraft and crew costs only (pilots and mechanics), based on 1,200 annual revenue operating hours.

6. Mature operating costs, in which the aircraft has been in service 4-5 years.

Bell 214B has payload of 8,000 pounds, although a turn of logs will typically run about 4,500 pounds. It is available in cargo or passenger configurations for a variety of uses. Photo courtesy Bell Helicopter Textron

which affect the operation. In the higher elevations of mountainous regions during the heat of the summer, the payload of a given aircraft may decrease by as much as 25 percent of its rated lift capacity at sea level. For example, in the interior of Oregon and in Idaho rising temperatures cause loss of payload capacity, and afternoon production therefore suffers. Typically, temperatures vary from 65 degrees Fahrenheit in the morning to 90 degrees in the late afternoon. The same does not hold true in coastal Oregon and Washington, where altitude and temperature differences have no significant influence on helicopter operations. Figures 15.1 and 15.2 illustrate the relationship between altitude, temperature, and payload.

Descent rate

Changes in elevation (altitude) also affect cycle time since helicopters have a maximum descent rate. Yarding speed depends on aircraft velocity and yarding distance. As yarding distance increases, with constant helicopter speed, cycle time increases proportionately. However, in steep terrain, where a helicopter is most likely to be used, there is a definite relationship between speed and maximum descent rate. Since the descent rate is already fixed, it is necessary to choose a forward speed that will make it possible to achieve the minimum inhaul time.

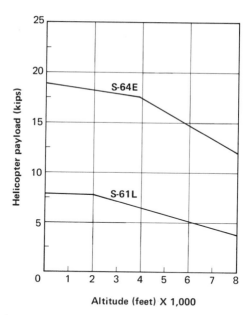

Figure 15.1. *Lift capacity of Sikorsky helicopters in logging operations versus altitude at 80 degrees Fahrenheit.*

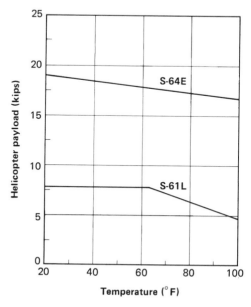

Figure 15.2. *Lift capacity of Sikorsky helicopters in logging operations versus temperature, operating at 4,000 feet altitude.*

Figure 15.3 can be used to illustrate the problem. Assume that the elevation (H) is 2,000 feet; the yarding distance (L) is 4,000 feet; and the glide angle or descent angle is 50 percent. The maximum rate of descent is 2,000 feet per minute. The helicopter must travel 4,000 horizontal feet and 2,000 vertical feet to arrive at the landing. If the maximum vertical descent speed is 2,000 feet per minute, it will take one minute to travel from the yarding area to the landing.

If the helicopter went faster it would have too much altitude when it arrived at the landing and would have to hover to descend, thus increasing cycle time. If the trip were made in more than a minute the helicopter would reach ground level before arriving at the landing. In the example, the shortest inhaul time is one minute, and it is set by the change in elevation.

If the yarding distance were increased the inhaul could be made in the same time by increasing aircraft speed. For example, if the landing were 8,000 feet from the yarding area the helicopter could still descend 2,000 vertical feet in one minute. The forward speed would be 8,000 feet per mintue or 90 miles per hour. If the landing were located still farther away from the landing area the forward speed necessary to make the descent in one minute might exceed the machine's capability. In short, with an elevation difference, landing location is not critical so long as the landing is not too far from the yarding area.

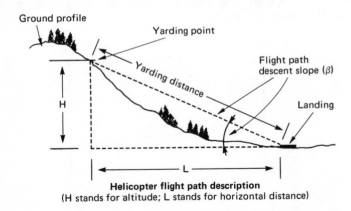

Helicopter flight path description
(H stands for altitude; L stands for horizontal distance)

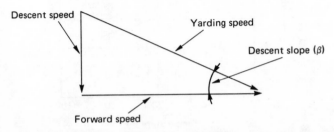

Helicopter speed diagram

Figure 15.3. *Descent rate, yarding speed, and landing location relationships.*

Helicopter velocity depends, to a large extent, on the average slope within the yarding area and the maximum descent rate. On moderate slopes, up to around 20 percent, the aircraft will operate very close to maximum cruise speed of 80 to 100 miles per hour. As the slope becomes steeper—35 to 40 percent—speed must be reduced to the 40 to 60 miles per hour range.

Volume per turn

The productivity of a helicopter is directly proportional to the volume per turn. Like any other primary transportation function, maximum payload is essential in aerial logging. In fact, because of high operating costs, maximum payload is all the more important when helicopters are being used.

The helicopter is weight constrained. Load volume has little to do with scaled volume per se, but rather with wood density. In order to ensure maximum payloads, wood density must be estimated or even tested for each species to be yarded from a particular show, and log lengths must be cut accordingly. Figure 15.4 is a typical log scaling versus helicopter load diagram.

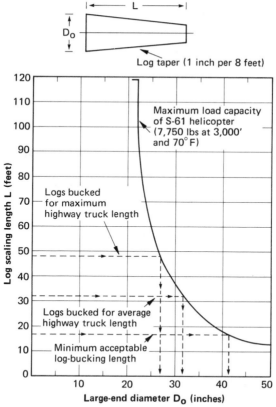

Figure 15.4. *Log scaling length versus large-end diameter.*

A problem that occurs when logs are bucked for optimum load length, is that buckers must ignore log grade. Figure 15.5 gives an example of the value loss in a medium-sized Douglas fir tree with a density of 50 pounds/cubic foot. Figure 15.6 estimates the value losses resulting from bucking for load length with three helicopters of different capacities. On the one hand, the lower the load capacity of the aircraft the higher the potential value loss, especially as scaling diameters increase. On the other hand, with a heavy lift machine like the Sikorsky S-64E the losses tend to be lower unless log diameters are very large. The fact remains, however, value losses in large timber through degrada-

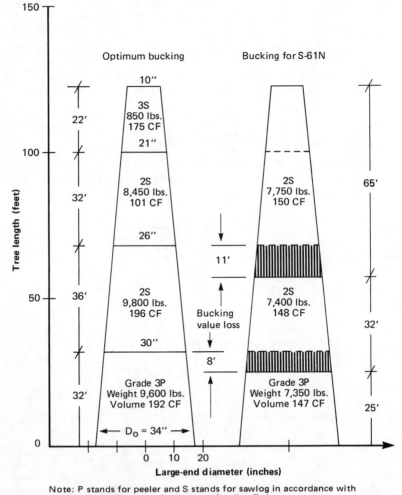

Note: P stands for peeler and S stands for sawlog in accordance with the Columbia River Log Scaling and Grading Rules.

Figure 15.5. *Illustration of bucking value loss of medium-size tree–Douglas fir (density 50 pounds per cubic foot).*

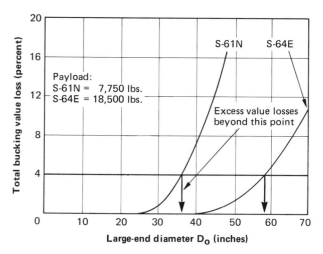

Figure 15.6. *Estimate of bucking value losses for different helicopter payload capacities and large-end tree diameters (Douglas fir).*

tion are very real and should be considered when helicopter logging is a possible alternative.

Weather

The productivity of any system is heavily dependent on availability. Helicopter yarding is greatly influenced by the weather—rain, fog, snow, wind, and, as has already been mentioned, temperature. In the Pacific Northwest the system is especially sensitive to rain, fog, and snow. The flight availability data given in Figure 15.7 indicate that in the Snoqualmie region of Washington State a logging helicopter will be able to work only six months of the year—with an even chance of being rained out. The seasonal average is only 65 percent availability. The season in the Northwest extends from mid-April through late October.

Weather graphs, such as Figure 15.7, are invaluable for planning helicopter operations. They will, of course, vary with the region. Thirty-five percent downtime in the Pacific Northwest or anywhere else is enough to have a significant effect on operating profits.

Winds and gusts also have a significant effect on helicopter logging. The wind both affects lift characteristics and creates hazardous conditions for the woods crew during hooking operations. When logs are being removed, the helicopter hovers in position with a 100- to 250-foot tagline hanging below the aircraft. The choker eyes are hooked to the tagline by a crew member called a *hooker*. It is the hooker's proximity to the tagline, while the aircraft is being buffeted by the wind, which causes the hazard. The situation is not unlike that of a highlead choker setter standing next to the running lines when a turn is moving to the landing. Winds or gusts of 20 to 25 miles per hour are sufficient

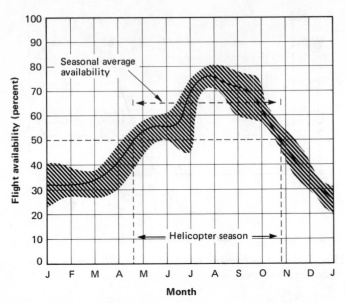

Figure 15.7. *Flight availability for helicopter yarding in the Snoqualmie, Washington, area in the 500- to 5,000-foot altitude range. Chart is based on maximum ceiling and minimum precipitation.*

to create a hazard. The hazard is further multiplied by a tendency to operate with as short a tagline as possible in the belief it will decrease hookup time.

Other variables

The several variables already discussed in this chapter are related, directly or indirectly, to operating variables discussed in previous chapters in relation to the logging systems described. In fact, all skidding and yarding variables affect aerial logging. However, aerial logging is, if anything, more sensitive to them because of the large capital investment and the emphasis on high production made necessary by such an investment.

Crewing

Depending on the equipment being used, a helicopter logging side may require from 14 to 28 men. The typical personnel requirements for an S-61 or S-64 yarding show, including everything from cutting the timber to loading out the trucks, are:

Woods crew   3-5 faller and buckers
                                    1-2 turn markers
                                    2 hookers
                                    4 choker setters

Landing crew	2-3 chasers and knot bumpers
	1 loader operator
Aircraft crew	4-5 pilots
	4-5 maintenance mechanics
Supervision	1 project supervisor
Total	22-28 men

The cutting crew, of course, is required whatever logging system is used. The equivalent of a rigging crew is also required on a conventional yarding side. In the case of helicopters, three to four choker setters are used with one lead man called a *bull choker*. Unlike turns in conventional systems, all turns for the helicopter are marked in advance to assist the hooker in locating turns and to assure maximum loading. In addition, the limits of the yarding strip are marked with colored ribbon, which delineates the top, sides, and bottom of the strips. Either ribbon or tape is used to mark the turns. The men who do the marking are appropriately called *turn markers*.

The hooker's job has already been briefly described. Two men are required because of the demands of the job. The two men work as a team, trading off after each six or so turns.

The landing crew is composed of chasers, knot bumpers, and the loader operator. The large crew is required because of the great volume of wood handled in a day. The helicopter can land a turn at, on the average, intervals of two to two and a half minutes. Each turn consists of one to five logs. The chokers must be unhooked and sent back to the choker-setting crew without causing delays in the yarding operation. Also, if the helicopter is yarding tree-length material, the trees or segments will have to be cleaned up (the knots bucked off) or rebucked into proper lengths. The job of the loader operator is equally demanding, for he must move the logs and get them loaded onto the haul trucks.

Typically, a working helicopter has two pilots and two copilots, who work in two five-hour shifts, each spending two and a half hours in the pilot's seat. Copilots are responsible for engine controls. In many operations an auxiliary helicopter is often used to fly the woods crew in and out (cutters as well as choker setters) and to return chokers to the woods. However, chokers may also be dropped 'hot' by the working helicopter, depending on size of the aircraft and general conditions. At any rate, the presence of an auxiliary aircraft makes necessary a fifth pilot.

Maintenance is generally done during nonyarding hours. Routine inspections are made during yarding, refueling, or lunch breaks. At the minimum two mechanics are required for each aircraft flying.

Auxiliary equipment

Part of the cost of any logging operation is the necessary auxiliary equipment, such as landing saws, crew buses, fuel storage trailers, fire equipment, and a water truck. Generally fuel is delivered to a landing on a scheduled basis.

However, with a helicopter show the fuel truck is auxiliary equipment.

The aircraft require several other pieces of equipment not normally found on a conventional logging operation. Because maintenance is critical a shop trailer with spare parts is located on site. A hoist is needed for engine and rotor maintenance. Auxiliary power units are necessary for starting the aircraft and as a source of electrical power.

There are, of course, other equipment requirements not normal for conventional operations. For instance, when chokers are preset, as is often the case, up to 150 twenty- to twenty-five-foot chokers are needed. At times even more chokers would be helpful, but at a certain point cost becomes prohibitive. The investment in auxiliary equipment on a helicopter operation is high—much higher than on conventional operations. Most operators figure the extra investment at between 15 and 20 percent of the helicopter purchase price.

The yarding cycle

The helicopter yarding cycle is a serial operation which involves hooking the turn, yarding it to the landing, releasing it, and returning for the next turn. As has been mentioned, chokers may be set 'hot,' just as in conventional cable systems, or they may be preset. However, presetting seems to make the most sense because of the necessary orientation toward high production. At hourly rates up to $2,948 it makes good sense to uncouple the choker setting from the yarding element.

The basic elements of helicopter yarding, excluding choker setting, are hooking, yarding, release, and haulback. The time distribution of the elements, as well as the range of percent distribution, is illustrated in Figure 15.8. Hookup and release time are fairly consistent. However, the percent distribution will vary with flight distance from the yarding area to the landing. As flight distance increases, the percent of time spent hooking and releasing decreases. This same relationship exists among the work elements of conventional systems. However, there can be some variations in hooking time depending on size of timber, makeup of loads, and terrain conditions.

Hookup

The hooker is a key man in the helicopter operation. His job is not unlike that of a 'Cat' hooker in a tractor operation but is considered more demanding and perhaps a bit more hazardous. He is responsible for, among other things, finding and hooking marked turns, picking up unchoked merch logs, and directing the flight of the helicopter when hazards or hang-ups occur. In addition, he must reset chokers when the turn is too heavy or rehook a turn that for some reason is aborted by the pilot. Finally, the hooker is responsible, to a large extent, for the safety of the rest of the woods crew—the other hooker, the choker setters, and the cutting crew if they are working nearby.

A turn might include anywhere from one to five chokers depending on the log size and weight. The chokers will probably be preset, but if the load marker

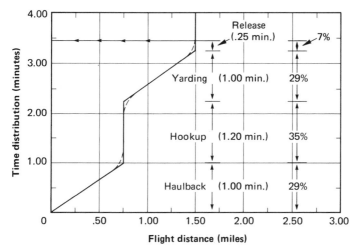

Figure 15.8. *Typical time distributions for the various yarding phases for an S-61L helicopter in logging operations in a clearcut area.*

has underestimated the weight, or if merchantable trees are left unchoked, the hooker must correct the problem. As the number of chokers increases to five the hooking time is likely to increase by 25 to 30 percent. This increase in time is caused by several factors. One factor is the difficulty experienced in hooking five choker eyes into a limited hook space. Another is that as the number of logs per load increases so does the number of aborts due to overload.

Productivity is directly proportional to volume per turn. Just as it is in any other logging system, maximum load size is important. Although it is against Federal Aviation Agency (FAA) regulations, overloads of 10 to 20 percent can be yarded but not without added expense. Continuous overloading definitely increases aircraft wear and maintenance costs. Perhaps even more important is the safety factor associated with overloads.

The turn is hooked to a tagline normally between 100 and 150 feet in length. Longer taglines are required under certain conditions, such as in a partial cut where the residual stand must be protected. In this case the tagline might be between 200 and 250 feet long.

The helicopter tagline assembly also has a load cell which monitors load weight and indicates overload conditions when they exist. The chokers are attached to a load release hook, which releases the load electrically when the turn is at the landing. In addition, there is an emergency release hook, which can be operated either electrically or manually. Figure 15.9 illustrates the tagline system.

The hooker works around hazards similar to those experienced by the choker setters. He works downward from the top of a hill so that he is never working under felled and bucked timber. In steep terrain, in fact under any conditions, he must move into the clear before a turn can be removed. However, there are additional hazards associated with helicopter logging to

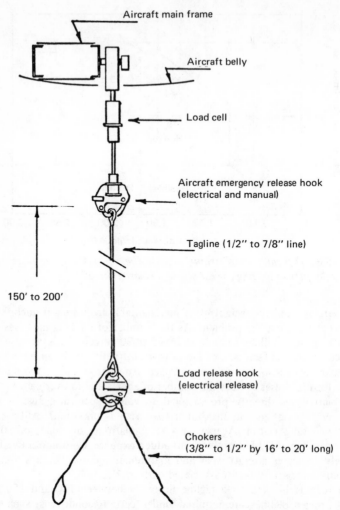

Figure 15.9. *Representation of the helicopter hook and tagline system.*

which the hooker is particularly vulnerable.

One hazard is created by the downwash caused by the helicopter's rotor. The downwash results in buffeting, causing loose limbs and rotten tops to fall out of standing timber. The buffeting also causes heavy dust swirls directly under the aircraft. Since the hooker must work in these conditions, he must do so with care. Not much can be done about the dust swirls, but in partial cuts the helicopter often will fly over the timber, before hooking begins, in an attempt to dislodge any loose limbs or tops.

The brush crew, both hooker and choker setters, must always be aware of the flight path of the helicopter. The pilot should never fly over the ground

Close-up of the load cell and attachment to the main aircraft frame. Load cell is rated at 20,000 pounds and is accurate within 100 pounds. A digital readout of the load is used by the pilots.

crew. Though the hooker attempts to direct the helicopter crew away from the ground crew, the ground crew should be alert nonetheless.

There is one special hazard for the hooker—that is the possibility of electric shock off the tagline. The electrical shock is caused by static buildup in flight, weather conditions, and poor grounding of the hook and electrical cables.

Close-up of the tagline release hook. Button at the center of the hook is the manual release which is operated from the cockpit or by the ground crew. Hook is rated at 30,000 pounds. This hook design appears to work very well.

Close-up of the load release hook. It is manually opened by the choker setter attaching a load and electrically by the pilot on load release at the landing. Guard ring protects the hook and helps the hooktender grab it. It may become charged with static electricity during a flight and shock the hooktender unless discharged by allowing the hook to hit the ground first.

Because of the potential for shock many hookers wear linemen's gloves when working around the hook.

In addition to his normal hooking responsibilities the hooker can be of enormous assistance to the helicopter crew and can significantly affect cycle time by helping to locate the turn in the yarding area. The hooker will always, or should always, wear brightly colored clothes, generally fluorescent orange. Good radio communication is also helpful. In some cases simple clicks on the radio are adequate. When the helicopter is outbound for the yarding area the hooker should move around, since a moving figure is easier to pick up visually. In some cases it may be necessary for him to attach a colored ribbon or tape to a choker and wave it in the air.

The hooker must be constantly aware of departure routes and be ready to react quickly in the event of an abort or hang-up. When the turn is clear of the area the hooker uses the radio to advise the helicopter crew of the next turn's location. The hooker can also use the radio to warn the pilots of obstacles in the flight path, coupled logs (Russian couplings),* and hang-ups (such as chokers winding around standing trees).

Hot versus preset chokers

While many operations use preset chokers, setting chokers hot is an acceptable method in some situations. Hot choking ordinarily takes place in a green partial cut, when the hooker is working at least a day behind the choker setters. Chokers are set in strips marked for location. However, despite good communication there are times when the hooker cannot find the strips. When this happens, or if the helicopter is working in small timber, some chokers may be set hot in order to avoid yarding delays.

The choker-setting crew can normally set a five-log turn in five minutes, but the helicopter can consume chokers faster than the men can set them. The choker inventory could be increased, but normally it is fixed. Under these conditions it is just a matter of time before the operation goes hot; that is, before chokers are being set in the hooking area.

One serious disadvantage of hot choker setting is the risk of accident. There will naturally be more men working around the yarding operation. Flexibility is also a problem. When working hot there is no place to go if the wind shifts in midday and makes yarding impossible. With preset chokers there is always another place to go, or at least there should be. An advantage of setting chokers hot, however, is that there is less chance of losing them.

The preset method allows for much more flexibility, since the chokers are set anywhere from four to eight hours ahead of hooking. If the wind shifts, the entire operation can move to another strip and production can continue almost uninterrupted. When the chokers are being preset the choker-setting crew has the easiest job on the show. Nevertheless, the crew can have a significant impact on production through poor turn makeup, poor choker setting, and the choking of obvious culls.

* A Russian coupling is a condition where two logs are attached to each other because they were not bucked clean.

Yarding time and distance

The time consumed in yarding depends to a large extent on yarding distance, slope (descent rate was discussed earlier), and machine cruising speed. As it turns out, yarding velocity increases proportionately with yarding distance. Therefore, when slopes are involved inhaul cycle time remains independent of yarding distance until the distance from the landing to the yarding point is so great that the helicopter is operating at maximum speed. Any further increase beyond this point will increase total cycle time linearly with yarding distance. On level ground, however, inhaul time varies directly with yarding distance.

Yarding distances generally range from a half mile to three miles. Yarding cost, to a large extent, is related to yarding distance. The upper boundary is set by ascent and descent rates and the altitude difference. The lower boundary, on level ground, is set by maximum air speed. An air speed of 80 miles per hour is considered reasonable with a heavy load. In the Northwest, yarding costs range from $107 to $303 per MBF for a half-mile to three-mile range. Beyond three miles, with some exceptions, costs increase to the point where helicopters are not feasible.

The inhaul element is basically a pilot show. He chooses the route and therefore directly affects flight time from the yarding area to the landing. Good radio communication is essential, especially during the yarding cycle. The pilot must inform the landing crew when he is coming in with a critical load; for instance, a Russian coupling. He may also, by request from the landing crew, land a load in a particular location. Constant communication is necessary for a great many reasons, but one of the most important of these reasons is to maintain crew safety.

Choker release and landing

The release element involves spotting the load in the proper place and releasing it from the tagline. This is accomplished electrically, from the helicopter. If an auxiliary helicopter is being used to ferry the chokers back to the brush crew the working helicopter will leave immediately after releasing the load. The chaser or chasers will then unhook the chokers and bundle them for pickup and delivery to the brush. In some cases only one aircraft is being used, and it will have to pick up the prepared bundles and deliver them before beginning the hooking element.

The landing area (see Figure 15.10) must be sufficient to accommodate load release, log decking, and truck loading. These activities require an area of approximately 200 by 300 feet. In addition, if the landing area is also used for refueling or other activities it must be proportionately larger. The landing should be located to accommodate the aircraft approach and haulback pattern. This means minimizing obstructions in the flight path and orientation with prevailing winds. Finally, the landing area should be oiled prior to use and should be wet down more or less continuously while in use, to minimize the dust problem.

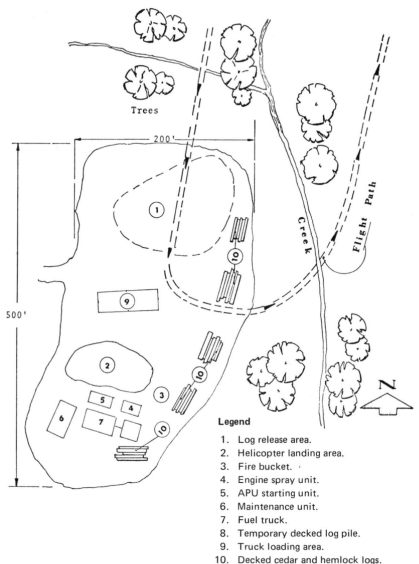

Figure 15.10. *Layout of a typical helicopter landing and log release area.*

Legend

1. Log release area.
2. Helicopter landing area.
3. Fire bucket.
4. Engine spray unit.
5. APU starting unit.
6. Maintenance unit.
7. Fuel truck.
8. Temporary decked log pile.
9. Truck loading area.
10. Decked cedar and hemlock logs.

Haulback time

The haulback element is uncomplicated. As a rule, a direct return is negotiated at maximum speed. Return time is almost totally dependent on yarding distance. If the yarding aircraft is also responsible for returning the chokers the haulback time is slightly longer, as is the total yarding cycle.

Aerial logging   301

When the helicopter begins the haulback element the hooker will move around to aid the pilot in perceiving the turn location. Bright clothing and good radio communication are also helpful.

Delays

Net yarding time in a helicopter operation is generally expressed as total available time minus time required for aircraft refueling. Four minutes for refueling and 45 minutes between refuelings is considered reasonable. However, refueling is just one of several delays affecting aircraft availability. Other delays are those attributable to weather conditions, aircraft maintenance, and other productive and nonproductive conditions. The distribution of delays might be as follows:

Weather time loss	35%
Aircraft maintenance	4%
Other mechanical equipment	3%
Delays	6%
	48%

The time lost due to various delays leaves only 52 percent of total available time for yarding. This percentage may vary somewhat, but in the coastal region of the Pacific Northwest 70 percent availability is considered the maximum, with the average around 50 percent. Causes of other delays are:

1. Chokers are not bundled at the landing, resulting in delay in pickup.
2. Hang-ups in the brush delay inhaul time.
3. Overweight loads result in abort of flight.
4. Poor choker setting causes the aircraft to set the load down for rehooking.
5. Inability of pilot to find hooker in return element results in increase of hooking time.
6. Sometimes the hook fails to release at the landing. The chaser must manually release the hook and reset it.
7. If the hook catch tongue is left open it must be reset by the hooker before the turn can be hooked.
8. An inexperienced crew will not only cause delays but will not achieve maximum production until it becomes more proficient.
9. Difficulty in positioning the hook causes an increase in hooking time.
10. Hook malfunction occurs and requires a new hook and and cable set. (Two or three hook-and-cable sets are kept at the landing for just such a contingency.)
11. Bundles of chokers are sometimes lost in the outbound flight and must be found. This occurs by accident when the pilot inadvertently releases the hook in flight (referred to as a *punch-off*).

Downtime and delays are serious, perhaps more serious in helicopter operations than in conventional systems because of the high cost of operation. With between 31 and 42 minutes out of each potential operating hour actually available for productive work it stands to reason these delays have a dramatic effect on costs. It naturally behooves the operator to minimize delays in every way possible. Good planning is helpful but is not a solution to the weather problem. Maintenance, however, can in most cases be scheduled so that it has the least negative impact on operations. As for the "other delays" just listed, simply being aware of them is helpful. Some can be avoided with good supervision, and the effects of others can be minimized with proper planning and preparation.

The place of helicopter logging

Helicopter logging can be very expensive. Even considering the favorable impact on the environment resulting from less road building and less soil damage, the use of helicopters may not be economically feasible.

This does not mean that helicopters do not have a place in the logging industry. However, it is clear from what has been said that the variables should be examined very carefully, as should the alternatives to helicopter logging— highlead, slackline, etc. In general, for helicopter logging to be feasible the value of the timber on a unit basis must be high. Road access must be extraordinarily high-priced to help offset yarding costs. Finally, if roads can be built, but the area is subject to sensitive environmental conditions (such as poor soil stability), the helicopter may be the answer. In any event the total cost-benefit equation must be examined carefully or financial disaster can easily be the result—regardless of how good the job looks.

Balloon logging

Balloon logging, unlike helicopter logging, is not a true aerial logging system. It is basically a lift-augmented skyline that combines the lifting capability of the balloon with the characteristics of certain conventional cable systems. The objective is to fly the logs, clear of the ground when necessary, from the yarding area to the landing. In addition to its aerial capabilities the balloon system also extends the reach of conventional systems far beyond their normal capabilities. Figure 15.11 illustrates the three basic skyline modes used in balloon logging.

The aerial logging capability and extended reach are major advantages of the balloon system, allowing the harvest of timber from extremely difficult terrain. Such terrain is defined by long, unbroken slopes and steep, mountainous features, where sufficient deflection is not available for conventional skyline systems. Properly used, the balloon can result in reduced road requirements. In particular, it can reduce the necessity for mid-slope roads, which are major environmental offenders.

Balloon yarding began in the late 1950s, with the first attempts taking place in Sweden followed by further experiments in British Columbia in 1963. In both instances barrage balloons were used, and it was found that they had insufficient lift capacity for the job.

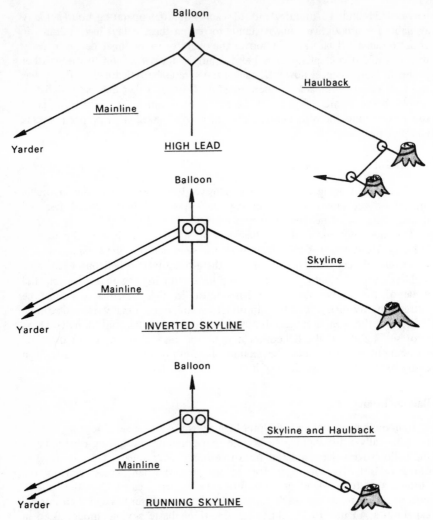

Figure 15.11. *Balloon logging rigging systems (Peters 1973b, p. 579).*

A successful balloon system must have a balloon with a 15- to 20-thousand-pound lifting capacity. In addition, the system must have a yarder that is matched to balloon size, as well as to terrain features in the area of operation. The Bohemia Lumber Company of Eugene, Oregon, pioneered a balloon system featuring a natural- or onion-shaped balloon and a Washington 608 Aero-yarder with the capabilities required. The only commercially available balloon logging system, it has been used by several companies in the north-western United States since being marketed by Flying Scotsman, Inc., also of Eugene.

The barrage balloon was used in the first balloon yarding experiments.

The balloon

The Raven balloon developed by Bohemia Lumber is helium-filled (530,000 cubic feet) and has a static lift of 23,000 pounds. The balloon is also called a *natural-shaped* balloon. The lower, conical section acts as an expansion

The experimental kite or delta-wing balloon (U.S. Forest Service).

The helium-filled, natural-shaped Raven balloon.   *Photo courtesy* Forest Industries

chamber for the gas which expands when heated by the sun. The expansion chamber prevents skin rupture caused by too much pressure.

The natural-shaped balloon carries the load through a series of steel cables embedded in the coated nylon fabric making up the balloon's skin. The cables are attached to a steel ring at the top, collected at the lower end, and fixed to the tether line. A series of tiedown ropes are also attached to the balloon. These ropes are allowed to hang loose during yarding or may be collected and attached to the tether line. The tiedown ropes are used to secure the balloon at the bedding area when the balloon is not being used.

The bedding area is between one and two acres in size and must be well sheltered from the wind. If stumps are not available the tiedown is made to a deadman, buried about 30 inches deep for the center tiedowns and 24 inches for the outside tiedowns. The tiedown area is used for initial inflation, for shelter, for storage during the off-season, as well as for routine maintenance and repair activities.

The yarder

The yarder should be a two-drum, continuously interlocked machine that allows maximum power transfer to the drums as well as maximum control. The Washington 608 Aero-yarder utilizes a hydraulically operated planetary inter-lock system which is substantially the same system used in Skylok yarders. The yarder has a level-wind system to ensure proper winding of the 1 1/8-inch lines onto the drums. The machine can carry 5,500 feet of mainline and 7,000 feet of haulback. Average line speed is between 1,200 and 1,600 feet per minute. The maximum line pull is 80,000 pounds on the mainline and 40,000 pounds on the haulback. The Washington 608 Aero-yarder is capable of working over distances up to 5,000 feet, depending, of course, on the way in which the system is rigged.

These photos show the Raven natural-shaped balloon during inflation and inspection at the bedding-down area. (1) Inflating the balloon. (2) The balloon completely pulled down at the bedding-down area. Balloon is now held by the bedding-down straps. It is ready for inspection and storage. (3) The fully inflated balloon. (4) The balloon being inspected.

A later development by Flying Scotsman, Inc., termed the "yo-yo" system, substituted two single-drum Smith-Berger yarders for the multidrum machine. Ideally, one machine would be placed at each end of the yarding strip or, if terrain is a limiting factor, in the most favorable spot to shorten the haulback line as much as possible. This offers better control of the balloon than a haulback line extending from the yarder to the back end of the setting and back again to the yarder. This system is also capable of reaching to a maximum distance of 5,000 feet from the landing.

Operating variables

Like the helicopter system, the balloon logging system is subject to the same variables as conventional cable systems. In the case of the balloon, the most important variables to be considered are probably wind, snow, turn volume, and yarding distance.

Balloons are relatively insensitive to rain and fog, but are very sensitive to wind and snow. The airborne survival velocity is 50 miles per hour and yarding operations must cease altogether when winds exceed 20 to 25 miles per hour. Snow, particularly wet snow, has a tendency to accumulate on top of the balloon. When the accumulation becomes heavy enough the balloon will tip, allowing the snow to fall. The sudden load reduction results in a phenomenon known as *impact loading* and can cause the skin to tear at the places that the load lines enter the fabric.

Turn volume and yarding distance

Turn volume and yarding distance are closely related in a balloon system. With a balloon, load-carrying capacity is directly proportional to maximum lifting capability. The net lift depends on the weight of the cable system attached to the balloon. As yarding distance increases the net load-carrying capability decreases.

Figure 15.12 illustrates the relationship between turn volume and yarding distance. The upper curve in the graph shows the maximum turn volume and indicates that turn volume decreases with yarding distance. The reason for this decrease is the weight of the additional line that must be carried by the balloon as yarding distance increases. In order to maintain control and to carry the logs free of the ground, the maximum load cannot be carried. A practical turn volume, approximately 80 percent of the system's potential, is shown by the second curve in the graph. This volume allows the balloon effectively to resist side wind forces along the yarding road, assures proper control, and allows the turn to be flown clear of the ground.

In order to achieve maximum practical turn volume the logs must be bucked for weight. It follows that grade, per se, is a secondary consideration, if it is considered at all. The available data suggest, however, that value losses with the Raven balloon become significant only in large timber (butt diameters exceeding 50 inches). In 50-inch timber (and less) the value loss should amount to no more than 2 or 3 percent of the potential. The volume capability of the system is between 3 and 3.4 cunits per turn. Providing the logs are bucked properly, and this should be carefully monitored, such turn size will result in minimum losses in timber under the 50-inch limit.

Productivity and yarding distance

Balloon yarding production, measured in cunits per hour, is the product of turn volume and turns per hour. Since turn volume is directly related to yarding distance, productivity must also be directly related to yarding distance,

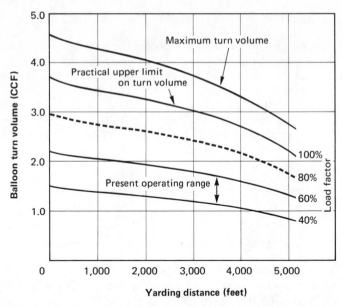

Figure 15.12. *Typical balloon yarding turn volume with the Raven 500,000-cubic-foot balloon for Douglas fir (density of 50 pounds per cubic foot) versus yarding distance. (Graph assumes yarding with a Washington 608 Aero-Yarder operating in the highlead mode out to 3,000 feet and with the inverted skyline mode out to 5,000 feet.)*

as it is in any conventional system. Figure 15.13 displays the productivity-yarding distance relationship. As the reader would expect, productivity is high with relatively short yarding distances. The number of cunits per hour produced peaks at a yarding distance of slightly more than 1,200 feet and then falls off significantly. The potential productivity is displayed on the upper curve and is based on minimum cycle times and maximum turn volumes. Experience indicates that productivity is far short of the system's capability, with load factors between 40 and 60 percent of the practical maximum. With the balloon adapted to a highlead system (highlead mode), and with yarding distances between 500 and 2,500 feet, productivity in the range of 10 to 15 cunits per hour can be expected. Beyond 2,500 feet productivity drops rapidly, because of lower turn volume and increased cycle time.

At distances between 500 and 2,500 feet, cycle time is unlikely to change substantially. Any productivity improvement will have to come through efforts to increase turn volume. Perhaps balloon loggers should steal a page from helicopter loggers. In the helicopter system, buckers use weight tables and actually mark the weight on each log. This helps the hooker to make up maximum turns. Such a method used with the balloon system could result in a 15 to 20 percent productivity increase under favorable volume per acre and piece size conditions. Load factors would then be 70 to 85 percent.

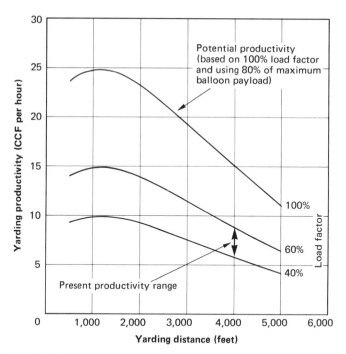

Figure 15.13. *Balloon yarding system productivity versus yarding distance for the Raven 530,000-cubic-foot balloon with the Washington Aero-Yarder.*

Yarding cycle time

Yarding cycle elements are the same for the balloon as for the helicopter. The elements are haulback, hookup, yarding, and release. The major differences, aside from equipment used, are that the chokers, for the most part, are set hot and there is no automatic release at the landing (a chaser must be used). Otherwise the systems are very similar. Chokers are attached to a hook which is, in turn, attached to the tether.

Haulback and yarding times, also termed *flight time*, are limited by line speeds and yarding distances. Operator control also can have an impact on the two elements. Hookup time depends on terrain characteristics, brush conditions, volume per acre, piece size, and cutting practices. Hookup time runs between two and five minutes when the chokers are set hot. Hookup time could be reduced by presetting the chokers and establishing a bunching and marking program that allows the fewest number of logs per maximum turn volume.

Release time ranges from a half to one minute and depends mainly on conditions at the landing. This time could be reduced if the chaser merely removed the chokers from the hook and replaced them with a second set. After haulback cycle began the chokers could be unhooked from the logs.

Figure 15.14 shows typical cycle time for a Washington 608 Aero-yarder and Raven balloon. The lower dotted line represents flight time with yarding speed and line speed equal and constant—the ideal condition. For distances up to 1,500 feet the average yarding speed is less than the potential line speed. This phenomenon is the result of the balloon flying in a higher arc from the landing up to about 1,500 feet, thus requiring increased flight time, as shown by the upper dotted curve. If the balloon were flown in a shallower arc, like it is for longer yarding distances, the two lines would more closely approximate each other.

System operation

The operation of a balloon logging system is almost identical to that of a conventional skyline system. Planning and layout are extremely important, with special attention given to landing selection and the selection of anchor stumps and tail hold stumps, which are used for corner blocks and sucker-down blocks (explained in Figure 15.16).

The landing area in a balloon show is slightly larger than that of a conventional system. More space is required because of the higher speed at which the logs arrive at the landing, the poorer landing accuracy, and the greater problem of crew safety. Logs are not generally loaded hot as they are landed. Instead they are swung to the loading machine because of the possibility of damage to the loader when it is placed too close to the landing impact area.

The landing area should be unobstructed. Though narrow drainages sometimes make this difficult, the extended reach capability of the balloon allows for over-flying a logged-off area to approach a landing. This sometimes makes landing selection somewhat easier.

Crew size and equipment requirements, except for the balloon and yarder, are the same as they are for a conventional system. They tend to vary depending on production, number of chokers used, and landing requirements.

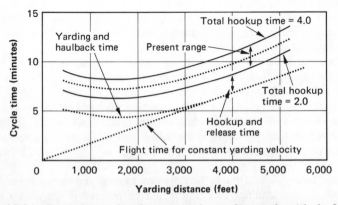

Figure 15.14. *Typical time distributions in the yarding cycle with the Washington 608 Aero-Yarder and the Raven 530,000-cubic-foot balloon.*

Balloon yarding systems are more precisely defined as lift-augmented sky-line systems. They extend the reach of conventional systems and improve the lifting capability. Balloon systems are operated in three basic modes: highlead, inverted skyline, and running skyline.

Highlead

The highlead mode is regarded as the conventional balloon system and is illustrated in Figure 15.15. The layout will look familiar, since only the balloon and extra brush blocks are additional. The balloon gives the extra lift which makes this system ideally suited for downhill logging.

The highlead system utilizes the familiar fan-shaped layout, and the yarding cycle is the same as it is in the conventional system. The cycle begins with the outhaul, when the balloon and butt rigging are at the yarder. The main line is slacked, allowing the balloon to rise and lift the tagline from the ground. With the balloon airborne, the haulback is engaged and the rigging is pulled (along with the balloon) out to the yarding zone. Once in the yarding area the rigging is stopped and spotted over the turn. Here a slight change from normal operation occurs.

In order to slack the rigging in a conventional system the running lines, either the mainline or haulback, are slacked. In a balloon system, lift is obtained by slacking one or the other of the running lines, depending on whether the system is hauling in or hauling out. Slack is obtained by tightening

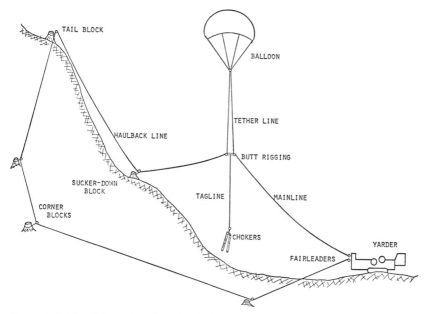

Figure 15.15. *Conventional method of balloon yarding.*

the lines. The balloon is positioned over the roadline by the sucker-down block, shown in Figure 15.16.

The hookup element begins once the rigging is in position over the turn. To position the rigging, the mainline is stopped and the balloon is 'sucked' or pulled down by pulling in on the haulback. After the logs are attached to the tagline, the operation is reversed. The haulback is released, allowing the balloon to rise, thus lifting the turn free of the ground. The mainline is then engaged to move the turn to the landing. At the landing the haulback is braked and the mainline, still engaged, sucks the balloon down until the logs touch ground.

Like conventional yarding, balloon yarding begins at or near the landing—wherever the logs begin. It then proceeds from the back to the front of the setting. Road changes are made by changing the position of the sucker-stump. Stumps are chosen in advance of moves. During moves the balloon is tethered.

In the highlead mode the maximum yarding distance is approximately 3,000 feet. The limit is the line capacity of the haulback drum. Yarding from behind rock outcroppings and ridges can be accomplished but is limited by circumstances which may affect the ability to pull the balloon low enough for the

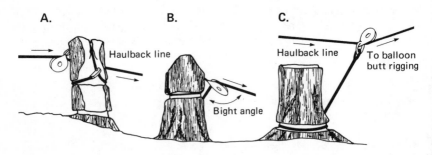

Figure 15.16. *General arrangements of blocks and stumps. A and B show the preparation of an intermediate support block for balloon yarding. The overhung block arrangement is used, whenever needed, to keep the haulback line off the ground as well as to keep it from rubbing nearby trees. The lower cut (ground level) notch in the stump (A) is made so that the stump can later be used as the sucker-down stump. It is important to keep the line bight angle (B) as large as possible (140 to 160 degrees) in order to reduce the net force on the stump. The overhung stump is subjected to a large twisting force, which increases the danger of tearing the stump loose. C shows the general arrangement for the sucker-down block. This is the last block carrying the haulback line to the balloon butt rigging. The tension in the sucker-down block may be very high, and it is therefore important that this be a solidly anchored stump. The line from the block to the stump is generally a 1½-inch line. The sucker-down block is often 14 inches in diameter and the intermediate blocks are lighter 12-inch blocks. Blocks are either carried or flown out to the field by a light-duty helicopter.*

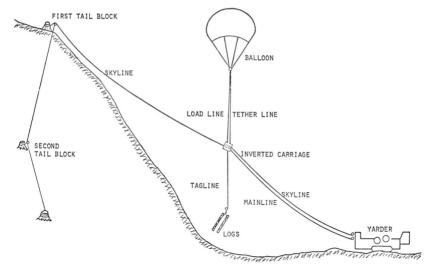

Figure 15.17. *Inverted skyline method of balloon yarding.*

tagline to reach the ground. Also, the haulback must be kept off the ground. Under such conditions, planning the layout is very important.

The inverted skyline

The inverted skyline, shown in Figure 15.17 is used when reaches greater than 3,000 feet are necessary. Yarding distance is limited by the line size and capacity of the main drum—about 5,000 feet.

With the inverted skyline the haulback becomes the skyline. The butt rigging is replaced by a carriage that rides the haulback. The mainline is attached to the carriage. The balloon pulls the mainline, rigging, and carriage out to the yarding area—a sort of gravity system in reverse. The rigging is positioned over the turn by tightening the haulback, which is now a skyline.

The load is lifted clear of the ground by slacking the haulback. When the load is clear it is pulled to the landing by the mainline.

The running skyline

This skyline mode requires a two-drum machine and functions just like the running skyline discussed in Chapter 14. The haulback doubles back on itself, serving both as a skyline and a haulback (see Figure 15.18). This system allows for greater pull-down capacity and there is less tension on the lines. However, there is a greater potential for line breakage caused by the lines sawing on each other.

Because of the added weight of the lines attached or connected in some way to the tether, both the inverted skyline and running skyline modes have a lower turn volume than the highlead mode.

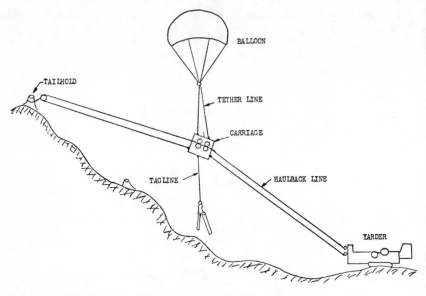

Figure 15.18. *Running skyline method of balloon yarding.*

Cost and operating comparisons

Studies of balloon systems now in operation indicate productivity in the range of 10 to 13 cunits per hour with the highlead mode and 8 to 10 cunits per hour with the inverted skyline. Up to 3,000-foot yarding distances, yarding costs are substantially less than helicopter yarding costs. It is estimated that a 3,000- to 4,000-foot balloon system can operate at less than 50 percent the cost of a helicopter system on the same or similar terrain.

The yarding cost stated in terms of unit cost for volume removed depends on total operating costs and productivity. However, certain added factors such as availability and utilization also have a significant effect on cost. Availability is expressed as the ratio of total hours available for yarding and total hours available in a season. Utilization is expressed as the ratio of actual hours of production and total hours available for yarding. Yarding efficiency is the product of both availability and utilization.

Availability depends on the weather factors discussed earlier and on equipment uptime. Utilization depends on performance of the system defined by the efficiency of the layout, the ability to minimize delays and downtime, and the quality of bucking and load marking.

Total hourly costs depend on the length of the operating season. In the coastal region of the Pacific Northwest a 9- to 10-month season can reasonably be expected. This translates into between 1,500 and 1,600 total available hours. Hourly operating costs with a seven-man crew are estimated at $400.

Based on current cost and on the state of the art, balloon logging certainly appears to be more feasible than helicopter logging—at least economically.

When compared with conventional systems such as highlead or slackline, even the balloon, since it costs two or three times as much as these systems, is apt to look unattractive. However, conventional systems and balloon systems are not strictly comparable when operating ranges vary between 600 to 1,200 feet and 3,000 feet. A realistic comparison should consider the total cost, including yarding and road building.

With its superior reach the balloon system, like the helicopter system, is able to operate at very low road density. The range is between 1 to 2 miles per section for the balloon system, and 4 to 12 miles per section with conventional highlead. Another advantage of the balloon over conventional cable systems is that it can be used for steep, mountainous terrain, especially with long unbroken slopes. Under such conditions there may not be sufficient deflection to operate any sort of skyline system. However, since the balloon creates its own deflection, it offers an attractive alternative.

Overall, the balloon does offer a feasible alternative to conventional systems. Because it is simply a modification of the conventional systems, loggers do not have any problem working with it. One major drawback is the high capital cost. Including the yarder, the balloon, and the support equipment, the balloon system costs about $1.5 million—well above the price of conventional systems. However, operating costs are realistic considering the range of yarding distances within which the system can work. The extended reach capability and reduced road requirements make this system a feasible alternative under the right conditions.

Hydraulic loaders are widely used. This one is trailer mounted.

Section Five

16.

Loading and Unloading

In an earlier chapter, logging was described as being predominantly a transportation function. Once the tree is reduced into manageable lengths all subsequent operations are performed with the objective of moving the forest products closer to an end-use point. The primary efforts—ground skidding, cable yarding, and aerial logging—are aimed at gathering and concentrating logs or bolts. Secondary transportation is concerned with moving them from the woods to an end-use point for manufacture or sale.

Secondary transportation is ordinarily associated only with the movement of forest products by truck, rail, or water. However, these products must be loaded on the primary conveyance, and when they arrive at their destination they must be unloaded. Loading and unloading are interfacing components closely related to secondary transportation. Loading is the link between logging and transportation, and unloading is the link between transportation and downstream activities such as log sorting, storage, and manufacture or sale. The components of loading and unloading have the greatest impact on log movement between the landing and the end-use point. For this reason, in this text they will be considered a part of secondary transportation.

Log loading

Log loading is a key component in any logging system, since it is the means by which forest products—tree-length stems, logs, or bolts—are transferred from the ground to some form of conveyance which will complete the transportation cycle. Usually loading takes place at the landing, but in some cases it takes place at the stump, or at a distant transfer or reload point.

Loading is not so difficult to explain as other components—for instance, cable yarding. It is nevertheless a complex function. It can be accomplished in any one of several places, and equipment ranges from a hand-loading pulp hook to an expensive, self-propelled machine. The type and size of machine used will depend on where loading is to take place, piece size, and volumes to be loaded. Another variable which must be considered is the productive capacity of the logging system, since it is of little value to have a loading machine that can handle 20 loads per day and a logging system that can produce only 10 loads. While it does not pay to have a loading capacity greater than the production capacity of the logging system being used, it also does not pay to have a loading capacity less than the production capacity.

Of even greater importance is the relationship between loading and hauling. Since hauling is the most costly component in the total harvesting system, it makes sense not to have trucks waiting at the landing for loads. If the loading operation is slow, because of machine or operator capability, the trucks lose time. Of course, if loading capacity is greater than hauling capacity, then the machine and crew will be idle. Most operators, however, would rather have excess loading capacity than idle trucks or idle logging equipment and crew. The ideal condition—the key to efficient overall operation—is a balance of log movement from the stump through hauling.

Specifically, woods loading involves placing some form of tree segment onto a conveyance, generally a truck bed or trailer, in preparation for hauling to some second concentration point or final destination.

Hand loading

The author, while working in a small pulpwood operation in Vermont, had the doubtful pleasure of loading 48-inch bolts by hand. The bolts were first loaded on a wooden dray at the stump. Then the loaded dray was skidded to the roadside by a small tractor. At roadside the contents of the dray had to be transferred to the haul truck. The handling of the bolts was aided by the use of a pulp hook, a small, hand-held steel hook. The hook is driven into the side of the bolt near the end, and the bolt is lifted onto the truck bed and stacked. When bolts are very large two men may be required to lift them. Or, the bolts may be loaded one end at a time.

Hand loading is not efficient, given today's labor market, and can be accomplished only with relatively small pieces. Despite the inefficiency and physical difficulty, however, some one- or two-man operations in the pulpwood regions are still obliged to load by hand.

A variation of hand loading that is used with larger logs is *gravity or roll-way* loading. The truck is positioned on the downhill side of the log decks, and the logs are rolled onto the truck bed by using a peavy. The stakes on the truck are dropped and two small logs are positioned from the ground to the truck bed to form a makeshift runway. Sometimes a log cribbing, called a *roll-way*, is constructed, as illustrated in Figure 16.1. Roll-way loading is not much of an improvement over hand loading, and load size is limited.

Powered systems

While the powered systems discussed in this section are not much of an improvement, in terms of efficiency, over hand-loading systems they at least

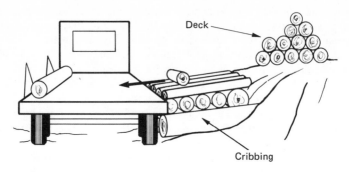

Figure 16.1. *Illustration of roll-way loading.*

replace manpower with machine or animal power. As is true for hand loading, some of these systems are still used, but they have no place in modern logging systems.

The *cross-haul* or *parbuckle loading* system is a variation of roll-way loading. A tractor or horse can be used for power and a light cable is used to pull the logs onto the truck. Because mechanical or animal power is used, rather than man power, the truck can be positioned on flat ground or even above the logs to be loaded.

A cable is attached to the truck. The end is run under the piece to be loaded and back over the truck to a tractor. The tractor pulls against the cable and rolls the log up onto the truck bed. A roll-way can be used if the loading is being done in one position, or skid logs can be placed between the ground and the truck bed so that the logs are rolled up the skids and onto the truck. By reversing the lines this system can also be used to unload.

Any sort of block hung in a tree, gin pole or A-frame can be used to load logs. The *stationary block* method involves hanging a block in a tree and positioning the truck below it so that the block is midway over the truck bed. A cable is run from a tractor winch, through the block, and hangs to the ground, where the log deck is located directly against the tree. A crotch line is shackled to an eye in the cable. On each end of the crotch line is an end hook. The end hooks are jabbed into the ends of the log and the operator engages the winch, picking up the log. When the shackle hits the block, with the winch still engaged, the log cannot be raised any further and will be pulled out away from the tree and over the truck bed. One man on each end of the log is required to position it on the truck and to release the end hooks. The sequence of activity is repeated until the load is completed.

The gin pole and A-frame are simple improvements over the system just described. A *gin pole* is an erect pole between 30 and 50 feet tall. It is rigged with two or three guylines and leans far out so that the loading block is positioned directly over where the truck will be spotted. From this point on, the stationary block procedure is used, except that the load is lowered onto the truck rather than pulled over and lowered. The *A-frame*, which can be mounted on the truck or on the ground, is essentially the same as the gin pole except there are two poles rather than one. The frame is the same as that used on a jammer, which is used both for yarding and loading.

The big-stick loader

Equipment and techniques used for loading have improved over the years just as have equipment and techniques used in other logging system components. Perhaps the improvements began when some logger discovered he could mount an A-frame on the back of a truck and move the machine to the logs rather than deck the logs beneath the loader. Indeed, one of the first mobile loaders was called the Logger's Dream and was manufactured by the Taylor Machine Works in Mississippi. The boom was a modified A-frame fabricated out of tubular steel and about 25 feet high. The boom was guyed and loading was done by using a cable powered by a hoist mounted on the truck frame.

As late as the middle 1960s about 80 percent of the loading was done by hand, and improvements were obviously necessary. However they came about, these improvements did occur. Today, logging operations in North America utilize a variety of mobile loading equipment, which requires skilled operators and adds a great deal of efficiency to logging operations.

Loaders range from the very expensive to the very inexpensive. The *big-stick loader* is on the inexpensive side but represents a real improvement in small pulpwood operations. The big-stick loader is a very simple piece of equipment

Bobtail pulpwood truck with a big-stick loader. The loader boom on this truck is attached to a center post midway on the truck bed.

which, in general, replaces hand loading, especially in the southern and south-eastern United States.

The big-stick loader has a steel frame located either midway on the bed of a bobtail pulpwood truck or directly behind the cab. Attached to a center post mounted on the frame is a 4-foot horizontal boom capable of swinging 360 degrees. A small-diameter cable runs from a powered winch mounted on the truck frame, through sheaves in the boom, and out to the pulpwood bolts. Several bolts may be strapped together by a length of cable and loaded, or one bolt at a time may be handled. Bundles or individual pieces may be pulled to the truck from 50 or more feet away. A full load is about three cords.

Self-loading trucks

The big-stick loader is one example of a self-loading truck, although most operators who use it probably do not think of it in those terms. In the first place, big-stick loaders are limited to bolts while self-loading trucks can handle anything from bolts to 100-foot poles. In addition, big-stick loaders utilize only a simple boom and powered winch (some of the winches are manually operated). The most widely used loading device found on self-loading trucks is the hydraulic knuckle boom.

The knuckle boom is permanently mounted on the haul truck immediately behind the truck cab. The operator's seat and controls are located directly on top of the cab. The hydraulics are operated through a power takeoff on the truck.

In operation the loader is positioned between 6 and 12 feet from the log deck, which is parallel to the road. In this position a full load can be handled without relocating the truck—providing there is a full load in the deck. Each loader is equipped with hydraulic outriggers or stabilizers. The stabilizers take the strain off the truck during loading and keep the truck stable.

At the end of the boom are hydraulically operated grapples. The boom is lowered to the level of the logs in the deck, the grapple grasps a log, and, finally, the log is swung around into loading position over the bed of the truck and then lowered into place.

The self-loading truck is a versatile piece of equipment. It can be used to pick up partial loads on completed landings or logs that have been lost along the road, or it can be used in small operations where the investment in a loader, apart from the truck, is not feasible. Loading requires no help, the truck driver doubles as the loader operator, and it is not difficult to learn how to operate the equipment. Fifteen or twenty minutes of instruction will enable a man to operate the equipment. Depending on the man, moderate skill can be achieved after four or five days of operation.

Self-loading equipment, despite its advantages, has some drawbacks. Because of size limitations it is best suited to small logs—up to 20 feet in length. In addition, the weight of the loader reduces the payload of the truck. For instance, a 16,000-pound-capacity loader weighs sightly more than 7,000 pounds, depending on the manufacturer. The 7,000 pounds must be subtracted from the load capacity of the truck. This reduction in capacity can be

Self-loading truck with a Ramey loader. Note the controls over the cab.

especially critical when the truck is running on public roads where load weight limits are enforced.

The loader used on the truck can also be used to unload once the truck has reached its destination. Furthermore, unloading need not be to the ground. The machine can unload into a deck or into some other carrier such as a rail car. It is also used on crawler tractors and wheeled equipment for preloading.

Tree-length spruce loading on a fifth-wheel trailer in Ontario, Canada. A front-end loader is being used.

Pallets

The pallet system, mentioned in an earlier chapter, eliminates both manual loading and mechanical loading. The reduction in loading time has a significant effect on truck productivity. Only 6 to 8 minutes are required to unload an empty pallet and pick up a full one. Compared with hand loading, the pallet system can double the volume hauled.

A pallet, which may have anywhere from a one-cord to a six-cord capacity, is preloaded in the woods either manually or with a self-loader mounted on a tractor or wheeled machine. The pallets are then forwarded to the landing for loading on a haul truck. A truck, depending on its own size and the size of the pallet, may haul anywhere from one to five pallets in a load.

The trucks used for hauling pallets are specially made with L-shaped rails, instead of a solid bed, into which the pallets fit. Of course, a solid bed can be used if the pallets are loaded with a crane. The truck backs up to the loaded pallet, and a skid frame, which reaches from the truck bed to the ground, is put into position. The pallets are then pulled up the skid ramp and onto the truck by a winch and cable mounted on the truck. For unloading at the destination, the pallets can be removed from the truck and stored or can be unloaded without removing pallets from the truck at all.

Some trucks have a steel tipping frame fastened to a hydraulic hoist behind the cab. The frame is hinged at the rear of the truck chassis and extends over the rear of the chassis so that when it is raised the tail end reaches the ground. The tipping frame eliminates the need for a skid frame but a winch is still used to pull the pallet to a carry position.

Front-end loaders

It has already been suggested that tractors and wheeled machines with knuckle-boom attachments can be used for loading both in forwarder operations and at the landing. A second loading tool, which may be more common, is the forklift used on front-end loaders.

Many of the front-end loaders used in the woods began as earth-moving machines that used buckets. By replacing the bucket with forks the machine can be used to lift such objects as logs, lumber, or crates. Other front-end loading machines were designed especially for loading logs.

The following explains how the front-end loader is used. The forks are lowered to the ground and run beneath the logs. The head and forks are then tipped back allowing the logs to roll back on the forks. The curved top clamps that form the top of the head are then closed to retain the load while it is moved. The loaders can either pick up and lift several logs at a time or pick up and lift only one log. If the logs are relatively small, the forks can be used to retain and lift by closing the tips of the clamps on the log, which is resting on the tip of the forks.

To unload onto the truck the head is tipped forward and the clamp is raised, thus allowing the logs to roll onto the truck bed. The tractor is then backed away from the truck and travels to the next load. Figure 16.2 illustrates the procedure used for loading logs with a front-end loader.

Logs to be loaded by a front-end loader should be decked perpendicular to

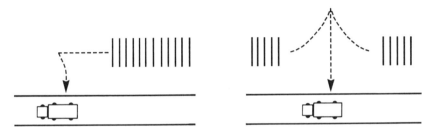

Figure 16.2. *Top view of loading logs with a front-end loader. Dashed lines indicate the path of the loader.*

Loading redwood with a front-end loader.

the truck's loading position. The loader moves in to pick up a load and backs out of the pile—generally while lifting the load. The machine is backed out in an arc. At the end of the arc the load is parallel to the truck trailer or bed. As the machine moves forward to place the load on the truck the load is lifted high enough to clear the stakes on the truck and any logs which have already been loaded. Several logs at a time may be loaded, depending on the capacity of the machine. However, as the load grows higher fewer logs can be loaded at a time because it is difficult to control their position on the truck bed. It may be necessary to place the last few pieces on the truck one at a time in order to build a full-capacity load.

In most cases logs can be loaded efficiently on a flat landing, with the machine lifting the logs high enough to reach the top of the load. Logs to be placed on the side of the truck opposite the loader will have to be rolled into place or loaded from that side. Some loggers use a tractor to dig what amounts to a loading pit, or a ramp from which the logs can be placed on the truck. The bed of the truck will then be lower than ground level, allowing for greater visibility and generally higher loads than can be accomplished with the same machine loading on level ground.

Safety is a serious consideration. With a front-end loader moving in a fairly large area the operator must be especially sure that the area is clear of any personnel. Needless to say, nobody should be standing or walking on the side of the truck when it is being loaded. For one thing, the operator cannot see over the load. In addition, because the operator does not have positive control, some logs are apt to roll off on the side opposite the loading machine.

Front-end loaders are versatile machines which can be used for loading, forwarding, swinging, or, with a bucket to replace the forks, road and landing construction. Some operators install grapples at the rear of wheeled equipment (the same thing would work with tracked equipment) and use the machine for skidding as well as loading.

For all the advantages of the front-end loader, some disadvantages are also evident. Loading with a front-end loader requires a fairly large landing—larger than that required by a stationary loader. If an operator has a machine large enough to load off-highway trucks it cannot also be used efficiently as a skidder. On the other hand, a machine that is of ideal skidder size may not be an efficient loader. The primary difficulty is the height the machine can reach. However, if it cannot reach high enough to build a capacity load, the operator can always resort to the ramp or loading pit.

Loading cranes

Cranes such as those used in the construction business as earth-movers or as heavy lift equipment are equally well adapted to loading logs. Loading cranes range in size from 1/2 yard to 2 1/2 yards, a lift capacity rating that relates to the size of the bucket the machine is designed to handle.

Most loading cranes are self-propelled units mounted on tracks or rubber tires. Some of the smaller machines, such as the knuckle-boom loaders, may be mounted on a motor truck—that is, the truck bed provides a platform for the engine, hydraulic motors, cab, and boom. Some loaders are still air-controlled. However, most use cables and hoisting drums or hydraulics for lifting. The rubber-tired rigs, as well as truck-mounted machines, have hydraulic outriggers which keep the machine stable during loading operations and keep the weight off the tires.

Booms

There are three types of booms used for loading logs—the gooseneck boom, the hinge-type heel boom, and the straight boom. The *gooseneck boom* is a curved boom constructed of two segments, the lower of which is hinged to a machinery platform. At the curve in the boom is a gantry to which are attached the lines that raise and lower the boom. At the upper end there is a fairlead. If tongs are used, only one sheave is necessary. If grapples are used, two sheaves are necessary.

The *hinge-type boom* has no curve, being straight instead. It still has a gantry, however, which serves the same purpose as the gantry on a gooseneck boom. The sheaves at the end of the boom, like the sheaves in the gooseneck, carry the cables for closing grapples and for lifting. In the case of tongs only lifting capacity is required.

In use, with either the gooseneck or hinge-type boom, the logs are heeled into the boom. The log is grasped off center, closer to the end of the log than the middle. As the log is lifted it is heeled against the underside of the boom. The lifting line or tong line continues to lift until the log is nearly parallel to the ground. One end of the log is free and the other is supported by the underside of the boom. Once the log is suspended, it is swung into position over a truck trailer or bed and lowered into position. As the log is lowered the unsupported end descends first, followed by the supported end. Because of the position and weight distribution of a log after it is heeled, a hinge-type boom has a greater lifting capacity than a gooseneck boom. All other things being

Crane with a gooseneck boom picking up log.

equal, the log is heeled in closer to the machine with a hinge-type boom.

A third type of boom is the *straight boom*. It is of the familiar lattice construction and is most frequently used as heavy lift equipment in ship building, heavy construction, and cargo handling. Nevertheless, the straight boom has been applied to log handling and is frequently used in log yards for building high log-storage decks. However, it is also used for log loading in the woods.

Like other loading machines, the straight boom can be used for both yarding and loading. For example, in one application the operator casts the grapples up to 150 feet below the road, catches a log, and yards it into the landing. If a truck is present the log might go right onto the load. If not, it will be decked, waiting for the next truck to arrive. The boom is about 60 feet long and is truck-mounted for greater mobility.

Grapples

For many years logs were loaded with tongs. The tongs, a giant-sized version of those used by the iceman years ago, are made of forged steel. Rather than a handle the tongs have two clevises on the upper end, for one each leg. A head loader drives one of the points into a log and then signals the loader operator to go ahead on the lifting line, which causes the points to dig into the wood on both sides of the log. The log is lifted, heeled, swung around on the boom, and loaded on the truck. A second man, appropriately called the *second loader*, assists the loader operator in positioning the log correctly and shakes the tongs

Loading logs with straight boom and a Bucyrus Erie crane.

loose after the log has been positioned. For logs too large for the tongs, a strap is used.

Grapples have replaced tongs in many cases, although tongs are still used in some regions as well as for yarder-loader operations. The grapples used are similar to those used for ground skidding, except that they are cable driven.

There are three main drums on a grapple machine and one smaller drum called the *tagline drum.* One of the main drums is the hoist used to lower and raise the boom; it is called the *boom hoist.* The second main drum opens the grapple. The third main drum closes the grapple and lifts the load after the grapple is closed around the log. The tagline is attached directly to a grapple arm by a small shackle. It is used to swing the grapple, along with the boom, so that it can be cast beyond the end of the boom. The tagline is first spooled in, pulling the grapple toward the machine, and then released, allowing the grapple to swing like a pendulum. When the grapple is over a log, it is dropped on the log and closed.

Preloading

Preloading has already been mentioned several times, especially in conjunction with forwarding systems, pallets, and certain of the mechanized cutters. In each case the load was moved from the woods to the landing and then loaded on some sort of fifth-wheel trailer. This operation is quite uncomplicated and reduces landing and loading delays in the hauling operation.

Assuming that the set-out trailer is already loaded when the haul truck

When the logs are too large for the grapples a cable sling must be used.

Photo courtesy Clyde Equipment Co.

returns, the operation is as follows. The truck drops the empty trailer in the designated position, backs the fifth-wheel attachment under the loaded trailer, lifts the standards which were holding up the front end of the trailer, hooks up air and water for the trailer brakes, and the truck is then ready to leave with a load. Total elapsed time for the operation is between six and eight minutes.

There are two specialized devices used for preloading and storing logs for pickup. The first is called a Batson Bunk and is no longer manufactured, although there are quite a few of them still in use. The second is called a Lever Loader and is a relative newcomer to the industry.

Both machines are used with the pole trailers commonly found in the Northwest. The loads are strapped, wrapped, and weighed in advance of the truck's return. The loads can be built with any of the loading machines described thus far—front-end loader, crane, and so on. Loading can also be accomplished with a skidder equipped with a swinging boom, thus eliminating the need for a loading machine. Of course, the skidder could be used to load only with smaller timber.

The Lever Loader is equipped with bunks. Once the load is prepared and the truck is ready, the rear of the load is lifted by two hydraulic cylinders. Before the cylinders are engaged a cable is wrapped around the load behind the cylinders. When the load is lifted, the cable holds down the back of the load, forcing the hydraulic cylinders to lever the front end of the load into the air so

A load of pulpwood waiting for a haul truck. The load was built with a Buschcombine. The cable visible at the front of the trailer secures the load.

that the trailer can be backed under it. With the load on the trailer the cylinders are retracted, allowing the load to settle onto the trailer. The hold-down cable is released and the truck is ready to leave.

The loading can be done anywhere from the landing to a wide spot in the road. Several preload bunks at the landing allow for a certain amount of sorting, which is beneficial to downstream operations. Also, since the load is weighed there is no reason for any truck to leave the landing with less than a full load.

Work elements

There are nine basic elements involved in loading. Several of those elements are repeated when there is no truck at the landing and the machine is sorting logs for loading. The common elements are unload trailer, truck delay, swing out (repeated), pickup log (repeated), swing log into position (repeated), load log, production delay, and breakdown. Another element, called *tailing*, is involved in both loading and sorting functions.

These elements can be categorized as fixed or variable in relation to the loading operation. When loading, all elements involving truck preparation—unload trailer and truck delay—are fixed. They occur with each load and require approximately the same amount of time. The other elements, except breakdown, are variable and depend on the number of logs to be loaded. Since

Lever Loader—the load is ready for loading and the trailer is being backed under the load. Note that the cable at the rear of the load levers it up so the trailer can be backed under it.   *Photo courtesy Logging Specialties*

it is not predictable, breakdown is a random variable. In terms of log hauling, all loading functions are necessary delays, but delays nonetheless.

Unload trailer: Unload begins when the truck is in position under the loading boom and ends when the trailer reach has been attached to the truck and the safety chains have been hooked.

Truck delay: Truck delay begins the instant the safety chains are hooked and ends when the truck driver closes his truck's door after entering the cab. This element includes hooking up air and water.

Swing out: Swing out begins with the boom in position over the truck trailer either for unloading the trailer or immediately after loading a log. The element ends when the grapples have been attached to a log and the lifting cycle begins.

Pickup log: Pickup begins when the lifting cycle begins and ends when the log is heeled against the boom or when the boom begins to swing back toward the truck. The pickup element may include delays if the wrong log is selected and the operator elects to discard one log in preference for another.

Swing in: Swing in begins the instant the boom begins to move with the log heeled or partially heeled and ends with the log over the bed of the truck or positioned over the trailer.

Load log: Loading involves positioning and repositioning the log on the bed or trailer and ends when the grapples are finally disengaged.

Load preparation: Load preparation begins when the last log is loaded and the driver is signaled or the boom turns away from the load the last time.

Sometimes, as a safety precaution, the loader operator will leave the boom over the load with the grapples still attached to the last log loaded.

Productive delays: Productive delays include such things as backing the truck during loading, checking the weight scales during loading, or releasing the bunk pins.

Breakdown: In this context breakdown includes mechanical failure of either the truck or the loading machine. Specifically, this might involve the freezing of brakes on the truck or the breaking of a loading cable.

There are other delay elements ranging from no logs available to an inefficient operator. For instance, if the operator has had time to sort logs and deck them in preparation for loading, but has not done so, this would constitute a delay in the loading element. The operator would then have to tail logs around before he could choose the best logs to load. On the other hand the operator might be required to take a little time from loading to assist the chaser in unhooking a turn. This type of occurrence is not uncommon when the ground is steep and the landing small.

In West Coast operations, and in others as well, the loading machine is frequently idle. If a single loader is stationed at one landing with only one cable machine feeding the landing it is quite possible that the loading machine might be idle up to 50 percent of the time.

The time distribution (percentage of total cycle time) for loading an average of 12 logs (Douglas fir) with a loading crane equipped with a gooseneck boom and grapples is:

Element	Time (%)
Unload trailer	8
Prepare trailer	6
Swing out	9
Pickup log	19
Swing in	11
Load log	19
Load preparation	28

The time would be approximately the same with a straight boom, assuming the operator was skillful. Ordinarily, for a load of 12 logs at least 20 minutes would be required with an average loader operator.

Since the elements and time distribution are for a crane loading operation, they would not be applicable to other loading methods, such as front-end loading. The elements would be similar, however, with some adjustments for the method used. Swing in and swing out, for instance, would become travel-to-log-deck and travel-from-log-deck when a front-end loader is used. If a trailer is not used, the unload-trailer element and the prepare-trailer element would not occur. Load preparation time would be similar, though perhaps a bit shorter.

Log loading operations

In operations where a front-end loader or a loading crane is used the task of loading is fairly similar, at least conceptually. There are some differences in element descriptions, but some of the fixed elements (unload trailer, prepare trailer, and load preparation) are the same.

In the Northwest, log-hauling trucks use pole trailers, which are stored on the truck. Whichever loading method is used, the trailer must be removed from the truck before loading begins. The trailer has a lifting eye located in the center of the bunk, and this device is used to remove the trailer when a crane is used. When the trailer is unloaded, the stinger must be hooked to the truck. Once the trailer is hooked, the air and water lines are coupled and the truck is ready to begin loading.

The loading process for front-end loaders was briefly described earlier in the chapter. For a loading crane, loading is a matter of swinging the boom over the log deck, picking up a log, swinging back to position the log over the load, and placing the log on the truck. The first logs loaded are the bunk logs and must be long enough to extend a short distance over the front bunks and also overhang the trailer bunks. If the logs are extended too far over the front bunks they will rub on the truck and bind the trailer in a turn.

A second loader, a holdover from the tong loading days, stands on a platform built on top of the truck's cab. The loader operator signals the second loader to back the truck or move it ahead as needed to position the truck for loading. The second loader, in turn, signals the truck driver by tapping signals on the truck or by giving voice signals. One tap with the branding hammer means move ahead. One tap stops the truck. And two taps means back up.

After the bunk's bottom is filled, the next logs loaded are the wing logs. Wing logs are placed against the top of the bunk stakes on each side of the load. Additional logs are then placed between the wing logs. When the bunks are filled to the top of the stakes all additional logs are placed in the saddles which develop between two adjacent logs. Each saddle is filled, building the load in pyramid fashion until only one saddle remains—at the top of the load. The log on the very top of the pyramid is called the *peaker* (Figure 16.3).

After the last log has been loaded the driver is signaled, with a series of taps, to pull ahead. Once the truck is pulled out from under the loading boom at

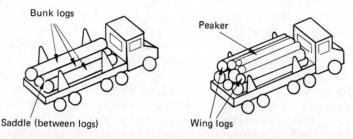

Figure 16.3. *Technique for building a load.*

least two wrappers are put around the load depending upon load size. Wrappers are lengths of cable with a short length of chain attached to each end. In some cases a length of chain is used rather than cable. Whichever the case, each wrapper must be long enough to reach all the way around the load. The wrappers are secured with binders, one wrapper to each binder.

Generally the wrappers are thrown over the load by the truck driver or the second loader, if he is helping. When the loads are very high, as might be the case when off-highway trucks are used, the crane will be used to lift the wrappers over the load.

It is extremely dangerous to walk around either a truck being loaded or an unsecured load. A person on the side of the truck opposite that from which the wrappers are being thrown can be hit. Of even greater concern is the possibility a log will roll off the load before it is secured. This is a common cause of accidents in logging operations.

Delays and efficiency

The time required to load a truck, whatever equipment is being used, depends on log size and log availability. The time might be anywhere from 10 minutes to one hour—10 minutes for a five-log load and one hour for a 50-log load. These average times give some idea of the effect of log size.

Very large logs and very small logs offer special problems. Some logs are so large the grapples are not able to grip them. When this happens a strap (a short length of cable with eyes spliced into both ends) is used to pick up the log. Continued attempts to pick up a large log when the grapples are too small causes damage to the log because the grapple ends dig into the sides of the log and are then ripped loose. Some logs are so large the loader cannot lift them. In such cases the log is loaded one end at a time.

When the logs are small—short—they must be loaded between the wing logs. If they are short enough they will be double ended; that is, two logs will be placed end-to-end in the same saddle. Log lengths are important in any operation. Too many short logs will result in longer than normal loading time and smaller than average loads. Smaller loads mean a higher unit cost for hauling. When a reach is used, as with the pole trailers used in the Northwest, the logs must be long enough to reach from one bunk to another. There must be at least enough bunk logs and wing logs to build a load.

Logs should be decked on both sides of the road if the road is being used as a landing. When this is the case, the loader can load from either side of the road. When cable logging, and using a tower, the loader generally loads the logs 'hot' or soon after they are landed—perhaps as soon as they are landed. Also, the landings on cable sides always tend to be too small. Hot loading from a small landing can and often does slow down the loading process. If possible the logs to be loaded should be sorted and decked beside the loading machine before the trucks arrive.

The last problem to be discussed is waiting time, which trucks experience at the landing. This is a major cause of truck delay and is caused by one or more of several factors.

In many cases several trucks will be hauling from the same landing. The number of trucks may vary from two to six, depending on the production expected. The trucks assigned to one landing are called a *string*, and all too often they arrive either all at once or at very short intervals. This means that while the first truck to arrive is being loaded the others must wait. If 20 minutes is required to load each truck and there are three trucks in the string, each arriving 5 minutes after the last, the following delays will be experienced. The first truck does not wait at all. The second truck waits 15 minutes and the third truck waits 30 minutes.

It is bad enough if the delays occur only once during a day. However, this is generally not the case. Because of the timing of downstream operations the string will often come back at the same arrival interval for each trip—up to three or four trips per day. This means in one day, on one landing, there is an automatic accumulation of nonproductive delay time totaling between 135 and 180 minutes, depending on the number of trips made. This type of delay is usually caused by a problem in truck scheduling. The problem can be remedied by starting the trucks out on longer intervals or by staggering the intervals throughout the day.

A second cause of truck delay at the landing is the result of running out of logs. Each day the foreman or his designate will estimate how many loads will be available on the next day, and the trucks will be scheduled accordingly. As sometimes happens, however, the estimate may be incorrect or a breakdown may result in less production than was expected. When the trucks arrive there are no logs. Often, when communications are poor, the trucks simply wait until logs are available or until someone discovers the problem—whichever comes first. The loader operator should always communicate with his foreman when logs are not available. If there is another landing available, the trucks should be rerouted. The same advice pertains whether availability is the result of no logs, yarder breakdown, or loader breakdown. The effect is the same in each case, no logs can be loaded.

Log unloading

The old adage is, "what goes up must come down." In the logging business the adage is changed to "what goes on must come off." The truckloads of logs which originate at a woods landing or elsewhere must be unloaded someplace. Where, of course, depends on the location, downstream facilities, and the final destination of the load. In some operations the logs must be loaded and unloaded several times before they finally make it to the mill pond or log yard.

In one operation, which moves logs from the interior of British Columbia to the coast, the logs are transferred three times before they finally arrive at the mill. First they are loaded and banded with steel straps in the woods. The load is hauled and dumped into a lake where the bundled logs are rafted and towed 65 miles, lifted from the water, and loaded back onto trucks. The trucks haul the logs over a mountain pass to another lake where the logs are dumped again for another 65-mile water trip. For the second time the logs are lifted out of the water and loaded back onto trucks. The truck then hauls the logs, still bundled, to the mill yard where they are unloaded for the last time.*

The example is an extreme case, but there are a great many instances where logs are either reloaded to other trucks or transferred to some other transportation mode, such as rail or water. In each instance the logs must be handled, unloaded, and reloaded.

Log unloading is accomplished using a variety of machines. Some, like the A-frame dump and parbuckle dump, have been used for many years. Heavy lift cranes, log stackers, and various types of overhead cranes are also used and are more recent arrivals. The unloading system used depends, to a large extent, on the facilities available at the unloading area, whether the load is to be just unloaded or reloaded, and whether the loads are being unloaded into the water or onto dry land.

A-frame dumps

The A-frame dump is used for unloading from truck to water. It is so named because of the two poles utilized to lift the load off the truck and swing it out over the water. The unloading line runs from a drum on the unloading machine up to the apex of the A-frame, through a block and down to the spreader bar. A second line runs from another drum out to the spreader bar and is used to position it over the truck and to retrieve it after the load has been dumped.

Four lines are attached to the spreader bar, one on each corner opposite each other. After a truck is positioned under the A-frame, between the top log of the bulkhead, called the *brow log*, and the unloading machine, the spreader bar is lowered. The longest cables, which are always on the water side, are then placed between the reach of the truck and the load. These cables, equipped with nubbins, are then attached to special hooks which hang from the two shorter cables.

When the load is strapped, it is picked up until the dead weight results in the straps holding the load securely. Then the wrappers are removed. Wrappers should never be removed until the straps are secured and the loading line is carrying the weight of the load. After the wrappers are removed the load is lifted until it clears the truck stakes. Then it is allowed to swing out over the water while the loading line is being slacked. The load swings out over the water because the A-frame poles are set at an angle, with the top being over the water. When the load is released it swings out like a pendulum.

When the load hits the water the straps become slack and a trip mechanism in the hook releases the straps. The hooks always release when tension is removed. The logs are pushed away from the dumping area by a boom boat.

A-frame unloading is fairly efficient and is still commonly used on the West Coast. The use of the system is under attack, however, because environmentalists charge that it contributes to water pollution. The charge results from the fact the system is what is called a *hard dump*. That is, dropping the logs into the water results in loose bark and other debris getting into streams and

* The company that manages the harvesting operation described is Eurocan. The logs are transported from the interior of British Columbia, around Ootsa Lake, to the town of Kitimat on the coast.

A-frame dump in Alaska. Note spreader bar hanging in front of the A-frame. The logs are dumped into the water and rafted for further transportation.

waterways. This material and the tannic acid which leaches from the bark of some species cause some pollution and unsightly floating accumulations.

Heavy lift cranes

Lattice-boom cranes, bridge cranes, and gantry cranes with a lift capacity of up to 100,000 pounds are used in many areas where large volumes of logs are delivered every day. These types of systems, especially gantry cranes, are used extensively in Europe where the logs are fairly uniform in length and diameter. A fourth type, with a fixed boom revolving around a free-standing column, is used for unloading and storing small logs, mainly in southern operations. This system unloads logs from either rail cars or trucks and stores the logs in two rows beneath the boom.

Bridge cranes and gantry cranes are both elevated. Bridge cranes run on elevated tracks, while gantry cranes are elevated by supports and also run on

Top: Kockums K-100 log boom does all log handling at this southern chip mill.
Bottom: Grapple of Heede portal crane removes truckload of logs to storage.
Photos courtesy Kockums Industries and Heede International

Loading and unloading   341

This 50-foot straight boom is unloading and decking the logs. This type of operation is common where there is a need to build the decks high and fully utilize the space in the log yard.

tracks. The cranes are of box or stress construction and are electrically operated. There are several types of gantry crane, including revolving and nonrevolving, cantilevered and circular. These machines can offload onto dry land or water and are fixed installations.

Grapples are commonly used to remove entire rail-car loads or truckloads of logs. The load vehicle is positioned for unloading, the grapples are spotted over the load, and the load is picked up. Following pickup, the boom, bridge, or grapples are moved with the load, and the logs are taken to storage—either in cold decks up to 50 feet high or into water storage. The equipment is capable of unloading 10 to 15 cars or trucks per hour.

A major disadvantage of the overhead crane design is capital cost, which can easily exceed 1.5 million dollars. Operating costs are generally lower than those for other unloading systems. One of the advantages of the equipment is the reduction in breakage. Most unloading systems incur significant handling breakage, but overhead systems minimize this problem. Also, when they are used for water storage, there is much less debris since the loads can be lowered more gently than they can with the A-frame.

Log stackers

Log stackers, heavy lift machines with up to 60-ton lifting capacity, have been used in the logging industry since the middle 1950s. The machines operate similarly to the front-end loaders already described but have a much greater lifting capacity. There are only three manufacturers of log stackers. Two employ diesel-hydraulic systems; the third employs electric systems.

Machine heads are equipped with forks, clamps, and kick-off arms or kickers. When unloading a truck the forks are placed between the stringer and the load, and the load is clamped. Only after the load is clamped can the binders be removed. If rail cars are being unloaded, no binders will be present. After the binders are removed the load is picked up and carried to storage. Logs to be water-stored can be placed in the water by the stacker with a

The Raygo Wagner stacker unloads long logs at a southern operation.

Loading and unloading   343

minimum of disturbance—certainly an advantage over the A-frame.

The big stackers can ordinarily lift and carry an entire load. When smaller forklifts are used, or with the occasional oversized load, the load is split. The forks are forced into a part of the load, which is then removed. The remainder is removed with a second bite.

Pulpwood handling

Long logs delivered to a pulp yard can, of course, be handled by log stackers. A good deal of the wood produced is either long-log or tree-length. The operation in such a case is not any different from that already described. Short-wood, however, is handled somewhat differently.

Many pulp yards use a specialized machine, a commercial lift truck with a pulpwood handling attachment replacing the bucket or forks. Instead of forks, the front frame carries two slings. The slings are placed around the load, up to two cords at a time, and the boom is raised to unload the truck. The load can then be stored on racks on the ground or transferred directly to a waiting rail car, which will take the load to the pulp mill yard.

Short-wood rail cars used in the South and Southeast are normally loaded with two rows of pulpwood, and the rows must be even. The tilting boom on the lift truck is built with a gridded frame used to 'bump' the load and push in any protruding pulp sticks. This makes the load secure and wedges the sticks into place. Both sides of the load must be trimmed in this manner.

Rail cars are unloaded with loading cranes equipped with lattice booms and grapples. The lattice booms are used because they can build higher piles of pulp inventory, thus better utilizing log yard space.

Unloading delays

The actual unloading of trucks and rail cars takes little time. A large log stacker can approach a truck, clamp a load, and lift it clear of the bunks in about 30 seconds. The problem is not in unloading, per se; but rather in some of the other activities, productive and nonproductive, which take place at a delivery point.

Before a load of logs even approaches an unloading point, whether a water dump or a sort yard, it must be measured for volume either by scaling or by weighing. This function is critical to the business because it is the means for determining the value of the load. It can, however, be time consuming.

Let us look, for instance, at the job of scaling and grading a load of logs. The truck stops at a scaling ramp, which may be able to accommodate up to four trucks at a time depending on the size of the ramp and the number of log scalers available. The work involved with scaling and grading a load may take between 15 and 20 minutes and is a necessary delay. What happens, however, when more trucks are present than can be handled by the four scaling positions at the ramp? The overflow must wait, just like it does at a landing. The amount of waiting time depends on the skill and speed of the scalers, the size of the loads (in terms of the pieces), and the number of trucks in the queue.

A load of pulpwood being bumped up to even ends and wedge pieces into load.

One might suggest that the solution to the waiting problem is simple—larger ramps, more ramps, and more scalers. This, however, is not the case. The ramps are manned to accommodate the loads per day expected. Most of the time the real problem lies in truck flow. Trucks arrive in surges during specific periods in the day. This situation is similar to the situation where the trucks arrive at the landing within five minutes of each other and continue to arrive at the same intervals throughout the day. At a scaling ramp, however, the string does not amount to three or four trucks. Instead, there may be a dozen or more trucks arriving from several different landings.

The same surge that affects waiting time at the scaling ramp also occurs at the dumping area. At one dump observed by the author the surge occurred three times per day, at 9:00 A.M., 11:00 A.M., and 2:30 P.M. During these times there were sometimes as many as 30 trucks waiting to be dumped. It does not make any difference what sort of unloading machine is used. The same surge phenomenon is observed in a dry-land sort yard where both cranes and log stackers are used.

In each case the same possible solution might apply—more men and equipment. This is a solution to be sure, but another problem is created simultaneously. Between the surges trucks arrive at intervals that are not entirely predictable. Anywhere between 5 and 20 or more minutes might elapse between truck arrivals. During these intervals there are no trucks to unload and both men and equipment are idle.

In a log sort yard the stackers perform many functions, of which unloading is only one. Logs are placed in inventory and removed from inventory; they are sorted for grade; and, in some cases, rebucked or remanufactured. The stackers also load logs out of inventory onto trucks and rail cars for further transportation. The stackers may also be used to feed a mill or place logs into a mill pond.

Each of these functions is more or less scheduled to utilize the equipment most efficiently. However, when a truck or a train arrives, the emphasis is generally on moving that vehicle out of the yard as quickly as possible. The trouble is that unloading is not the only function to be performed. The load is unloaded and must then be carried to a storage, sorting, or scaling area if a roll-out scale is used. Following load storage the stacker travels empty back to the truck unloading area. In all, perhaps only 20 percent of the time is actually spent unloading trucks. The rest of the time is spent traveling, either empty or loaded, and placing loads in bunks, scaling bays, or inventory decks.

Once again, the solution is in the efficient scheduling of trucks and the right balance of equipment available to handle the incoming traffic. This is no simple task. The traffic and functions performed in a large sort yard are overlapping and complicated. The flow of traffic through a sort yard can influence the speed of that traffic and the time required for unloading.

Finally, after the truck is unloaded it can return for another load. If a trailer is used, however, it must be reloaded before the truck returns to the brush. This function is fairly simple and is performed either by the unloading machine or by a special piece of equipment. While the function is simple it does require time—a necessary productive delay. However, the surge problem also affects the time required for reloading trailers.

Reloads and transfers

When logs are hauled on public highways there are weight, width, and length restrictions which constrain load size. If logs are being hauled relatively long distances over private lands, where such restrictions do not apply, it pays the operator to use off-highway trucks which can haul a larger load than trucks on public roads. In addition, several modes of log transportation may be required to move forest products to the final destination.

Reloading generally applies to removing the loads from trucks, storing them for a brief period, and then reloading them on rail cars or some other mode of transportation. If off-highway trucks are the primary mode, then the load size must be reduced for highway trucks. Log cars, on the other hand, may have a larger capacity than trucks, and so larger loads are built. A transfer is a load shift from one mode to another. For instance, truckloads of logs may be transferred directly to rail cars when both carry loads of the same size.

Reloading the pole trailer for another trip to the woods.

A rail transfer may be accomplished by a heavy lift stacker, which can handle entire truckloads, or by a type of crane powered by a yarder. The transfer is located adjacent to the rails and a string of empty cars is spotted along the rail spur. A rail car is spotted under the crane and loaded as the trucks become available. Each new car is spotted by a line controlled by the yarder. Small forklifts or specially designed spotter vehicles also may be used. The rail spur on which the transfer takes place is generally constructed on a slight slope so that the train can be moved with a gravity assist.

The problem of delays is substantially the same in reloads and transfers as it is in larger sort yards, although perhaps on a smaller scale. The trucks still have to wait for unloading during surge periods and the trailers still have to be loaded. While the reload and transfer equipment does not have a great many functions to perform, it can still handle only one truck at a time.

17.

Log Transportation

There are two good reasons for emphasizing the secondary transportation functions. The first, and most obvious reason, is that transportation is crucial to the overall harvesting system. Prior to log transportation, by whatever mode, there are no profits earned; only costs are incurred. It is not until the forest products are delivered to some final destination that a profit is actually made. The second reason is that transportation is the largest single expense associated with logging production. The estimates vary, but a transportation cost of 50 to 60 percent of total operating costs seems to be acceptable. The importance of transportation is further emphasized when one realizes that most operators measure production in terms of loads produced per day.

In addition to the direct cost of transportation—truck hauling, rail, barge, raft, or river drive—other costs of the system must also be considered. Trucks require roads, bridges, and culverts. And the roads must be maintained. Railroads run on tracks. Rafting requires booming grounds and storage areas. In other words, whatever the mode, support equipment and facilities are necessary, and the cost is high.

For instance, in 1970, truck road costs in the Northwest amounted to about 34 percent of stumpage costs. In northern California, during the same year, timber purchasers spent $23 million for roughly 1,800 miles of permanent road (Dennison 1971, p. 15). With this rough idea of costs consider the enormous capital investment in the national forests, where approximately 60 percent of the United States' commercial timber is found. Within the national forests there is a 200,000-mile road system. The U.S. Forest Service builds roughly 800 miles of new road each year and rebuilds some 400 miles per year. Timber purchasers build 3,500 miles new road and rebuild 1,300 miles each year (Caterpillar Tractor Co. 1970, p. 2).

Transportation of logs by truck began around 1913 or 1914, when names like Federal, White, and Diamond T were common. Over the years trucks have become the dominant transportation mode and, in some cases, the only transportation mode used for moving forest products. In other cases, trucks are used in conjunction with some of the other modes mentioned earlier.

Both before and for a time even after trucks became established, the logging railroads were king. In 1931, a survey conducted by *The Timberman* revealed there were more than 7,200 miles of logging railroads in service (*The Timberman* 1949, p. 63). Some are still in service, in eastern Canada as well as in the western United States, but the days of the privately owned logging railroad are past. Common-carrier railroads, however, are extremely important in wood movement in both the United States and Canada. In eastern Canada (east of the Rocky Mountains) about 16 percent of all pulpwood deliveries are made by rail (Martin 1971*a*, pp. 31-40). It was estimated that nearly 40 percent of pulpwood requirements in the southern United States during 1974 would be delivered by rail (Lilley 1973, pp. 6-7).

Prior to the days of railroads and trucks, forest products were floated or driven on rivers and lakes from the woods to the mill. Today, river driving, barging, and rafting are still important transportation modes in some areas, but they retain only a shadow of their former importance. The last white-water sawlog drive in the United States occurred in 1971 on the Clearwater River in Idaho. Water driving of pulpwood is still common in eastern Canada, but is on the decline.

In eastern Canada, for example, driving accounted for 50 percent of all pulpwood transportation during the 1961-1962 season. In the 1969-1970 season the share for river driving had dropped to 40 percent. Most of that share was lost to truck transportation.

Log-carrying barges and rafts are another form of water transportation used in some areas. Barges holding up to 150,000 board feet of logs are loaded and towed by tugboat over canals, lakes, and rivers. This mode is especially common on Lake Superior from both the United States and Canadian sides. Barging is also common on the coast of British Columbia. As a rule, barging is the final mode of transportation and is generally associated with trucks.

The use of log rafts of various types is common in many areas of the United States and Canada. Nearly all logs produced in Alaska, for instance, are trucked to a dump on the water, rafted, and towed to the mills. Rafting differs from driving in that the logs are confined by boom sticks bound together by chain and cable. It is generally associated with large rivers and waterways where, like barges, the rafts are towed by tugboats. In the early days, the early 1900s, ocean-going rafts 600 to 900 feet long and 50 feet wide were used to transport four to five million board feet of Douglas fir logs (at a time) from ports in Washington and Oregon to San Francisco and other port cities in California.

There are some other types of log transportation, such as the chute and the flume, which are no longer or very little used. When they stood they were monuments to the logger's ingenuity. The only flume ever seen by the author is located on the Columbia River a short distance from Portland, Oregon. It is still used to transport lumber to a finishing mill, a distance of about nine miles. However, some flumes are still active in eastern Canadian pulpwood movement.

In a flume, a log rides a sheet of water at speeds up to 100 miles per hour.

The distances varied from a few hundred yards to 22 miles. Sometimes they were on ground level and sometimes they towered up to 100 feet above the roads, rivers, and valleys.

All the transportation modes mentioned will be discussed in this chapter. However, truck transportation, because it is the most commonly used mode of transportation, will receive the most detailed treatment.

Truck transportation

Trucks used in logging vary widely in size and load-carrying capability, depending on such variables as topography, climate, size of operation, haul distances, volumes available, and the product to be hauled. They vary from single-axle vehicles, in the 100- to 135-horsepower range, that carry a payload of up to 18,000 pounds, to multiple-axle vehicles that can pull a load in excess of 200,000 pounds and are powered by 400- to 500-horsepower engines. Engines with even higher horsepower ratings are used to haul the largest loads possible over very steep terrain. Such vehicles are used in the interior of British Columbia to haul 100-ton loads. The smaller trucks may be flat-bed trucks adapted to pulpwood hauling.

Trucks also vary as to the volume of wood that can be carried. The bobtail truck mentioned in an earlier chapter is a single-axle vehicle which carries up to three cords or approximately 2.58 cunits.* A tandem-axle truck can accommodate up to five cords or about 4.3 cunits. Large tractors can haul up to 10-cord loads.

A truck with two powered rear axles (four sets of wheels powered) is called a *tandem-axle truck*. Live tandems, where both axles are powered, are fairly common in pulpwood and longwood operations. They generally have better weight distribution and increased tractive capability. When only one of the rear axles is powered it is called a *dead tandem* (three axles, four-wheel drive). With a dead tandem the load is distributed on three axles but only one axle is active when the truck is loaded.

Truck tractors or tractors are used to pull semi-trailers and the pole trailers used extensively in the Northwest. Some tractors are equipped with a fifth-wheel attachment, a connection located over the tandem driving axles on the tractor. These vehicles are used with semi-trailers equipped with fifth-wheels. A pole trailer has two axles and is attached to the tractor with a steel tube called a *stinger*. The stinger is telescoping and can be adjusted for length. In either case, semi-trailer or pole trailer, the tractor carries part of the load. A full trailer or pup trailer, by contrast, is pulled by a truck and none of the load is supported.

Log-hauling trucks and trailers are equipped with bunks and stakes which help contain a roundwood load. The bunks are generally attached to the truck and trailer frame. On a flat-bed truck both the plank bed and bunks are attached to the frame. Stakes made from steel, wood, or some other suitable

---

* The cunit estimate is based upon an 86 cubic feet per standard cord. A standard cord is 4 feet high, 4 feet wide, and 8 feet long.

A one-log load on a highway truck. The second loader is getting ready to throw the wrappers over the load. This is a pole trailer truck, which is commonly used in the Northwest.

material are on the ends of each bunk. On pole-trailer rigs, the bunks are designed to swivel up to 360 degrees. When the trailer is not loaded a pin is used to keep the bunk stationary. When the bunk is loaded the pins are removed so the load will turn and the trailer will track properly. Bunks used for hauling on public roads are 8 feet wide to the outside of the stakes. Other bunks, on specially built trucks, are up to 16 feet wide with stakes as high as 7 feet. These large rigs are capable of carrying up to 30-cunit loads depending on the size of the logs.

A standard 8-foot bunk carries around 7 to 8 cunits while 12- and 15-foot bunks can carry up to 15 and 22 cunits respectively. In general, trucks that haul loads not exceeding legal highway limits are *on highway* or simply *highway trucks*. The large vehicles are classed as *off-highway trucks*.

Spotter trucks

At times trucks are used in a sort of relay to move wood from a landing. This is likely to occur when the roads are muddy or there are severe adverse grades over a significant part of the haul from the landing. At times, a relay is also used with preload trailers. The trailers are picked up at the loading area and pulled to an intermediate concentration point. The regular haul trucks arrive at the concentration point, drop their trailers, and pick up a new load. The spotter truck drops its full load, picks up an empty trailer, and returns to the landing.

When working on extremely muddy ground, as is found in some areas of the South and Southeast, a large, high-powered truck or a converted skidder is used to spot the trailers at the landing and to pull loaded trailers to the main haul road where they will be picked up by a regular highway truck to complete the haul.

A similar condition exists in very steep terrain where road grades are always at the maximum acceptable limit. A large tractor that can negotiate the grades fully loaded is used to relay the trailers up the steeper grades to the regular haul vehicles. This system allows the use of just a few larger tractors, which are more expensive, rather than requiring a whole fleet of large tractors.

The trailer

The flat-bed trailers and solid-frame trailers with bunks are fairly common for hauling pulpwood and long wood in the United States and Canada. Long wood, of course, is loaded lengthwise on the bed. In some cases short pulpwood is loaded in the same manner. In the latter case the truck bed will be loaded with several tiers of short logs parallel to the truck frame. In the Lake states, as well as in other areas, 100-inch pulpwood is loaded crosswise on the truck, with the load slightly higher in the front than in the back. The stakes are located at the front and rear of the truck bed or trailer bed and the load is secured with two or more cables or chains. Loading the bolts in a tapered-down

A fifth-wheel pulpwood trailer. The trailer is sectioned to facilitate unloading. Note the short-wood storage.

352   Logging practices

fashion, from front to rear, places more weight on the front axle and less on the rear. In some states, where there is no weight limit on front axles, this loading method offers distinct advantages. The load on the rear axle is controlled to within the legal weight limit and the front axle may be overloaded.

When a haul vehicle is carrying tree-length wood the stems are usually loaded with the butt forward. The bunks are built so they are wider between the stakes at the front and narrower at the rear. The arrangement causes the small ends of the stems to be pinched together in the rear bunk and allows the load to be built quite evenly along the length of the trailer. Depending on the length of the trailer and length of the stems, some tops may overhang the trailer. The logs are loaded in the manner described because of the natural taper in the trees and a butt orientation that facilitates unloading and storage.

Some truckers, in order to carry a larger payload, use a slightly different method. They load a tier of tree-length stems with butts forward and then lay a second tier on top of the first with butts back. The bunks in this case are the same width. The advantage here is a larger overall load making full use of stake height and bunk space. The two tiers lie on the truck and trailer like two wedges lying on top of each other with the wider tops in opposite positions.

Some trucks are equipped with dumping platforms rather than a stationary bed. To unload the pulpwood the platform is raised hydraulically and the wood simply rolls off the end of the truck. This method is particularly handy for unloading into water. When dump platforms are used, the rear stakes are released or tripped. The tripping lever is located at the side of the truck and at the rear. It is connected to the stakes by a rod that runs the width of the truck bed or else a chain-release mechanism is used.

The same quick-release device is used for side stakes as well. Quick-release side stakes are located on the side of the truck from which the logs are off-loaded. The stakes are hinged and rest in a pocket. When they are released

The logs are being pushed off the rail cars by the unloading machine. The stakes are tripped by one of the hands from the other side of the car.

the weight of the load pushes them out of their pockets and they swing down against the side of the truck. This same type of system is used to release the stakes on rail cars being unloaded by a jilpoke, a hydraulically operated ram that pushes the logs off a car.

A second, more effective, quick-release system allows the bottoms of the stakes to swing out. The tops of the stakes are chained to the stakes on the opposite side of the trailer. When these stakes are tripped the logs roll out from under them. With this system there is less chance of damaging the stakes, since the logs go under rather than over them. With hinged stakes, if they do not rest against the side of the truck properly, unloading may bend or break them.

Usually there are two side stakes on each side of the truck and/or trailer when the logs are being loaded lengthwise. When logs are being loaded cross-wise on the truck bed, the stakes are located on either end of the truck or trailer bed. Some trucks and trailers have three or more sets of stakes set at intervals. If so, the truck is loaded in units which are more suitable for unloading. When pallets are used, of course, there is no need for stakes.

When a pole trailer is carried on the tractor, the stakes (tractor stakes) are between the rear wheels, and the stinger rests in a notched receptacle located on the top of the truck cab. Pull trailers or pups are loaded piggyback on the truck bed, if there is one; otherwise they are towed. Carrying trailers piggyback on the return trip reduces tire wear and allows a bit more maneuverability.

Weight scales

Measuring a haul vehicle's capacity on board feet or cunits really makes no sense in terms of operating efficiency. These units are used because they are common volumetric measures and can be equated to load weight. While costs may be determined on a volume unit measure, capacity is really measured by weight.

The amount of weight that can be hauled depends on whether the truck is a highway or an off-highway truck. For off-highway trucks the weight hauled is a matter of common sense. How much weight can be hauled depends on the safety factor and the capacity of the truck. If trucks are consistently over-loaded the maintenance costs will increase, as will mechanical downtime. Haul time also has a tendency to increase since a prudent driver will generally pull an overloaded trailer a little slower than one loaded normally.

Highway trucks are regulated by legal weight limits, which are fixed by the federal government at a gross combination weight of 80,000 pounds. Some states in the West allow greater loads under permit, and some states in the South have more restrictive weight limits (see Table 17.1). These regulations have a significant impact on log-hauling costs because a great many log trucks haul over public roads. In the western United States, for instance, it is esti-mated that 75 percent of all logs are hauled on public highways (Hammack 1966, p. 36).

In order to insure maximum loading, many haul trucks are equipped with some type of scales. Not only do drivers want a full load, but they also want to make sure they are not overloaded, because if they are caught overloading

Table 17.1 *Maximum Gross Combination Load Limits in Selected Timber-Producing States (in pounds)*

State	Without permit	With permit
California	80,000[1]	--
Idaho	80,000[1]	105,500
Montana	80,000	105,500
Oregon	80,000[1]	105,500
Washington	80,000[1]	105,500
Alabama	80,000[1]	92,400
Arkansas	73,280	--
Florida	80,000[1]	--
Georgia	80,000[1]	--
Louisiana	83,400	88,000
Mississippi	80,000[1]	--
North Carolina	79,800	--
Oklahoma	80,000[1]	90,000
Texas	80,000[1]	--
Michigan	80,000[1]	148,000
Minnesota	80,000[1]	--
Wisconsin	80,000[1]	--
New England states	80,000[1]	--

Source: Go West 1981, p. 35.

1. Limit set by Federal Highway Act of 1975.

means a fine. Basically, there are three types of scales: hydraulic, air, and electronic.

Hydraulic scales have been used the longest. The system measures the void between two steel plates, called a pad, located under a beam which supports the fifth wheel and/or the bunks. The weight is read on a gauge located close to the scale pad. With a hydraulic system the driver must get out of his cab to read the scales. This is done during the loading cycle and results in a delay. Sometimes the scales are checked several times during loading and logs may have to be removed and replaced, or repositioned on the truck to balance the axle loads. In addition, the scale is subject to errors resulting from torsional twists in the truck and trailer frame, temperature changes, and improper adjustment.

Air scale systems have also been used for a long time. The air system measures the air pressure in an enclosed chamber and registers the pressure, in terms of weight, on a gauge. The controls for the air scale are located in the cab. Through the use of selector valves each bunk is weighed one side at a time (or at once, on newer models). The two figures are added together to find the total weight on the bunk. Air scales are not subject to the same difficulties as the hydraulic system. Accuracy depends only on how closely one can read the gauge—generally to within 500 pounds, plus or minus 1/2 a percent.

Electronic scales are the newest addition to weight scale systems used in trucks. The system uses a strain gauge to measure changes in linear distance of a steel bar located under the bunks. The measurements, made by an electric current supplied by 12-volt truck voltage, are converted directly into actual pounds on a digital readout panel located in the truck's cab. Even on uneven ground the electronic system is accurate to within 1 1/2 percent. On level surfaces such scales are accurate to within plus or minus 1 percent.

Scales, especially the more accurate air and electronic types, are very effective cost-reduction tools. On public roads, however, drivers are apt to load light rather than risk a citation for an overweight load. This results in both the weight and the volume being lower than necessary. In the process, hauling efficiency is lost and unit costs increase. Experience has shown that the use of accurate, truck-mounted scales can increase load size by up to 10 percent and result in a substantial cost reduction on an annual basis. One operator reported a 5,258-pound increase in load size using an electronic system, while a second, independent operator reported an average of 3,413-pounds increase per load per trip (Arola 1972, p. 48). This amounts to nearly a cunit in the first case and slightly less than a cunit in the second case.

Operating variables

First and foremost, log hauling is tied inextricably to all other harvest system components. The timber cutters can and often do have a direct effect on hauling costs. In the Northwest, for instance, all they have to do to significantly decrease volume per load and increase hauling unit costs is to manufacture a preponderance of short logs. Skidding or yarding, whichever is used, is the point of production. No log can be moved from the landing until it is placed there through the efforts of the woods crew. As has been pointed out, logging production is variable and depends on a great many factors. The number of loads per day hauled varies with the effectiveness of preceding operations and the operating variables affecting those operations. The effects of loading and unloading, both elements of the hauling component, have already been discussed.

Assuming that a log supply is assured on a timely basis and can be loaded and unloaded in a relatively efficient manner there is left only one major variable, or rather family of variables, which has a direct and significant effect on log hauling—roads. Trucks are, to a degree, inflexible. They are confined and limited by the roads built for them to run on.

Travel time and haul costs are affected by road surface, gradient, road alignment, haul distance, and other specifications such as width and turnout locations. Of all the variables mentioned, distance is the most important factor. Some operators contend that with the right trucks—that is, enough horse-power—all variables except distance can be ignored.

Haul distances

Haul distances vary from region to region and from logging side to logging side. In some cases the haul might be only 10 or 12 miles, while in other cases it might be as much as 200 miles. One hundred to 150 miles, especially in some of Canada's crown limits, is not an unusual haul. The distance involved determines how many trips are possible and therefore how much volume can be hauled in a day. These two factors, in turn, relate to costs—total or unit.

Haul distance is important in any case, but is especially important to the independent contractor who works on a unit basis. He must choose the haul and calculate his costs accordingly if he is to make a profit every day. A larger firm, with its own wood supply, may haul its own logs. If the hauls are long, then the higher cost becomes a part of the cost of doing business. With its capital resources and equipment, a large company can view costs a bit differently than a small company.

At any rate, the cost of trucking logs varies with the length of haul. The longer the distance the higher the cost. Very short hauls cause the truck to spend more time being loaded, unloaded, and scaled or weighed. All of these activities are necessary, but they reduce productive hauling time. Also, in many areas a short haul means more time is spent on woods roads, with steeper grades, more curves, and narrower widths, than on public roads.

Haul distance has a big impact on unit cost. Fixed costs include depreciation, licenses, insurance, overhead, and return on investment. Variable costs, including fuel, oil, tires, and maintenance, are calculated on a per-mile basis. Drivers' wages are variable to the extent that they are paid only for time worked but in calculating trucking costs are considered fixed.

If a hauling contractor is paid on a unit basis—that is, on so much per cunit hauled—these variables can make a big difference when it comes to selecting a long haul versus a short haul.

Compare, for example, a 200-mile haul (400 miles round trip) with one trip per day with a 25-mile haul (50 miles round trip) and four trips per day. The total operating cost per day for the 200-mile trip is only 20 percent greater than the 25-mile trip. But on a cost-per-cunit basis, the long haul is nearly five times higher if variables are calculated correctly.

If the operator had calculated his costs on a per day basis and recognized no variable costs, the unit costs would have been lower for the long haul, and the operator would have lost money. The use of hourly rates does not help. Unfortunately straight day rates and hourly rates are commonly used to the disadvantage of the operator. However, the point is not how costs are calculated, but the fact that unit costs are generally higher on long-haul trips than on short-haul trips, given the same load size.

Fuel costs have taken on new significance since sharp increases began in 1973 and accelerated again in 1980 and 1981. At the beginning of 1981, for example, diesel at the pump in central Oregon was $1.10.9 and was being delivered in bulk at 99.4 cents per gallon, plus 4 cents road tax.

All costs have increased, however. Figures developed by the Oregon Log Truckers Association showed the following average percentage increases between 1977 and 1980:

Fuel	113.4%
Repair and maintenance	30.7

Table 17.2 *Truck Speed Road Class (Graveled Road Only)*

	Speed					
	Loaded	*Unloaded*	*Loaded*	*Unloaded*	*Loaded*	*Unloaded*
Road features	*20 mph*	*35 mph*	*10 mph*	*18 mph*	*7 mph*	*11 mph*
Road width (ft)		26		22		18
Gravel depth (in)		6		4		0-3
Sight distance on curves (ft)		160		120		80
Maximum grade (%)		10		12		15

Source: from a study by W. E. McGraw, Canada Department of Forestry, August 1963.

Tires	36.4
Drivers' salaries	33.3
Equipment costs	33.4
Insurance	65.3%

An Oregon Log Truckers Association survey in 1977 showed an average operating cost per hour per truck of $23.97 among both owner-operators and multiple-truck operators. The average annual operating cost per truck was $44,416.

Class of road

Classes of roads were described in Chapter 5. Depending on location, these descriptions vary. In general, road class describes such things as surfacing, road width, maximum grades, and sight distance on curves. Truck speeds, loaded and unloaded, are generally higher on roads with good running surfaces, minimum grades, and good alignment. For instance, Table 17.2, adapted from a Canadian study, shows relative travel speeds on three classes of gravel roads.

In the state of Washington hauling costs are based on a combination of factors, which include a basic charge (for loading, unloading, and waiting time), a charge per log hauled, the amount of defect, and finally, a variable charge depending on the class of road. The variable rate, or transportation charge, is stated in cents per mile per thousand board feet hauled. The lowest rate is applied to a class A road (the best road) and the highest rate is allowed on a class E road (the poorest road).

Class A roads are paved or macadamized, with grades of 6 percent or less. Class E roads are very rough, with grades exceeding 22 percent; in some cases a truck may not be able to run under its own power. The rates used reflect the ability of a truck to haul over a road. Paved roads with minimum grades are much faster roads than gravel roads with the same average grades. These differences are reflected in the transportation rates.

Effect of curves

An abundance of curves on a road slows down traffic. If the line of sight at a curve allows the driver to see oncoming traffic, the road is not too bad. However, on single-lane roads with short-radius curves, round-trip time is reduced and driving can be hazardous.

Figure 17.1 shows the relationship between round-trip time, curve radius, and the number of curves per mile of road. On the one hand, the longer the radius of the curve, with the number of curves constant, the shorter the round-trip time. On the other hand, if the number of curves per mile increases, with the same average radius, round-trip time will increase. The relationship between time and curve radius is much more critical than the relationship between time and the number of curves.

Road grade

On any class of road, speed will be reduced when grades are adverse and steep. On adverse grades the truck must be run in lower gears. If a truck is purchased to run consistently on steep grades it will have to have higher horsepower and will cost more than a truck of lower horsepower. Favorable grades, especially at higher percentages, are not particularly advantageous. A truck running consistently on steep favorable grades needs a greater braking capacity. The use of conventional brakes on long grades can lead to brake failure or a decrease in braking efficiency, because the brakes get hot from absorbing energy from the moving vehicle. Retarding, or engine, brakes can

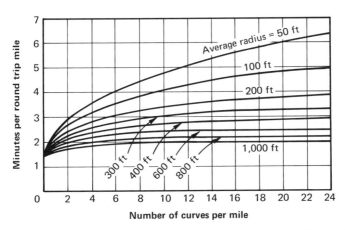

Figure 17.1. *Graph showing the effect of curves and curve radius on loaded truck travel (Byrne, Nelson and Googins 1969, p. 13).*

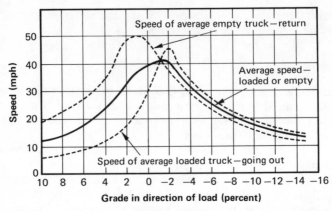

Figure 17.2. *Graph showing logging truck speed versus road grade for gravel roads (Byrne, Nelson, and Googins 1969, p. 16).*

slow a vehicle down and are very effective at low speeds. Figure 17.2 illustrates the relationship between speed and grade for loaded and empty trucks.

Other variables

While there are other variables that can affect hauling operations, we will only discuss two of them—legal restrictions and truck size.

Legal restrictions differ from state to state and sometimes within states. Because they are so variable no attempt will be made to delineate specific legal requirements. Broadly speaking, however, each state in some way controls length, height, and width of loads hauled over public highways. There are also laws concerning axle weights, braking systems, mud guards, and so forth.

Legal restrictions can have much to do with the size of truck used and how that truck is equipped if it is to be driven on public roads. Sometimes, when a firm owns its own timber it will find it economical to build and use private roads, even if those roads run adjacent to public roads. On private roads larger loads can be hauled, and in general, the larger the load carried the lower the cost.

Of course, as the vehicle gets larger, costs increase, but few costs increase proportionately with truck size. Capital costs for the larger equipment are higher, as are fuel costs and tire costs. However, the increase in productivity, in most cases, will more than offset the small increase in costs. One Canadian firm, which runs trucks with 15- and 12-foot bunks, reported that while costs increased 18 percent overall with the larger trucks, they were able to haul 47 percent more wood (Trebett 1966, p. 86).

Haul elements

A truck's hauling cycle, stated in minutes, will vary depending on the hauling distance, condition of the road, size of the log which must be loaded,

Pulling up a relatively steep grade slows down the haul cycle.

and many other variables. However, despite the time variations, the distribution stated as a percent of total time available should be fairly constant. In chapter 16, on loading and unloading, it was suggested that each of those components are really elements of the log transportation system—that they constitute necessary productive delays in the hauling cycle. In this section hauling will be further related to loading and unloading, which are identified as haul elements. In addition, there are three other elements: warm-up, haul-out, and haul-in.

Warm-up: Each morning the driver of a particular truck goes through a routine that involves checking oil, water, battery, and brakes. In addition, he will check the inflation of each tire. The truck will be started and allowed to run so that the oil will start circulating in the engine. The total time required for warm-up is around 10 minutes depending on circumstances. For instance, if the truck needs to be fueled more time will be needed. However, refueling generally takes place the preceding evening to avoid delay at the beginning of a new day.

Haul-out: Haul-out begins when the truck leaves the parking area, wherever that is, and starts for the landing. During the day, haul-out time will vary if the truck is directed to different landings. The second occurrence during the day will vary from the first if the starting point is other than the dumping area. Ordinarily haul-out begins at the dumping or unload area and ends when the truck arrives at the landing.

Starting from the unloading area and hauling from the same landing, haul-out time will vary but little from day to day. That is, if it takes an hour to make the trip to a certain landing during one cycle, then it should take about the same amount during every other cycle, barring some unforeseen condition. The variation should be no more than about 15 percent total for all trucks hauling from the same landing.

Haul-in: The haul-in element begins when the truck leaves the landing and ends when it arrives at the unloading area. The time required to haul in a load is naturally longer than the haul-out element, generally about 20 percent longer, with about the same variation between trucks that is experienced in the haul-out element. The main reason, of course, is that the vehicle is carrying a load and cannot move as fast as it can when empty. Also, road variables such as alignment and grade have a noticeable effect on a loaded vehicle, while the effect on an unloaded vehicle might be negligible.

During the haul-in element the truck may have to be stopped one or more times. For instance, if binders are used the driver will ordinarily stop to tighten them after the load has had a chance to settle. There are also various delays associated with hauling, which will be discussed as a separate element.

Loading: This element has already been described in detail in Chapter 6. It begins when the truck arrives at the landing and ends when it leaves with a full load. In terms of delays, this element shows the greatest time loss.

Loading involves positioning the truck for loading, preparing the truck or trailer for loading, loading the logs, and preparing the load for hauling. Time for loading varies mainly with the size of the logs and the skill of the loader operator. The variation in loading time may be wide—anywhere from 20 to 60 minutes.

Unloading: Like the loading element, the unloading element may involve a great deal of delay time. The element begins when the truck arrives at an unloading area such as a reload or transfer point, or some other concentration point. The element actually involves, in many cases, several subelements, such as weighing or scaling, reloading the trailer, unloading, and waiting time.

Another variable which affects unloading is the type of machine used and its availability. If a stacker is used in a log yard, for instance, and that machine is also used to perform other log handling functions, it may not be available or may not be in position to unload when the truck arrives. In contrast, an A-frame dump is used only for unloading. Unless there is a queue, an arriving truck can unload immediately.

Delays

Delays in the trucking operation may vary from no logs at the landing to personal delays, both necessary and unnecessary, on the part of the driver. As a rule, trucking operations are, to some extent, better controlled than most other harvesting components. However, delays still occur and are quite serious.

A load of tree-length wood being unloaded. Note large ends are all forward.

Photo courtesy Howard Cooper Corp.

Delays account for 10 to 20 percent of the total cycle time beginning with the haul-out element and ending with the unloading element. The actual percentage will vary with the haul distance, primarily because most of the delays occur during the loading and unloading elements. If haul distance is short then the haul-in and haul-out elements will be correspondingly shorter. But unloading time and loading time are both relatively fixed for any one landing, so their relative percentage as a part of a total distribution will increase.

Waiting for loading and unloading is by far the major cause of truck delay time. In one western operation involving crane loading and A-frame unloading, waiting time accounted for over 50 percent of the total delay time. If waiting time was reduced by half at both the landing and unloading area the delay time as a percentage of total cycle time would be reduced accordingly.

As defined here, waiting time at both the landing and unloading area is the result of queueing. This phenomenon was described in Chapter 16 and can be brought under control by scheduling the morning start-up at staggered intervals. Staggering will not eliminate queueing but will tend to decrease it dramatically. A second alternative, to be used along with staggered start-up intervals, is to dispatch trucks to different landings throughout the day. This also has a tendency to break up the queues—or at least help to shorten them.

The solutions just suggested can be implemented best when a fleet operation is used. If a landing is being serviced by several contract haulers it will be difficult to stop queueing at the landing.

If all the trucks servicing a landing are also hauling to the same unloading area, as is often the case, then the queues follow from one element to the next.

Queues at the unloading area are further complicated when trucks from many landings are feeding the same facility. The waiting time attributable to queuing at the unloading area is generally higher than at the landings simply because there are more trucks involved.

Another occurrence which results in delays is not having logs to load. This also was discussed in Chapter 16 and no more will be said about it here.

Mechanical downtime involving both the truck and the loading machine is the third most serious cause of delays. If the truck is down there is little to do except dispatch a mechanic or tow it to the shop. A preventive maintenance program will help reduce unexpected truck downtime but certainly will not eliminate the problem. When the loader breaks, the answer is the same as that which applies to the problem of no logs. Communicate the problem as soon as possible and get the trucks rerouted if there is an alternative landing to go to. Of course, if the unloading machine is down, the trucks will simply have to wait unless there is a backup machine.

The delays described thus far have been attributable to the interface of trucking with other elements. There are also some difficulties which may be experienced within the trucking organization itself. One problem is that drivers, like other workmen, will sometimes take advantage of the situation if they are not properly supervised. The result can be unnecessary personal delays, too much visiting, and even running errands for some logging foreman. The drivers may also stretch the normal round-trip time in order to pick up a little premium time. This, of course, does not happen when drivers are hauling for themselves or when they are being paid on a trip basis. Nor does it occur all the time with hourly workers. But the foreman should certainly be well aware of the possibility.

One tool the foreman has to help control an operation is the tachograph. This device is found, when it it used, in the cab of the truck, and records on a circular graph the time the truck is moving or not moving. By carefully studying the graphs at the end of each shift the foreman can tell how much time was spent for haul-in, haul-out, loading, and unloading. He can also tell how many times the truck was stopped during a haul and how fast it traveled. This device is an extremely good tool to monitor trip time and operating effectiveness, and can be useful in determining problem areas.

Radio communication

Radio communication is another effective way to help control hauling operations. A dispatcher or the truck foreman (probably both) can be in constant contact with the landings and the trucks to direct movements and give assistance when it is needed. A radio system is especially helpful when the trucks can haul from several landings. As loads become available trucks can be dispatched. When there are problems, such as being out of logs, or mechanical delays, trucks can be rerouted en route thus saving valuable time. Those operators who use two-way radios in their trucks report they are very effective. The savings come in terms of cost reductions, an increase in trips, and greater profitability.

Scheduling

Scheduling is another activity, along with radio communication, which can reduce delays and improve productivity. But scheduling means much more than simply assigning trucks to the various landings each day. A working schedule is based on the trucks available, hauling distances, and log availability. Each afternoon, after the last string is out, the loader operator should call the truck foreman or dispatcher to tell him how many loads are available for the first string in the morning. If he will need fewer trucks or more, he should be forewarned. Trucking must always be balanced with logging production, or vice versa since trucking provides the paydays.

When an operator is hauling from several landings the haul distances can often be balanced to maximize loads and minimize costs. For instance, a truck on a long haul (say, two trips per day) might be sent to a nearby landing to pick up an extra load.

In many cases, when a fleet is being operated, each truck is assigned to a landing and will continue to service that landing until the logging is completed. If the cycle time per trip is 3.6 hours and three trips per day per truck are required, each truck driver would have to work 10.8 hours, not counting his lunch break. That is 2.8 hours of premium time each day if the driver is an hourly worker. However, if three loads are required per truck the foreman may assume that the premium time is simply part of the cost. This need not be so, however—at least, the cost need not be so high.

Let us assume there is another landing with a shorter cycle time—say, only 2.5 hours. Drivers on that landing can haul three loads in about 7.5 hours. The drivers would undoubtedly arrange their time to get in their 8 hours. Occasionally a driver may haul four loads, the last one on premium time. In any case, they would get 8 hours a day and would be paid for some work unnecessarily. This is where the problem lies, since on one landing the drivers are likely to be overpaid and on the other they are working overtime to get the required production.

One solution is scheduling the trucks so that one or two that are hauling from the farthest landing haul at least once from the closest landing and the same number of trucks from the closest landing haul one trip from the farthest landing. Let us call the farthest landing A and the closest landing B. We will assume that only one truck is being used at each landing, in order to keep the example simple.

If one truck hauls two loads from landing A and one from landing B, the total time will be 9.7 hours. If the second truck hauls two loads from B and one from A, the total time will be 8.6 hours. Under the old schedule the truck driver hauling from B would have been paid for 8 hours in any case and the driver hauling from A would have been paid for 10.8 hours, the last 2.8 hours at premium time. Under the new schedule the total amount of time is 18.3 hours, rather than 18.8 hours. This reduction in time spent results in a savings of 1/2 hour of premium pay.

This is not a dramatic example, but it illustrates the point. Scheduling can save money. In many cases the result will also be increased productivity and more trips per day. For adequate scheduling the following truck cycle time distribution* must be considered:

Element	Time distribution (%)
Haul-out	28.0
Loading	12.7
Haul-in	35.5
Unload	7.1
Delay	16.7

Railroad transportation

Railroads are still used in the Northwest to deliver logs to their final destination. However, unlike the logging railroads of old, few logging railroads today are privately owned, and those that are operate, for the most part, over common-carrier tracks. Also, railroads today do not go to the source of the wood. Instead, the logs are moved in stages: first from the woods landing to a transfer point; and then, after the load is transferred, on to a final destination by rail car.

In spite of private railroads having all but ceased to exist, common-carrier rail transportation of forest products is still important, especially in the pulpwood regions of the northeastern, eastern, and southern United States and of Canada.

Railroad hauling is best suited to high volumes, high tonnage, and hauls exceeding 300 to 400 miles. Shorter hauls are acceptable if solid trains of pulpwood cars can be put together and if there is minimum amount of switching involved. In some southern states, where these conditions exist, hauls may be less than 150 miles. Nevertheless, some hauls are substantially longer— ranging up to 800 or even 1,000 miles.

Under the right conditions rail hauling is far superior, in terms of cost, to any other mode of transportation. This is especially true where long hauls are involved. While the railroads are extremely capital-intensive, railroad maintenance costs are lower per mile than those incurred on most haul roads. River driving, often considered the least costly transportation mode, has some hidden, or at least not so obvious costs which sometimes tend to make rail transport more attractive.

River driving is a seasonal operation; large inventories are built up and held ahead of the mills. In one example, described by an official of the Canadian Pacific Railroad, the hidden inventory cost amounted to about $8,000 per month over a six-month period (Masson 1971, pp. 68-76). In addition, log handling in and out of the river has become more expensive. Finally, there is

---

* The distribution shown is based on actual time studies of log hauling in a West Coast operation. The logs were loaded by loading cranes and unloaded at an A-frame dump. The use of other loading and unloading equipment would tend to change the distribution. However, the distribution does indicate the general relationships among the various elements in the haul cycle.

Rail transportation is still fairly common in the Northwest. This train is being used to haul logs from a transfer point to the mill.

the ever-present impact on the environment, which, along with dumping into the water, has already been mentioned.

In order to achieve minimum costs from stump to mill several conditions must be satisfied. Volumes must be guaranteed and handled in a uniform manner over the entire year. When volumes are moved in and out of concentration points, loading and unloading must be accomplished as quickly as possible. This has the effect of increasing cycle time per car, increasing car utilization, and reducing the number of cars required to handle the volumes. Naturally, all cars must be loaded to capacity, a condition that most shippers would attempt to ensure at any rate. Finally, the maximum volume of wood must be shipped from a minimum number of points.

In the southern United States, where most car deliveries originate in concentration yards maintained by pulp mill operators or pulpwood dealers, nearly all of the conditions mentioned are satisfied. Large volumes are shipped from a minimum number of points of origin. Because of the penalty for holding cars and because of a shortage of pulp cars, each one entering the yard is filled to capacity and moved as quickly as possible. Finally, in many cases, perhaps most cases, the cars are shipped as unit trains, not stopping until they reach their final destination.

Volumes carried

One of the reasons rails present a cost advantage over truck hauling is the tremendous volume they can carry. While trucks can carry 5 to 15 cords, the largest rail cars commonly used can haul up to 38 cords of short-wood or approximately 33 cunits of solid wood, if standard cords are used. Even some of the older cars carry 20 cords of wood. One 72-foot car which was being designed for lumber a few years ago, could carry up to 45 cords. The actual capacity would be limited by track curvature restrictions and wood handling methods (Masson 1971, p. 73).

Pulpwood cars

Pulpwood is classed as a low-grade rough freight commodity. As such it can be handled on any type of car. Before World War II, it indeed was hauled in any type of car—from stockcars to boxcars. Following the war gondolas and flatcars were modified for hauling pulpwood and added to the list. Today the cars are more specialized, with steel end-racks or bulkheads on each end for hauling wood stacked crosswise. Others have steel end frames as well as skeleton steel side frames. With the side frames, pulpwood is stacked lengthwise on the car.

When gondolas, gravel cars, and flatcars were first used the racks were relatively primitive and not very efficient. Loads shifted in transit and wood was lost. This created a hazard and also constituted an expensive loss of value. To overcome the problem of shift, 6-inch by 6-inch timbers were bolted along the edges of the cars to form an abbreviated side rail. When the short-wood bolts, between 48 to 63 inches long, are loaded, they cant to the center, creating a tighter load and alleviating the problem of shift.

Some standard 46-foot cars have been converted for hauling pallets by adding light steel bulkheads, cross rails, and tiedown devices. Special steel pallets, which conform to railroad clearance requirements, are used to contain the 4-foot wood. Each pallet has a five-cord capacity and five pallets are carried on each car.

This string of pulpwood cars is ready to head to the mill over 150 miles away. Note side stakes and chain wrappers. Car ends have steel bulkheads.

The Canadian National Railroad has designed 52-foot end bulkhead cars to carry long-wood (8 to 10 feet). The car measures 52½ feet inside the bulkheads and is equipped with side stakes. It can carry 33 cords of 100-inch pulpwood in six tiers. Some U.S. firms are also ordering such cars built because of the economies of handling 8-foot wood.

Cars used for hauling long logs (up to 40 feet) are found primarily in the Pacific Northwest. These cars vary from flatcars with platform decks to skeleton cars with steel or wooden bunks located over each axle. Some of the bunks are equipped with stakes similar to those used on trucks. Other log cars are equipped with short steel blocks called *cheese blocks*. They are located, like stakes, on either side of the bunks. The cheese blocks, or stakes, are tripped from the opposite side of the car for dumping or unloading with a jilpoke. The capacity of the cars varies with the size of car and the height of the stakes. They may carry anywhere from 8 to 12 cunits depending on whether they were transferred or reloaded and on railroad regulations.

Water transportation

River driving

Water transportation in the form of river driving is the oldest form of log transportation and has long been used in both Canada and the United States. In eastern Canada, pulpwood driving is still common, but volumes moved this way are declining. Some driving also takes place in the eastern United States, primarily in the New England states.

The bulk of the river-driven wood in eastern Canada is moved in a few major drives. For the most part, the drives are 300 miles or longer and involve large volumes of wood. One such drive, which is considered representative, covers nearly 300 miles of the upper Ottawa River in Quebec. The drive begins with 100,000 cunits and picks up wood along the way until it peaks at about 300,000 cunits (Martin 1971b, p. 33). The St. Maurice River, also in Quebec, carries more pulpwood annually than any other river in the world and has been driven for over a hundred years—since 1828. The drive now involves approximately 850,000 cunits per year.

The shorter drives, up to about 100 miles, are or have been replaced by truck hauling, generally for economic reasons. Truck hauling has increased in eastern Canada as river driving has decreased. Most of the 33 percent decrease of volume driven since 1951 is now hauled by truck, although some of the volume was captured by barging and towing. The railroads are simply not able to compete over such short hauls.

River driving operations

For purposes of driving, a river is divided into three sections: upper, middle, and lower. The upper stretches of the rivers, where the drives generally originate, are the most expensive in terms of improvement costs. Because the upper stretches can be driven only during spring runoff, if at all, dams called *splash dams* are constructed to hold back the water so that normal flow can be increased sufficiently for driving.

All stretches of the river require some work to make driving possible. Stream channels must be cleared of obstacles, such as rocks and debris, which can cause jams. When such obstacles cannot be removed, then side piers are constructed to act as sheers, directing the wood around the obstacles. In some cases the river channel itself might be moved to remove bends in the river that restrict driving.

River driving is an exciting business and generally requires very little time. The river drive down the Clearwater River in Idaho, when it was active, often negotiated the 90 miles of river in two days to a week. Of course, that was under ideal conditions. When the river fell because of changing weather conditions, it sometimes took the drive crew weeks or even months to retrieve the thousands of logs stranded along the banks.

If the logs are stored on the iced-over rivers of Canada they start floating toward their destination when the ice breaks up in the spring. If the logs are decked on the bank of the river they are tripped or watered when the water conditions are considered to be right for the trip. The logs or bolts are generally tripped by logging tractors or grapple cranes and make their own way to their destination. Logs that are left stranded along the river banks or in jams after the main drive is over are picked up by the rearing crew that follows the drive.

At some point the pulpwood bolts must be stored, since there generally is not room for a year's supply of wood at the mill site. As the bolts reach the lower river (sometimes they go only as far as the middle river) or a lake, they are bagged off in holding booms, which are used to encircle the sometimes massive volume of wood that arrives at the holding area. The booms are made of strings of floating, debarked logs called *boom sticks*. Each stick is attached to the next by a length of heavy chain called a *toggle chain*. The booms are anchored to a *dolphin*, which is several pilings driven together and lashed with cable or banding to make them secure, in a sort of tripod fashion.

On large, fast-flowing rivers the bag boom is not enough to hold the wood. In this case the log booms are built like raceways, with relative narrow compartments made of boom sticks and held together with piers and pilings. The piers are made of solid wood cribbing, which is layered like a log house and filled with rock. The holding booms are chained to hitching posts built into the middle of the piers.

There are several advantages to river driving; the principal one is low cost. However, if the driving is done primarily in the upper stretches of a river, it may be quite expensive. Often, when the upper stretches cannot be driven, the bolts are hauled by truck to the middle stretch and dumped into the water. The seasonality of river driving is its most important disadvantage. The mills must receive and hold up to a year's supply, all in just a few weeks. Two or three hundred thousand cunits of wood is a huge volume. The cost of holding inventory is quite high when you consider physical storage costs, storage facilities, maintenance, degradation, volume losses, and taxes. This cost is generally estimated at about 2 percent of the inventory value per month.

Hydroelectric dams constitute another problem. In some cases, the presence of large dams marked the end of driving the river on which the dam was constructed. This was the case with the Clearwater River drive in northern Idaho. The completion of Dworshak Dam on the North Fork of the Clearwater marked the end of river driving in that area. In some cases, flumes have been used to bypass dam locations.

Log rafting and towing

Free driving, which has just been discussed, involves dumping the logs or bolts into the river or stream and allowing them to float to the concentration point or mill, powered only by the flow of the water. Rafting, on the other hand, involves confining the logs or bolts with boom sticks connected by chains and towing them to a final destination or storage area. In some cases, as was described, free-driven bolts are captured in booms and then rafted the rest of the way to their destination. While driving can take place in smaller streams, rafting and towing are usually done on larger rivers and lakes.

Rafting is always done in conjunction with some other transportation mode, such as driving, trucking or railroading. In the case of both truck and rail, the logs are delivered to a dump, placed in the water, the rafts are built, and finally, the rafts are towed either to the mill or to storage.

The round-boom raft or bag-raft consists of two strings of boom sticks, one within the other, which confine the bolts much like a corral confines wild horses. The logs float free within the confines of the raft. Any sort of rough water will throw the bolts out of the raft and allow them to scatter. The round-boom raft is used extensively on the inland lakes of eastern Canada as well as on Lake Superior. Up to 12,000 cords at a time are rafted and towed by diesel tugboats from storage areas to the mills.

This tugboat is towing two six-section flat rafts—about one-half million feet of logs—to the log storage area.

Log booms used for storage. Note rafts tied up outside booms or log pens.

The rectangular raft is commonly used in the Northwest to float Douglas fir and Sitka spruce from the log dumps to mill storage. Like the round-boom raft, the rectangular boom is formed by boom sticks, which may be up to 84 feet long and between 30 and 60 feet wide. A hole is bored in the end of each boom stick and a chain and toggle is used to connect the sticks. When the logs are dumped, small, and very maneuverable boats called *boom boats* or *water broncs* are used to push the logs into the booms. This process, called *booming*, generally takes place hot—that is, as the logs are being dumped.

The booms are constructed of pilings and dolphins used to confine the string of boom sticks. Two head sticks form a V and the rest of the raft is rectangular. The end of the raft is opened and the logs are pushed into it in rows. A raft may be composed of between 6 and 12 sections depending on where it is being used. As each section is filled, *swifters* (pieces of small-diameter cable) are run over the raft and attached to the boom sticks on either side by dogs. The swifter also is dogged to some of the logs within the raft.

When individual logs are allowed to float within the boom sticks it is called a *flat raft.* Some logs, such as western hemlock, which are less buoyant and have a tendency to sink, are bundled by steel straps and rafted.* Small logs are also rafted in this manner, although for different reasons. Unless the small logs are bundle-rafted, the volume per raft will be very small and therefore more expensive to tow on a unit basis.

Rectangular rafts are used in rivers as well as the bays and inlets of Oregon, Washington, British Columbia, and Alaska. Almost all of the logs produced in

---

* Logs that sink are called *sinkers*, appropriately enough. Heavy-butted hemlock often sinks butt first with the top of the log protruding just above the water's surface. These logs are called *deadheads.*

Alaska are involved in rafting and towing since much of the logging is done on the islands of southeastern Alaska. Water transportation is the only economic method of moving the logs to the mills, which may be some distance away. Most of the timber in Alaska is moved in bundled rafts.

The round boom and rectangular boom are the most commonly used, although there are other types. For instance, a triangular boom, made of three boom sticks bolted together, is used in some parts of eastern Canada. There are also four other types of boom rafts and fence booms, which each have special applications.

Rafting and towing with tugboats (when it is possible) is less expensive than trucking. However, the method comes under the same attack from environmentalists as other water transportation systems, and regulations restricting this practice have been enacted in many areas. The primary concern is water pollution.

Barge transportation

With barge transportation pollution is less a problem. However, the method is restricted to waterways and navigable rivers on which a barge system can operate. Barging also is associated with other transportation modes, generally trucking or driving. The timber must be delivered to barge landings or loading areas and the mill is often located adjacent to the waterway or very close to it.

The barges are either flattopped or open. Flattopped barges are used in the southern United States to haul pulpwood. The wood is ricked or stacked on the deck and unloaded by cranes at the destination. Flattopped, self-dumping barges are commonly used on the coast of British Columbia and can carry up to 1.5 million board feet of logs. These barges are equipped with cranes used for either loading or unloading. However, logs are normally dumped by flooding tipping tanks located on one side of the vessel. When the barge is unloaded, the tanks automatically empty themselves and the vessel is leveled. Dumping takes up to one hour.

Pulpwood may be loaded by flume, jack ladder, or cranes. The larger barges used on Lake Superior carry approximately 2,300 cords of round wood, while smaller barges carry between 150 and 300 cords of pulpwood.

One barge operation in Canada uses two open barges with an overall length of 402 feet and a width of 75 feet. The barges carry about 4,000 cords of river-driven wood. The wood is stored behind a holding dam before being loaded. When the vessel is ready a flume is used to feed a steel loading plant, which, in turn, is equipped with a number of flumes which direct the wood into the barge. The barge is filled to about 70 percent capacity from one side before it is turned and loaded from the other side. At the mill site the barges are unloaded using a heavy lift crane equipped with a two-cord grapple, which dumps the wood into a conveyor that feeds debarking drums. Unloading takes between 24 and 40 hours (Williams 1971, pp. 94-99).

Barges are either pushed or towed conventionally by tugboats generating up to 2,000 horsepower. In some cases, with one tug pulling and the other pushing,

A self-propelled, self-dumping log carrier. *Photo courtesy* Forest Industries

up to three barges at a time are towed. When a barge is pushed, the tug fits into a V-shaped notch in the stern of the barge and is held in the notch by snubbing lines. Push towing is not used in rough water; instead, conventional towing is used. When towing conventionally, a main tow line linked to a towing winch is attached to a towing bridle or directly to the barge.

Unloading at the mill site is usually done with cranes as was just described. The cranes are equipped with lattice booms up to 80 feet long and use either orange-peel or clamshell buckets to pick the wood off the barge. The wood is unloaded into conveyors which take the wood either directly to the mill for debarking and subsequent use or to a storage area where it is simply piled into large, roughly conical-shaped decks or rows.

Section Six

18.
Safety Management

19.
Cost and Production Control

18.

Safety Management

Much emphasis has been placed on safety throughout this text. Safety is a major component of the logging business and can have significant impact on both production and cost. Generally, this reference to safety takes the form of a "do and don't" list: don't work below felled and bucked timber, do wear proper safety equipment, and don't work in the bight of the line.

However, there is much more to a safety program than simple lists. In fact, a successful safety program is not a program at all, it is an operating philosophy. It requires commitment from all levels in the organization. It requires good judgment, management controls, and excellent training.

Safety, according to Webster, is the condition of being safe from undergoing or causing hurt, injury, or loss. Most healthy people do not intentionally risk injury or death and do not purposefully place themselves in dangerous situations. Being injured or finding oneself in a dangerous situation is usually an unexpected or unplanned event: an accident.

If there is any deficiency in industrial safety programs it is that managers, supervisors, and workmen think accidents and injuries are synonymous. People often think of accidents in terms of injury—cuts, bruises, broken bones—or fatality. Injuries are not accidents but are the *results* of accidents.

Not every accident results in personal injury. For instance, suppose a skidder operator rolls his machine down a steep sidehill but rides out the incident without injury. An accident occurred, the machine was wrecked, but there was no personal injury.

The primary objective of any safety program should be accident prevention. It is only by reducing accidents that the incidence of serious personal injuries can be reduced. And, it is only by reducing accidents that the business losses that result from those accidents can be reduced.

The importance of an industrial safety program is obvious when a person is injured or killed. No one likes to see someone injured. But what of the accident's other effects: lost production, property damage, medical expenses, and workmen's compensation claims? These have a common denominator—dollars. A safety program is important because not having an effective program is expensive and wasteful in terms of both human misery and business losses.

If ever an industry had a solid rationale for a strong accident prevention effort, logging does. Logging is one of the most hazardous occupations and is one of six industries chosen by administrators of the Occupational Safety and Health Act for special attention. The six industries have been targeted because of high injury incident rates (Table 18.1).

OSHA incident rates are expressed as cases of days per 100 full-time employees, using 200,000 employee hours as the equivalent.

$$\text{Incident rate} = \frac{\text{Number of recordable incidents x 200,000 hours}}{\text{Manhours worked}}$$

A recordable incident is an occupational death, occupational illness, or occupational injury. Any injury that stems from a work accident or occurs in a work environment and that results in loss of consciousness, restriction of work or motion, transfer to another job, or medical treatment other than first aid is classified as an occupational injury.

Table 18.1 indicates that the incident rate for all private industry is 9.4 per 100 full-time workers. Lost work days, an indication of severity of injury and illness, amounted to 63.5 days. Experience in logging and logging contracting was 2.7 times greater in total cases and nearly five times more in lost work days.

The purpose of these injury statistics is not to conjure images of logging as a man-killing business, but to illustrate that it is hazardous. Injury experience has improved dramatically over the years, yet there is still much room for improvement. The business is not any less hazardous. Physical conditions that set the stage for accidents, injury and non-injury, are still present. The logger is still faced with the personal risks involved with felling, bucking, yarding, loading, and hauling logs or other timber products from the woods to the mill.

Table 18.1 *OSHA Target Industries*

Industry	Incidence rates per 100 full-time workers[1]		
	Total cases	*Lost workday cases*	*Lost workdays*
Mining	11.5	6.4	143.2
Roofing	22.5	11.7	212.7
Longshoring	18.0	Not available	Not available
Metal fabrication	19.3	8.0	112.4
Logging	25.9	15.6	316.2
Meat packing	28.4	13.2	168.9
All industry	9.4	4.1	63.5

1. U.S. Department of Labor 1980.

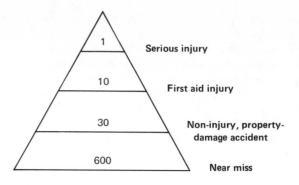

Figure 18.1. *Ratios between various types of accidents (Insurance Companies of North America study).*

Improvements in injury statistics have been gained as the result of training, experience, effort, and commitment. Management, labor unions, equipment manufacturers, and the workmen have all cooperated to make the woods a safer place to work.

Accident prevention

The most effective approach to a safer working environment is an accident prevention program. There is a distinct difference between accident prevention and injury prevention. In the former you work to avoid or eliminate accidents; in the latter you work to eliminate the cause of injuries. Unfortunately, the two are often confused, as indicated in an earlier paragraph. Since accidents are the cause of personal injuries, accident prevention is the logical beginning point of any industrial safety program.

The distinction between injuries and accidents is clarified by statistics developed by the Insurance Companies of North America (INA), based on nearly two million accidents. The INA accident ratio (Figure 18.1) shows the relationship between near-miss accidents (accidents that did not quite happen), non-injury property damage accidents, first aid injuries, and serious injuries. Obviously, if emphasis is placed on controlling those incidents at the bottom of the pyramid—the near-miss and property damage accidents—both serious and non-serious personal injury accidents will be reduced.

The cost of accidents

The rationale for approaching safety through accident prevention becomes even stronger when a manager views the direct and indirect costs of accidents. The avoidance of human misery should be reason enough to pursue an aggressive accident prevention program, but it is not the only reason. Altogether there are four good reasons for accident prevention programs:

1. People
2. Economic factors

3. Employee morale
4. Public relations

In round numbers, there are 14,000 on-the-job deaths and two million disabling injuries in the United States each year. In terms of lost wages alone, the cost is about $18.3 billion per year. The lost wages, however, are overshadowed by the loss of limbs and life, and in the case of fatalities, by what happens to the widows and children who are left to fend for themselves. The misery and grief, while real, are not measurable in dollars and cents.

The injured party and his or her family are not the only ones affected when an injury occurs. There is something about working in an environment where accidents and injuries are commonplace that affects the morale of all involved. Fear and anxiety are poor companions, especially at work. It is difficult to describe the feelings of a man or group of men who have just witnessed the injury or death of a friend and fellow workman. The emotional impact is severe and lasting.

The reputation of a company is based on many factors, not the least of which is the manner in which management looks after the welfare of its employees. There is no honor in being identified with an industry that is the target of the government agency responsible for the health and safety of America's workers. There is little to recommend an industry or a company with a high incident rate. In these days, when public opinion carries so much weight and the cost of government interference can be so great, it pays an industry to do everything in its power to improve its reputation for commitment to safety.

The brevity of this discourse about people, morale, and public relations is in no way intended to minimize the importance of these topics. That importance should be obvious without elaboration. Unfortunately, however, people react most often to those things that affect them economically. And accidents cost money, whether or not there is an injury.

The cost of lost wages associated with job-related fatalities and disabling injuries is about $18.3 billion per year, as indicated previously. In addition, industry spends about $8 billion in compensation resulting from job-related injuries and illnesses. Finally, the statistics tell us that the production loss is around $10 billion per year. This increases the total cost to about $36.3 billion annually, and it is unlikely that all the costs involved are accounted for.

Everyone loses with each injury. The workman loses wages, at the very least. The company loses because it must absorb additional compensation costs, loss of production, and the loss of material and equipment. The public loses because it must ultimately pay the price in the goods produced.

Compensation and medical costs are fairly straightforward. Most managers in the logging business are aware of the magnitude of these expenses. In fact, some logging operators have learned from sad experience that a high loss ratio can result in bankruptcy. With each lost-time accident the insurance premium increases until it can no longer be paid. Unfortunately, many loggers consider only the compensation and medical costs and therefore grossly underestimate the cost of accidents.

Remember the accident ratio triangle (Figure 18.1)? For every serious injury and every 10 first-aid injuries there are 30 non-injury, property damage accidents.

A log truck comes off a long, favorable grade in too high a gear and the transmission is damaged. The driver could have lost control of the truck with fatal results. Instead, the owner buys a new transmission. The operator of a loader, positioned too close to a deck of logs, swings it into the deck and caves in the side of the machine. The operator could have been injured but wasn't. A timber faller, working under the side-lean of a tree, cuts off his corner and the tree comes over sideways. The tree could have fallen on the cutter, but instead the saw is smashed. In every case there was property damage. In every case serious injury or a fatality could have resulted.

In one example familiar to the author, a mobile tower was pulled over. The accident was the result of the guylines being worn and too much tension being applied to the running lines during the yarding cycle. When the tower fell, the chaser was struck by a guyline and, luckily, sustained only serious bruises. The injury was serious enough to put him on light duty for a week. The tower was badly bent as a result of the fall. The following costs were associated with that accident:

Chaser:

Medical	$ 60.00
Lost time	75.00
Replacement	375.00
Total cost	$510.00

Equipment repair:

Tower	$15,000.00
Yarder	500.00
Total cost	$15,500.00

Production:

The yarding side lost three days of production while it was moving and rigging a wooden tree to take the place of the mobile yarder. One of the three days was lost because very little was done on the day of the accident.

The direct cost of the accident was simply the cost of the injury and replacing the chaser while he was on light duty because of his injury. The total cost of the accident was $16,010. The cost of repairing the tower and yarder could easily have been much higher. Indirect costs still not accounted for include supervisory and management time required to investigate the accident, and time required to fill out the accident report forms, medical forms, and reports required by the state accident commission. It should be obvious that the cost of compensation and medical expenses does not begin to cover the total cost of an accident.

The ratio of direct costs of injury accidents (lost wages, compensation, and medical expenses) to indirect costs is set anywhere from 2 to 1 to 50 to 1.

Alfred Lateiner, a safety expert, places the ratio at 4 to 1 (Lateiner 1965). According to one consulting firm, the actual cost of accidents in the Ontario construction business was conservatively placed at five times greater than compensation and medical costs *(Canadian Forest Industries* 1969, p. 50). It seems that the ratio really has not changed all that much over the years. More recent insurance company statistics suggest the ratio ranges between 5 to 1 and 50 to 1. That is, if a firm's cost of medical and compensation payments is represented by $1, then operations are experiencing direct losses to equipment, facilities, products, and production of between $5 and $50.

The point is that there is a direct correlation between accidents (injury or non-injury) and production costs. Productivity and costs tend to improve as accident prevention improves. The injuries alone cost money in terms of compensation and medical expenses. But accidents, whether or not there are injuries, are extremely costly when physical damage, lost production, and downtime are considered.

Management responsibility

There is also a direct correlation between good management and a good safety record. Management alone is responsible for employee training and for providing a safe working environment for the physical and mental fitness of employees. Like it or not, management is also responsible for control and minimization of physical and mechanical hazards and for the unsafe actions of employees. In logging, like any other business, the primary reason for a poor safety record is poor management and lack of adequate supervision.

Maintaining an aggressive accident prevention program is doubly difficult in a logging operation since the very environment in which the men must work is hazardous, and unsafe conditions are often difficult to avoid. This means there must be a strong commitment to *working* safely all the time on the part of each member of the logging crew. Commitment is the first requirement of a safety program.

Commitment begins with the boss, whether he is a corporate president or a working owner. And, it is much more than an expression of interest. Strong commitment is manifested by active leadership, sustained interest, total support and participation, and the delegation of responsibility with accountability. It is not enough to issue orders and demand results. You cannot delegate interest; you have to create it.

Picture the owner-manager of a small logging operation who just purchased a new tower-yarder combination for a half million dollars. This represents a major investment for him and he is going to jealously guard his interests. He will see to it that his key workers are fully trained and will be present when the factory representative does the training. In fact, he will probably participate himself. And when that tower is raised for the first time he will be there watching anxiously. He wants to make sure that his investment lives long enough to earn him a return.

The same is true of the corporate president who authorizes a multimillion dollar investment in land and equipment to open up a new logging operation.

He will actively participate in reviewing the financial analysis. He will make sure the best men available work on the planning for the project and will insist on being apprised of the progress. He will not actually run the project for them, but there is no doubt of his total interest, emphasis, and commitment to what they are doing.

The same kind of commitment and active show of interest is necessary if a safety program is to work successfully. However, neither the owner-manager nor the corporate president have the time to run the whole business alone. Even if they had the time they probably would not have all the skills required. An owner-manager does not have to be an expert in everything to run a business—even a logging business. All he needs is to have experts working for him—he must give them responsibility and hold them accountable. It is important not to forget, however, that the manager to whom the responsibility is delegated cannot be an expert in all things either. He is going to require some training, and it is the owner's responsibility to see that he gets that training.

All industrial safety programs depend on one thing for success or failure. Responsibility for safety and accident prevention must rest primarily with the supervisor who is in direct charge of the operation. The manager is ultimately responsible, but the line supervisor or foreman is the one who gets the job done with the crew. In other words, with the manager's support, the supervisor must leave no doubt as to his total commitment to, emphasis on, and interest in safety.

The emphasis on safety is manifested in the way the supervisor manages his business. Just as in logging production, the management of safety requires real effort. In logging production, that effort results in acceptable productivity levels, meeting production targets, and achieving targeted cost results. The effort in accident prevention results in fewer accidents and lower costs. That effort involves accident investigation, conducting safety meetings, training employees, and enforcing the use of protective equipment.

The foreman is the key to a good safety record—providing he is a good supervisor. If he cannot implement and maintain a good safety record with his crew he probably is not performing the rest of his job well either. He either needs training in supervisory skills or his attitude is wrong—in either case the deficiency will be reflected in total performance, not just safety.

The lack of supervisory skills is probably one of the most serious deficiencies in logging management. In too many cases a man is made a foreman simply because he is a good logger. With training he might make a good supervisor of men as well, but the training is not always forthcoming.

The following are examples of conditions resulting from the lack of good managerial or supervisory action:

1. Lack of an organized safety program.
2. Inadequate safety work.
3. Lack of management commitment.
4. Failure to provide safeguards, first-aid equipment, and safe tools.
5. Failure to screen new employees properly.
6. Failure to assign employees to a job they can physically perform safely.

7. Poor employee morale.
8. Lack of enforcement of safety rules.
9. Failure to place responsibility for accidents.

Though this list is incomplete, it is sufficient to illustrate where the responsibility lies. Each item on the list is the undisputed responsibility of management. If management doesn't see to it that it happens, it doesn't happen. If the foreman doesn't train the men in the crew, then who will? Is it reasonable to assume that the employees will provide their own first-aid kits and stretchers? If a machine, a skidder for instance, has defective brakes, who is responsible for getting the machine repaired?

Causes of accidents

So far in this chapter an effort has been made to establish three facts:

1. Safety in logging operations is a problem of great magnitude.
2. Accident costs are production costs, and the total cost of accidents, both injury and noninjury, is higher than any manager should be willing to accept.
3. Management is responsible for accidents and for accident prevention.

Now the question is, How can the accident rate in logging operations be improved? While the solutions may not be simple, the approach to identifying the solutions should be fairly obvious to any professional manager or supervisor. Correcting a safety problem is really no different from correcting a costly production problem.

Suppose that a mill owner who has bought a timber sale and contracted the logging complains to the manager of the logging company that there is insufficient trim allowance on a majority of the logs being delivered to his yard. The mill owner explains that the result of insufficient trim is short lumber and a waste of nearly 2 feet on every log delivered. Not only is there a loss in product value, but the logger is also losing revenue because the logs are all being docked 2 feet by the scaler.

The problem is insufficient trim allowance, resulting in loss of product value to the mill and loss of scale, and therefore revenue, to the logging company. If the logging manager is smart he is going to visit his cutting crew immediately—that is the only place where a great many short logs can originate, since no cutting is done at the landing. Talking with the crew and measuring some logs, he will find that the problem results from one of three causes:

1. The cutters are not measuring the logs at all, but simply eye-balling the length to save time and increase production.
2. The tape being used to measure the logs is inaccurate.
3. The cutters do not know how much trim allowance to add.

One way or the other, the logging manager is going to make clear what he wants the cutters to do. He will specify the length required; he should, but might not, explain the seriousness of the problem; and he will either instruct his bullbuck to tell the men (or he will tell the men himself if there is no bullbuck) to keep a close eye on trim allowance in the future. If the logging boss is a good manager he will tell his cutters when he notices an improvement, so as to reinforce the results of the first visit.

This approach to solving the mill owner's hypothetical problem has some gaps in it, but the necessary steps are covered. The problem is defined and the possible causes are identified. The solution, in this case, is to give the cutters some training—either to remind them of something they have forgotten or to teach them something they do not know. Finally, the manager arranges for follow-up and control by making sure that trim allowance is checked in the future and that the men are informed of their progress.

The same basic approach is appropriate to finding solutions for a safety problem. To prevent accidents and injuries, their causes must first be known. Remember, injuries are not accidents, they are the results of accidents. If a manager wants to prevent injuries he must learn to prevent accidents.

Every accident is a link in a chain of events. In about 2 percent of all accidents (including near-miss accidents) an injury will occur, and this injury will be the final link. In slightly more than 98 percent of all accidents no injury will occur, and the accident will be the final link. Figure 18.2 shows the links in the injury chain.

The injury chain contains the five factors most safety experts agree must be present for an injury to occur. In the following example each factor will be explained.

A hooktender, working on a highlead logging side is having marital problems (background) and reports to work distracted and emotionally upset (personal defect). He is stringing out a section of haywire for a road change and fails to notice that a turn of logs is being started to the landing, causing the backline, which he must cross, to flop up and down (unsafe condition). When he approaches the backline, pulling the haywire, he does not notice the haulback moving and begins to step over it (unsafe act). Just as the hooktender steps over the running line the yarder engineer is taking the slack out of the haulback in preparation for going ahead on the mainline, and the hooktender is thrown into the air (accident). The man sustains multiple bruises, a concussion, and a broken back (injury).

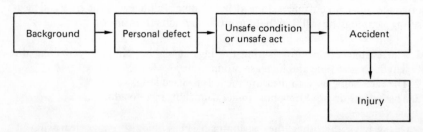

Figure 18.2. *The injury chain flow chart.*

This injury (which, by the way, is not at all unusual) was avoidable. The hooktender knew better than to be near running lines, especially when a turn is moving. He also knew that a man should never step over a running line, whether it is standing still or not.

Of the five events involved in the injury only two were direct causes. The accident and injury were results, the other three events were causes. Unfortunately, little can be done to control the home environment and the mental distraction caused by the home environment. Remedial action must therefore be directed toward the unsafe condition or the unsafe act. Since every accident is preceded by an unsafe condition or an unsafe act, controlling these two links in the chain can improve the accident rate.

Under normal circumstances unsafe conditions cause about 20 percent of all accidents; unsafe acts, 78 percent. The remaining 2 percent are attributable to acts of nature and cannot be controlled (Lateiner 1965, p. 81). In the injury-chain flow chart, acts and conditions are described as one factor. In woods operations it is difficult to imagine a logging show where unsafe conditions do not exist, by the very nature of the business.

For instance, widow-makers can be found hanging in trees whether a cutter is in the vicinity or not. Running lines always present a hazard—an unsafe condition. Felled and bucked timber can shift and roll—also an unsafe condition. For an unsafe condition to precipitate an accident, however, there must be a catalyst—an unsafe act. If a workman is aware of the unsafe conditions which exist and, despite this knowledge, provides the catalyst that makes the condition dangerous to man (an unsafe act), then an accident and maybe an injury will occur.

Some conditions, such as a worker's wearing unsafe clothing or a skidder's having defective brakes, can be remedied. Actually, these things are unsafe conditions caused by unsafe acts. Failure to repair the brakes on the skidder is indirectly caused by the failure to act—thus, in a manner of speaking, it is an unsafe act. Exceptions notwithstanding, the way to reduce accidents and injuries in logging operations is to remove or control unsafe acts—to act on the people component of the accident equation. To control the people component, management must be sure the men they hire are physically and emotionally able to perform the job, have the proper attitude, have knowledge of the potentially unsafe conditions, and are properly trained.

Training

Training and job knowledge are the two most important factors in an accident prevention program. Making sure a workman is fully aware of job hazards and training him to work around those hazards efficiently are the major preventive measures that management can take. Every new worker, regardless of experience, should be given some initial training. The totally inexperienced man should be given an explanation of the job functions, shown how to perform the functions, and then intensively trained on the job by experienced and responsible workmen. The men responsible for training the new man must, of course, be aware of potential hazards associated with the job and be able to teach the skills required.

On-the-job training (OJT) is the most common method used in the logging industry to train new men. Ordinarily, such training begins with an orientation by the foreman, who briefly explains both the purpose and the hazards of the job. Once the orientation is complete, the new man is handed over to a crew leader (a hooktender or rigging slinger in the case of a highlead operation), who will be responsible for the training. Training involves learning by doing. The trainee will work with the rigging slinger, say, and actually set chokers under his supervision. The assumption is that the trainer knows the job and the job hazards well. At times this has proven to be a bad assumption. There are some hazards the trainer may have underestimated or overlooked. Training is ultimately the responsibility of the foreman. To aid him in discharging his responsibilities, job safety analysis is very helpful.

Job safety analysis is a systematic approach to analyzing a job and indicating the skills required, the hazards present, and the experience necessary before a man can be considered fully trained. If the analysis is done properly, then the foreman can perform his training function very efficiently. Figure 18.3 is an

Responsibilities	Apparent hazard	Safe work practices and methods to avoid accidents
Setting chokers in a safe and efficient manner.	To self: Loose bark, lines, rolling logs, traps, jaggers. To others: Same.	Wear good caulk shoes, look over logs to be choked, watch for lines, stay out of bight of line.
Learn whistles.	To self: Rigging movement. To others: Same.	Learn whistles in order to know what is going to happen as signals are sent in.
Learn to recognize danger areas, such as bad ground, rolling rocks, root wads.	To self: Trips, falls. To others: Same.	Wear gloves, proper clothing. Stay in the clear—behind and on side of turn.
Follow rigging slinger's instructions.	To self: Possible accidents from not following instructions. To others: Subject to similar accidents.	Always follow the rigging slinger's instructions.

General safety precautions:

1. Wear good caulk boots, suspenders, stagged pants, gloves.
2. Be alert at all times, stay on your feet as the turn goes in.
3. Watch your footing at all times, look out for the other guy.

Figure 18.3. *Job safety analysis for a choker setter.*

example of a job safety analysis for a choker setter on a highlead side. Such an analysis should be performed for every job classification active on an operation.

The men who know their jobs best are the ones who work most efficiently and safely. If each workman is taught to work and think safely, he can eliminate the unsafe acts which result in injury and in physical damage to equipment. In short, he can minimize accidents. Even though unsafe conditions exist, as they often do, having well-trained workmen can at least reduce the chances of an accident occurring.

The safety program

To be successful in reducing accidents and injuries a safety program must be defined by the following five factors:

1. Strong and sustained management commitment.
2. Delegation of responsibility and accountability.
3. Supervisory training.
4. Employee job training.
5. Active participation of both management and employees.

The absence of any one of these factors will result in a program that does not meet its objective. We have already discussed the importance of the first four factors. The fifth factor, active participation, has not been discussed; yet it is probably the most critical. Unfortunately, in many cases it is the least understood of the factors listed and the most difficult to bring about.

The corporate president who is experiencing strong pressure from his board of directors, governmental agencies, and other external public groups does not find it very difficult to muster the required commitment. Recently this has been clearly demonstrated in some industries that have come under close scrutiny from OSHA inspectors and administrators. The manager or supervisor who is made aware of the alternatives also has no trouble in determining where to put the emphasis. However, for some reason the job is still not done.

There are many formal safety programs to be found in the woods industry. There are safety coordinators, safety committees, safety inspections and safety meetings. Some companies have even gone to the trouble of printing *SAFETY FIRST* across the front of paychecks. But buckers are still crushed by rolling logs, fallers are hit by widow-makers, and hooktenders, like the one in the example, are rendered useless by the haulback. Obviously, there is something wrong with the approach. The commitment is there, the accountability is there, responsibility is delegated, the training is done. The employees are involved—but, in all too many cases, they are not really participating.

Attitude is one area where something can be done. Attitude is the potential for action. A safety attitude is a readiness to respond effectively and safely under any set of conditions. The problem is that attitude is a learned response; it does not always occur naturally. Nobody wants to be injured—it is just that too many are of the opinion that "It can't happen to me." A bucker who is killed by a rolling log would never have worked on the downhill side of the log if he thought it would kill him. Until all workmen develop a new safety awareness the correct attitude will not be forthcoming.

Personalization and participation

Most companies have the tools for laying the foundation of a good safety attitude. They can get a short-term, but favorable response through posters, slogans, and movies. However, once a particular stimulus is experienced many times without positive reinforcement its effect becomes negligible. Much more can be accomplished when the experience associated with the first look at the movies or posters is personalized. Personalizing the attitude occurs when the employee identifies with that attitude, when he sees it as his own. Or put another way, the attitude will be retained only as long as it yields personal satisfaction. And personal satisfaction comes through participation. The attitude becomes important to the employee because he has taken part in its development, both in safety meetings and on the job. And he gains personal satisfaction from that participation.

Once the attitude is personalized, becomes his own, there must be constant reinforcement. The employee must be constantly reminded that safety is indeed first. Participation provides that reinforcement.

Without the positive and active participation of all the employees a "company safety program" is doomed to failure. Nonparticipative programs are unsuccessful because in most cases the workers do not identify with the company; instead, they identify with their own work group. For a program to be successful, the workers must identify it as their own.

Setting goals

In establishing a participative safety program, management is really providing an environment in which the employees can motivate themselves. Part of establishing that environment includes providing incentives to encourage the individual to achieve a preselected goal. Some of the incentives might come from the company in the form of rewards and praise; some will derive from the individual himself as he interacts with his group.

With the help of supervision or management the employees must set some goals which they accept as their own. The achievement of goals and objectives provides satisfaction, one form of incentive that should not be overlooked. The objectives of a safety program must necessarily involve reducing accidents. However, safety and production cannot and should not be separated. In those cases where safety is treated as separate from production, the results are damaging to both the employee and the company.

Once goals are set—and they may be slightly different in every case—achievement must be tested as a matter of course. This involves a reporting system that tracks performance against accepted goals. In addition, attitudes and interest must constantly be reinforced. This is the job of the supervisor who spends most of his time with the workmen. This close interaction allows the supervisor to demonstrate his interest by practicing what he preaches. He can further influence the attitude of his workers by engaging them in discussions and descriptions of personal experiences. This personal contact also allows the supervisor to look at workmanship and production without appearing to be snooping.

Incentives are an integral part of any safety program. Some of the incentives are self-derived, such as the sense of achievement. But direct, non-monetary incentives awarded by management for excellent safety performance tend to motivate and focus attention on the program.

One West Coast logging company paid off in Green Stamps redeemable for merchandise and a cash award to the employees in the amount of the rebate saved from workmen's compensation premiums for better safety performance. In other cases, the incentive award is a small prize such as a pocket knife or flashlight. Safety dinners are also effective—perhaps more so than gifts alone—since they provide an opportunity to communicate directly with the crews and can also involve the families.

Obviously, the incentive does not have to be large. A pocket knife is not terribly expensive. The personal recognition is probably as important as the award itself. More than anything else, the incentive demonstrates management's interest and commitment.

Competition

Another essential part of a participative program is competition. One thing many supervisors tend to overlook is the power of the group. The tendency is to deal with individuals or pairs of people, not with crews. The most powerful element in any social system is the small group. If the groups are pitted against each other in competition for status, for reward, or simply for recognition, the program will be stronger and more successful.

Examples of the effect of competition are all around us. Think of two highlead logging crews who are competing for top position in production. Both crews perform at a higher level than usual. In contrast, consider an isolated crew. Without the stimulus of competition they will not do nearly as well.

In a safety program, management can easily get the kind of active participation that is not possible in other circumstances. The reason, of course, is that the objectives are perceived by the employees as being their own. However, the result will be an increase in productivity as the members of each crew begin to interact and work together. The group interaction will not only make a safety program successful, but also bring about a secondary effect in the form of greater job interest. As a company's accident prevention program begins to work, production and costs will also be favorably affected.

The companies with the best safety records are well-managed companies with skilled, professional supervision and well-trained employees. Such companies have totally committed themselves to a sound accident-prevention program. They are also the companies earning the highest net profits.

Cost and Production Control

In each of the preceding chapters dealing with planning and operations a great deal of emphasis has been placed on costs and productivity. We saw, for example, that on steep ground productivity decreases and unit costs increase because the rigging crew finds it more difficult to move in and out of the turns. Poorly aligned roads slow down round-trip hauling time and result in a higher cost. Highly defective timber results in higher unit costs because the divisor, i.e., net volume produced, is lower. Each of the many examples and conditions was discussed to give the reader an insight into the realities of logging.

It is important to remember that cost minimization is a prime objective of the harvesting system. For the contract logger or producer the business of harvesting timber crops is undertaken to make a profit. A large, fully integrated corporation recognizes harvesting operations as a cost center; the profits are made downstream. In either case, the cost of producing the wood—logs, bolts, or what have you—has a direct and significant effect on the final profits, wherever they are made. In this chapter, cost is treated as a part of the profit equation, and it is assumed that a profit must be made from the logging operation. Conceptually, it makes little difference whether the profits are real or imaginary. If the operator, whether corporate manager or owner-operator, does not at least act like an entrepreneur his chances for success will certainly be decreased.

By the time a logging operation is undertaken the question of what the effort is worth is usually answered. The logging price should be calculated and set in advance. The logging price, less stumpage and costs, equals profit. Profits can be increased by producing more or by lowering costs. Unless an estimating error has been made, however, the volume of timber cannot be increased. The job of the logger is to make sure he gets all the volume that is economically available. The only thing most logging operators have to work with is cost. How successfully cost is controlled will determine the profitability of any particular setting or contract.

Cost and productivity

Costs are not controlled by looking at figures scribbled on the back of an envelope or carefully printed on the pages of a computerized report. Those figures represent what happened yesterday, last week, or last month. That is history—it can be a valuable guide for the future, but is not much help in today's operation. To control costs the operator must know what productivity is and what it should be. And he has to know at the end of the day, not at the end of the week or month.

Cost control means good records and believable production standards. On a daily basis, which is the time horizon considered to be appropriate, control does not get down to actual dollars. Rather, it is based on those immediate indicators that will determine costs in a predictable fashion. Cost control involves production, time, and cost rates for labor and equipment. On a daily basis, production and time are the key factors—the day-to-day control points. If they are not controlled in the short run, costs are not really controlled in the long run either.

Costs are not really controlled for the entire system whether the firm involved is a small independent firm or a large firm. Costs, like operations, are better broken down into manageable units. This need to deal with manageable units leads to the control center concept.

A control center is a group of related men and machines which can logically be treated together for purposes of analysis and control. The system components described in Chapter 4 are the common units used in logging. They are cutting, primary transportation, and secondary transportation. In most businesses loading is also treated as a separate unit. Unloading is often a control center in the manufacturing system. In a logging operation, then, the control centers would be timber cutting (which includes felling, bucking, limbing and topping); skidding or yarding and in some larger operations both; loading; and, last, transportation from the landing to the conversion point or some other concentration point.

Measures of productivity

Each of these control centers has some natural measure of productivity. *Productivity* in this chapter is defined as units of production per unit of time. Each center requires a productivity measure which reflects its effectiveness and can be made available on a timely basis. "Timely" means it will be available the same day or, at most, the following morning. Of course, the timeliness constraint may require some compromise. If volume figures are not available for the cutting crew on a daily basis, then cutting productivity may have to be based on stem count instead. Similarly, for skidding or yarding, a log count may be necessary rather than actual volume. Daily results for log transportation are best measured in ton-miles of volume hauled, but due to the difficulty in making the figures available daily, a load count may be used instead.

All of the units mentioned are or can be used: stem count, log count, and load count. In some cases, however, their use provides misleading and inaccurate information. In order to be used effectively the terms must be defined.

Stems must relate to some unit of volume, as should logs and loads, so they can be verified if necessary. Indeed, they should be verified as often as possible. Physical units, such as logs or piece count, are more easily verified as to count than are events such as trips or turns. In one operation three different individuals, the loader operator, the dispatcher, and the truck foreman, submitted daily reports containing load count as a measure of productivity. It was only infrequently that all three agreed on how many loads were delivered. At times their reports differed by as many as 10 to 15 loads, or about 20 percent of the production.

Productivity must relate to time, for which there are several units to choose from. Time can be measured in units produced per man-hour, per crew-hour, or per machine-hour. Work shifts is another commonly used unit and is, perhaps, the most meaningful to operators and workmen. The man-hour is the best unit for activities performed by individuals or variable-sized crews. The crew-hour, on the other hand, is more appropriate when a standard-size crew is used, as in a cable yarding operation. When the machine is clearly the dominant cost component, as it is with helicopter logging, the machine-hour may be the logical time unit.

Once the appropriate productivity unit has been decided on as a basis of control, it must then be reported on a timely basis. Such a report should contain pertinent information other than production. For instance, for a cable yarding show, a production report might include such information as amount of machine downtime, number of chokers used, crew size, and number of logs produced. Figure 19.1 shows a sample daily production report. An actual report might contain more information. However, it should not be too long, since loggers do not have any great love for filling out reports. Regardless of contents, assuming the report contains the necessary data and the supervisor has it in hand at the end of the day—so what? Do the figures mean that everything is in control, or that the operation is below par in terms of

| Date begin: 11/28 | | | | Location: Stub Creek | | Equipment no.: 108 |
| Date end: 12/2 | | | | | | |

| | | Crew | Log | Downtime | | |
Day	Hrs.	size	count	Yarder	Loader	Comments
11/28	8.0	8	165	–	–	Raining.
11/29	8.0	9	183	–	–	Rain in the P.M.
11/30	4.5	9	79	3 hrs.	–	Main drive chain. Talkie-Tooters missing signals. Run out for wind at 3:30.
12/1	8.0	9	173	–	–	No problems.
12/2	8.0	9	191	–	–	Clear day, good logging.

Figure 19.1. *Sample daily logging side report.*

production? In order to make this determination, the supervisor must have some standards against which he can measure actual production.

Production standards

Many logging supervisors and managers draw on past experience and rules of thumb to help set these standards. They compare conditions on a particular setting against conditions observed on similar settings in the past. Over time, an experienced and observant logging manager can build a fairly effective set of mental decision rules relating to production levels under varying conditions.

One example is the cutting foreman who, through experience, can relate heavy brush concentrations to approximately a 10 percent reduction in cutting productivity per set-day. That is, a set that would normally cut 50 to 60 thousand feet of timber in a day would probably get between 45 and 55 thousand feet per day in heavy brush concentrations. The same experience leads an experienced observer to predict that with 45 thousand board feet per acre, a piece volume of about 300 board feet, and an 800-foot average yarding distance a highlead yarding crew will produce about 50 thousand board feet per day.

Some men can estimate production very closely and fairly consistently. However, production estimates based on experience and good judgment are not always right or are sometimes right for the wrong reasons. The same experience and judgment that allowed a correct assessment of production on one setting may lead to errors on the next. Poor performance may be accepted as good, and good performance may be judged poor.

Records and past experience can be and are used, but they should be used carefully. Both are valuable, but they should be supplemented with, or replaced by, a test period early in the operation when production, time, and conditions are all carefully noted. This sort of actual experience will yield goals more applicable to specific conditions.

The goals or standards set for the control centers are not and should not be inflexible. If the goal is a good average figure, actual production will be higher than standard on about half the days and lower than standard on about half the days. The manager must decide at what production levels performance will be accepted without question and at what levels some corrective action must be taken. That "tolerable" performance level should be set at, say, 10 to 15 percent below the standard. If production falls below this level, corrective action is indicated. If production remains within this level the manager can spend his time on more pressing matters. This is the essence of management by exception, which means that the manager gives attention to those things definitely out of control and does not bother with what is reasonably in control.

Although some managers may not want to bother, a control chart such as that shown in Figure 19.2 can be helpful in tracking actual performance to standard. It is especially helpful if a manager is responsible for several control centers. When production falls below 270 logs in any one day or when production is consistently unfavorable, the manager must determine what is

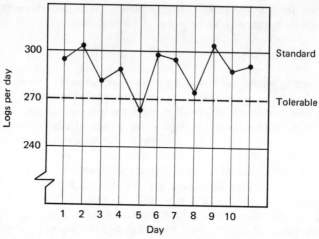

Figure 19.2. *Example of production control chart.*

causing the problem and take corrective action. This means, of course, that unknown causes must be identified and corrected. It does not mean that obvious operating problems are corrected only after they affect performance. The everyday problems must be solved—every day. When problems are not dramatic enough to call attention to themselves, the productivity figures may offer the first clue to their presence. When this occurs, as indicated on the control chart, the manager should take corrective action before costs get out of control.

Controlling productivity is one way to control costs—at least those costs controllable on a daily basis. However, there is much more to costs than the payroll for the crew. Expenditures for equipment, supplies, and taxes—anything that requires an outlay of money—must also be considered.

Fixed and variable costs

In order to make total cost information more meaningful, the accountants categorize costs into two types—fixed and variable. Variable costs or direct costs are incurred mainly as the result of activity. Labor dollars and fuel costs are variable costs. Fixed costs or period costs occur simply because time passes. The activity may cease, but fixed costs go on regardless.

Some examples of fixed costs are insurance costs, land and building occupancy costs, and property taxes. This general category is sometimes called *overhead costs* and is assumed to be noncontrollable in the short term. Variable costs, on the other hand, should be controllable in the short term even if they do not always vary as expected.

Not all costs fit into neat compartments such as fixed and variable. There is some fixed portion to nearly every variable cost, and fixed costs are never totally fixed. There is some question whether a particular cost should be attributed to use or to the passage of time. A prime example is depreciation.

Depreciation is an accounting convention adopted to spread the cost of a major capital purchase over its useful life. It is unreasonable to say that the cost of operating a business this year should include the entire purchase price of the new truck fleet or the new yarder. The trucks will last up to 10 years. The yarder, if it is well cared for, may last even longer. In general, however, logging equipment is more sensibly depreciated by use than by the passage of years. It is in that class of assets whose values decline more because of wear and tear than because of age.

Accounting records

The classification of depreciation, as well as the classification of some other costs, requires thought. The process is not automatic, and the value of the resulting accounting system will be a direct result of the thought and care exercised in developing and implementing the system.

There is, of course, something less than complete freedom for management to decide on accounting methods. The records must provide data for income tax and other tax purposes. Unfortunately, in many cases small and large businesses use accounting systems adopted for tax purposes only. In such cases, the revenue and expense accounts may be:

Revenue	Repairs
Cost of goods and operations	Insurance
Labor	Professional fees
Material and supplies	Fringe benefits
Depreciation	Interest of borrowings
Taxes	Bad debts
Rent	

In special situations other categories are added. For instance, a timber owner will account for depletion, which is a way of spreading out the cost of timber or minerals similar to equipment depreciation.

The actual recording and bookkeeping can be and certainly are accomplished in several ways. Larger businesses often have their own accounting and tax departments. Some smaller enterprises contract for accounting services or bookkeeping. Still other small operations with revenues exceeding the million-dollar mark and capital investments of over a million dollars insist on doing their own bookkeeping. One West Coast logger, who fits the latter description, had his wife, who had high-school bookkeeping experience, keep the company books. This approach to handling such a critical part of a logging business is sheer folly. The same operator would not dream of putting an inexperienced high-school graduate in charge of his logging crew and a half million dollars worth of logging equipment. But he does not even think twice about letting a person with the same sort of experience handle his accounting problems.

If a firm cannot afford to hire a full-time accountant, and many logging firms are not large enough to support such help on a full-time basis, there are many reputable firms which will provide the service. In some cases banks offer accounting services as a supplement to their major business.

The accounting and bookkeeping services may be generalized or they may be designed especially with the specific requirements of the firm in mind. Whichever the case, the manager should definitely spend some of his time with the accountants to be sure the information meets both legal and management requirements. In other words, the information developed must be understandable and in a form which will be useful in the management of the business.

The information most valuable for control of operating costs is not that which is needed for tax purposes or overall profit and loss determination. While a certain degree of compatibility is necessary, the needs are different. For instance, if maintenance costs are getting out of line management should be able to spot the problem. If yarding costs are too high the accounting system should highlight the problem and perhaps provide some clues as to the causes.

One Oregon operator known to the author discovered through a routine examination of his books that choker costs on one yarding side were exceptionally high. The ground was easy and there were relatively few hang-ups. A close inspection revealed that the swivels on the butt rigging were worn and causing abnormal wear on the chokers. The solution was simple enough once the problem was recognized. Choker costs came back into line, delays for changing chokers and picking up lost logs were reduced, and production increased slightly.

In another example it was noticed that the fuel bill for one of the yarders had increased substantially. The increase in fuel consumption had been accompanied by a decrease in production. The problem ordinarily would not have been spotted, since fuel costs are often lumped into one account. As it turned out, a leaking fuel pump was the cause of both the lost production and the increased fuel usage. The fuel was being pumped onto the ground, and since the yarder was not getting sufficient fuel it was not working to capacity.

The type of detail that provides information needed to highlight problems such as those described is not accumulated on the back of an envelope. It is provided through good, professional accounting.

The branch of accounting that furnishes operating cost data is known as *cost accounting.* Its basic function is to relate expenditures and noncash costs such as depreciation to some measure of business performance. The two main types of relationships are project costs and process costs.

Project costing involves collecting and summarizing all costs incurred for a particular undertaking, such as a construction job or a specific logging setting. Such costs are often tracked against a budget. *Process costing* involves collecting cost information attributable to a particular activity or type of operation. The process costs are matched with production to develop unit costs for that activity or operation. Examples of process costs are falling and bucking costs, yarding costs, and transportation costs.

Project costs are used extensively by contractors of various kinds in order to establish the profitability of a given contract. This approach can be successfully applied to logging as well as to road construction or any other type of activity administered on a basis of matching costs and revenues. It may be used anytime management wants to keep track of costs, whether or not profitability is a consideration.

Process costs are the equivalent in dollars of the productivity measures discussed earlier in the chapter. They are a more comprehensive measure than productivity because they include the dollar equivalent of labor hours as well as other direct costs, such as supplies expenses for equipment operation. Since daily costs are not ordinarily available, however, productivity is used for daily control.

Costs should be summarized at least monthly. This level of availability will allow the manager to become aware of any creeping conditions or trends which might not be otherwise apparent. Another benefit of process cost control is in planning. The knowledge of the unit costs of an activity under a certain set of conditions will be helpful in estimating the cost of similar operations.

Methods improvement applications

The cost of a similar project, or even accumulated cost experience to find an average unit cost, is often not sufficient. Such costs are helpful in the long run, but can be very inexact in specific situations. What a manager really needs is a tool to help him relate some standard performance levels to specific knowledge of conditions on a specific setting. With such a tool he could estimate the most likely overall cost of the operation. Process costs are helpful, but even at best there are certain imprecise elements involved, such as the mixture of constants and variables, machine types, and even inflation.

One way to make process costs more useful is to break the process being tracked into smaller subsystems or components. Performance measures, including piece rates and time requirements, can then be developed and a cost per unit of time can be applied to each of the subsystems or components in question.

In earlier chapters dealing with specific harvesting system components, the operations were broken down into elements. For instance, in Chapter 9 the skidding elements were defined and related to time (see Table 10.1). Those numbers were included to give the reader a better feeling of the relationships between the various elements and the total skidding cycle. Because the times per element are stated in ranges, any observer would have a difficult time relating those times to a specific set of conditions. If it was decided that more precise information was needed, some more detailed observations would be required.

The level of detail desired could be obtained by making the type of detailed observations made by industrial engineers. Those observations might point out what causes the range of time for a specific element and would allow the observer to arrive at a predicted average time per turn under a particular set of conditions.

If each of the logging components were treated in the manner just described a manager would eventually have a fairly detailed set of elemental times for logging operations where variations in conditions are critical. The type of information spoken of is available from several sources, but the most meaningful and most precise time data are developed within the operations where the information will be used. Each operation is subject to variation and peculiarities regarding operating methods and equipment usage. The data are most useful and effective when the user understands the variables. The desired level of understanding is attainable when the information is developed by the same organization which uses it.

There are several techniques for measuring time and work which have been used in logging operations. They range in precision from the application of motion-time data, stopwatch time study and work sampling, to actual gross time recording under controlled conditions.

Motion-time data

Motion-time data is time data for the classified motions of fingers, arms, and other body members. Its application produces highly consistent standards for manual operations of short cycle times and highly uniform motion patterns. It is a detailed and laborious technique for an analyst to use and is of most value to high-output, controlled activity. Motion-time data applications are not, for several reasons, a reasonable approach for measuring logging operations. First, logging operations are characterized by a high degree of variability in the cycles. Second, there are large delay elements. And third, a great deal of logging cycle times are machine controlled.

Stopwatch time study

A time study involves the use of a stopwatch to observe, measure, and record the length of time required to perform an activity. In addition, notations of such variables as distance traveled, pieces handled, and piece size are included. The product of these studies can be detailed tables or the synthesis of the individual elements to form broader standards of operation. Table 19.1, which deals with time required to load logs with a grapple loading crane, is an example of the time standards which can be developed through the use of a stopwatch study. The stopwatch time study is a practical method of measuring logging operations, particularly if the analyst is measuring relatively large elements.*

Work sampling

Work sampling is a statistically based tool in which the analyst, whether technician or logging foreman, observes an activity at random or at fixed intervals. Through a summarization of the spot observations the analyst is able to determine how much of the total work time has been spent in various states of activity or inactivity. Since the observations are taken at intervals rather

* An excellent source of additional information regarding time study and other measurement methods is the *Industrial Engineering Handbook* by H. B. Maynard (McGraw-Hill Book Company, 1963). A treatment of the subject as it relates to logging operations was written by the author and published in *Forest Industries* (Conway 1968*a*- 1968*d*).

Table 19.1 *Log Loading Time*

Logs per load	Loading time (minutes)	Logs per load	Loading time (minutes)
5	9.272	21	20.920
6	10.000	22	21.643
7	10.728	23	22.376
8	11.456	24	23.104
9	12.184	25	23.832
10	12.912	26	24.560
11	13.640	27	25.233
12	14.368	28	26.016
13	15.096	29	26.744
14	15.824	30	27.472
15	16.552	31	28.200
16	17.208	32	28.928
17	18.008	33	29.656
18	18.736	34	30.384
19	19.464	35	31.112
20	20.192	36	31.840

than continuously, the analyst must arrange some other means of determining pieces, distance, and volume. This can usually be accomplished with independent tallies or records.

Work sampling has many advantages. It is not a difficult procedure (it is rather like taking visual snapshots) and training an observer is simple. One observer can keep track of several operations concurrently without interfering with them, and the method seems to be well accepted by work crews—they do not appear to view it as a threat. Work sampling has been successfully used in a wide variety of operations and is easily adapted to logging operations.

Actual time recordings

When time per unit of work is actually measured by the machine operators the results are gross recordings taken over long intervals. It would not be sensible to ask a skidder operator, for instance, how much time was being spent on each of the elements listed in Table 10.1. Those numbers are better gathered by an analyst using either a continuous time study or work sampling. The most that could be hoped for is a record of major delays and perhaps a gross count of trips or pieces. However, if rough coverage is satisfactory and broad time categories are acceptable, the operators can do the job.

This technique, in which operators make the observations, is probably the most commonly used in the industry. With it logging managers keep track of daily piece counts, load counts, and downtime. And while the observations may be imperfect, they do offer some information which could not be gathered in a less complicated or less expensive manner. When a firm becomes serious about the use of time, standards, and cost control, however, it generally relies on the techniques of time study and work sampling.

Standard costing

Whichever technique is used to establish operating time estimates, these estimates can be combined with current man and machine hourly cost rates to set unit cost rates. The unit rates are called *standard costs* when they are analytically determined. The unit costs or standard costs are then used to estimate the total direct costs for a specified activity. To direct costs are added applicable fixed costs, a reasonable return on investment (or pretax profit if that method is preferred), and some amount for contingencies and risk. The result is an estimated cost for a project.

The following example takes the reader through the process of developing a standard unit cost for loading logs. The method used involves a continuous time study to set standard times for log loading with a grapple loader. The result is log loading costs. The example was adapted from a series of articles published in *Forest Industries* in 1968 (Conway 1968*b*, pp. 39-41).

First a continuous time study was taken using a set of work elements such as those described in Chapter 16 for heelboom loading. Measurements were taken with a decimal minute stopwatch to determine the amount of time required to perform each element of the operation. As each element was performed, the observer noted the time in minutes and decimal fractions as well as an assigned element number. Figure 19.3 is the list of elements and the numbers assigned to them. Figure 19.4 is a sample study sheet showing a partial loading cycle recorded.

When the time studies are completed the data are analyzed and standard times are developed. The standard times per element in log loading are shown in Table 19.2.

Figure 19.5 shows estimated hourly owning and operating costs. An additional cost of $0.11 per minute must be added for the second loader. The total rate for the machine, operator, and second loader is $0.509 per minute.

Number	Element
1	Unload trailer
2	Prepare trailer
3	Swing out for logs
4	Pick up log
5	Swing in with log
6	Load log
7	Tail logs or sort logs for loading
8	Production delays
9	Breakdowns

Figure 19.3. *Work elements with assigned numbers.*

Location: Eugene, Ore.				Study no.: 1
Operation: Log loading (Crane)				Date: 12/30/67
Analyst: S. Smith		Start: 8:30 A.M.		Finish: 4:00 P.M.

Element description		Time reading	Elapsed time	R	Normal time
	Start	0.00			
	1	1.06	1.06	–	1.06
	2	1.87	.81	–	.81
	3	2.02	.15	–	.15
Trouble getting logs.	4	3.64	1.62	5	.81
	5	3.86	.22	–	.22

Note: In the R column, each element must be given a rating based on 100 percent efficiency at the time of the observation. The 5 in this column means that, in the observer's estimation, the loader operator was working at 50 percent efficiency. Dashes indicate 100 percent efficiency.

Figure 19.4. *Sample continuous time study sheet (Conway 1968b p. 39).*

Now, knowing the standard normal time per log (1.102/log), the unit cost can be calculated using the rate data from Table 19.3.

If the manager is attempting to estimate the cost of loading logs for a particular setting, a standard cost for loading would be extremely helpful. Knowing the average log size and the approximate volume on the setting, he could easily compute approximate loading costs. For instance, if there were two million board feet on the setting and if the average log size were 330 board feet his total loading costs would be:

$$2,000 \times \$1.68 = \$3,360$$

Table 19.2 *Time Standard for Log Loading*

Element	Normal time (minutes)	Frequency of occurence[1]	Normal time per log (minutes)
Unload trailer	1.069	44/606	.078
Prepare trailer	0.882	44/606	.064
Swing out	0.103	1/1	.103
Pick up log	0.225	1/1	.225
Swing in	0.124	1/1	.124
Load log	0.222	1/1	.222
Load preparation	3.890	44/606	.286
Total normal minutes per log			1.102

1. 44/606 indicates the element occurred 44 times in 606 logs loaded — 1/1 means the element occurred on every log.

Depreciation period: 20,000 hours

Machine: Heelboom crane

Net depreciation value:[1] $200,000

Owning costs per hour (dollars)

$$\text{Depreciation:} \quad \frac{\text{Net depreciation value}}{\text{Depreciation period (hrs)}} = \frac{200,000}{20,000} = 10.00$$

$$\text{Taxes, interest, insurance:} \quad \frac{0.03[2]}{1,000} \times \text{Net depreciation value} = \frac{0.03 \times 200,000}{1,000} = 6.00$$

Total owning costs per hour 16.00

Operating costs (dollars)
Maintenance 6.25/hr
Supplies and expenses . 6.25/hr
Total operating costs 12.50/hr

Operator's loaded rate[3] 16.88/hr

Total owning and operating costs 45.38/hr

Cost per minute 0.756

---

1. Net depreciation value equals purchase price less salvage value.
2. Taxes, interest, and insurance are by rule of thumb $0.03 per hour per thousand of invested capital.
3. Operator's rate is loaded to include fringe benefits, social security, etc.

Figure 19.5. *Estimated hourly owning and operating costs.*

The same procedure would have to be repeated for the other harvest system components in order to arrive at standard costs per unit. However, once the procedure is developed, making cost estimates is much easier.

Break-even analysis

The detailed unit time and cost values developed through the standard costing procedure are useful for more than making cost estimates. They are equally useful as a measure of cost efficiency when standard costs are compared with actual costs after the job is underway. Perhaps these values are even more useful when used in an analytical tool called the break-even analysis.

The break-even concept is used in a great many industries. For instance, when an airplane manufacturer says he has a break-even point of 200 planes of a certain model, it means that when 200 planes have been sold, the excess revenues over cost will just offset fixed development costs incurred before any

Table 19.3 *Loading Cost per Thousand Board Feet*

Average volume per log (BF)	Logs per MBF	Time to load (minutes)	Loading cost per MBF ($)
500	2	2.204	1.67
330	3	3.306	2.50
200	5	5.510	4.17
100	10	11.020	8.33
50	20	22.040	16.67

sales were realized. If design and development costs were $200 million and the planes cost $5 million each to produce, then a sales price of $6 million per plane will allow the manufacturer to break even at 200 planes.

The application to logging, of course, does not involve costs of anywhere near the same magnitude, but the technique is useful nevertheless. It can be used for matching variable revenue against fixed plus variable costs, determining profitable production levels, and choosing among alternative logging systems.

Suppose, for example, that an independent logging operator can take a contract involving 2,000 cunits for a price of $75 per cunit. He calculates his direct costs to be $40 per cunit. It will take a month to complete the job. During that time the firm's fixed costs will be $45,000. Should he take the contract?

It does not take much analysis to calculate that there will be revenues of $150,000. From these revenues, $80,000 of direct costs and $45,000 of fixed costs are subtracted leaving a profit of $25,000. However, what if the weather turns bad and slows the operation? The overhead or fixed costs go on even if variable costs and revenues stop. A prudent man would ask what volume for the month represents a slip from a nice profit to a loss. In other words, at what volume will he break even? The dollars of revenue and cost per cunits of production are:

Cunits production	Revenues	Cost
0	0	45,000
1,000	75,000	95,000
1,500	112,500	105,000
2,000	150,000	125,000

These values can be used to construct a break-even chart as shown in Figure 19.6.

The chart tells us that the profit is gone if the volume falls slightly below 1,300 cunits. If the probability is high that something—bad weather, fire

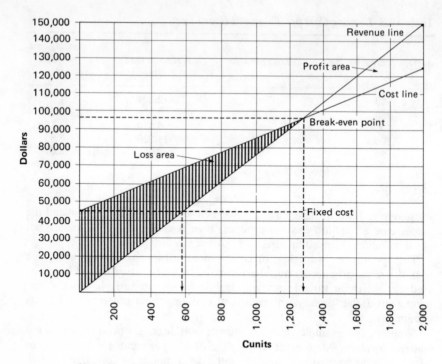

Note: To construct this graph, the points at which the cost and revenue lines cross the Y-axis must be known. Revenue will be zero when there is no production. The costs line intersects Y at 45,000 because fixed costs go on whether or not there is production. Only two points besides the Y-intersect are needed to construct the graph.

Figure 19.6. *Revenue-cost break-even chart.*

closure, a major mechanical failure, or labor problems, for instance—will keep the contractor from passing that 1,300-cunit level in the month, then he had better either reject the contract or ask for more money. He must produce at a level higher than that indicated by the break-even point in order to make any profit. At the break-even point, he will just pay his fixed and variable costs. Below the break-even point, between about 600 cunits and 1,300 cunits, the operator will pay all of his variable costs and all or part of his fixed costs. Below 600 cunits he will not even cover his fixed costs. Given no alternative, an operator might elect to operate below break-even since he will at least be contributing to his fixed costs. It would be a least-cost situation.

The graphic solution to the volume-profit relationship is probably the easiest to understand and is well suited to an operator's needs, since he can see at a glance what the situation is. However, the analysis can also be done algebraically, and for the type of problem that has just been solved, the arithmetic is fairly simple.

Only straight lines appear in the graph. Revenues will always be zero when there is no production. The point at which the expense line crosses the Y-axis, called the *Y-intersect*, will always be equal to the total fixed costs. Given this amount and the formula for a straight line, the volume-profit relationship can be solved algebraically.

The formula for a straight line is:

$$Y = a + bX$$

For the cost line, Y equals total dollars, a equals the Y-intersect or the fixed cost figure (45,000), b equals the variable operating cost ($40/cunit), and X is the number of cunits produced.

Since the revenue line starts at the origin, a equals zero. The formula for the revenue line then, is:

$$Y_r = bX$$

The r is a subscript meaning revenue and is there simply to differentiate between the revenue and cost. The b in the revenue equation equals the price paid per cunit ($75/cunit), and X is once again volume produced.

The solution for break-even point means, by definition, that total costs (fixed and variable) exactly equal revenues. Algebraically that statement is expressed as Y (total cost) equals Y_r (total revenues).

$$Y = Y_r$$

Using the equations for the two straight lines, which were just explained, the following equation will give the break-even point.

$$bX = a + bX$$
$$75(X) = 45,000 + 40(X)$$
$$35(X) = 45,000$$
$$X = 45,000/35$$
$$X = 1,285.7 \text{ cunits}$$

Another application of break-even analysis is to help choose between alternative systems on the same logging show. For example, assume that a logging contractor is trying to decide whether to use grapples or chokers on a logging show. Either arrangement will carry the same volume per turn (a bad assumption, made for simplicity's sake), but there are differences in line speed as well as in the time required to secure and release a turn. The comparative turn times are:*

Method	Travel time	Secure and release
Chokers	.28 min/sta	3.5 min/turn
Grapples	.48 min/sta	2.5 min/turn

* Note that travel time is average time per station (100 feet) for haul in and haul out.

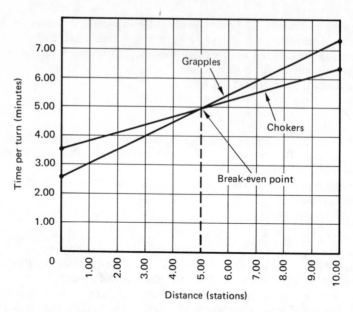

Figure 19.7. *Break-even chart—grapples versus chokers.*

Common sense dictates that the operator use the grapples for shorter yarding distances and the chokers for longer yarding distances. The question is, how long is long and how short is short? The break-even chart in Figure 19.7 provides a solution. The contractor should use both methods. Up to 500 feet, the break-even point, the grapples have an advantage. For longer distances, the chokers should be used.

Machine costs and logging costs

The cost of using machinery in logging operations is roughly comparable to the payroll and associated costs. Therefore, keeping this cost at its lowest practical level is quite important if overall costs are to be controlled. It may be obvious that the lowest purchase price does not necessarily lead to the lowest total costs. But from that point on there is not much about minimizing equipment costs that is obvious.

There are several approaches to the question of whether to repair or replace a particular machine, which do not require extensive cost records. None of these approaches will lead to a sound decision, but they are nevertheless used. One method, of course, is to run the machine out to its *service life*. That is, run the machine until it just cannot be restored to operating condition. This approach makes some sense only in a restricted class of equipment. A line truck, for instance, might be run out to its service life. Generally, the truck is used only infrequently and requires very little in the way of repairs. It is, however, not sensible to use the service-life approach on main operating machines.

Some firms use a *policy-life method* to determine when a machine should be traded for a new model. For instance, a manager might decide that log trucks will be traded in after 10 years—no more, no less—regardless of individual maintenance history. Or, he might decide that log stackers will be traded after 15,000 hours of service. The policy-life method may be satisfactory when there is thoughtful development of the guideline lives. However, the guidelines should not be applied automatically to every machine in a particular class.

The objective in equipment management should be to operate each machine in such a way as to minimize lifelong average total costs per unit of time or output. Total costs in this case are meant to be total average cumulative costs.

There are two basic elements in total costs. One is the cost of owning the machine and the other is the cost of operating it. Figure 19.5 gives examples of both costs. The cost of owning a machine is generally higher in the first year and less in each succeeding year. The reason is the loss of value over the years (economic depreciation) as measured by the decrease in market or trade-in value. Most authorities also maintain that interest on the market value is also a cost of owning, since the money could be used elsewhere if it were not tied up in the machine. The case is more clear when the machine is purchased with borrowed funds. Then the interest is obviously a cost of owning.

Operating costs, on the other hand, are ordinarily lowest during the early part of the machine's life and increase each year. For the purposes of the repair-replace analysis, only replacement parts, maintenance, and repairs are considered. Some firms also consider the loss of production occasioned by downtime on a machine. However, production is not really lost—the wood does not disappear—and although it may be considered a factor, it will not be included in this discussion. It is less complicated to treat only out-of-pocket costs.

The curves representing operating costs and owning costs in Figure 19.8 are cumulative costs per hour divided by cumulative hours. The total cost curve is the sum of the two other curves. When the total cost line is at its lowest, that is the time the machine should be traded. In the case of the example used in the figure, the machine should be traded at 10,000 hours and the total cost at that point is $49.75 per cumulative hour. Keeping the machine longer will result in still higher unit costs, probably higher than the costs of a new machine. Trading sooner will result in having a new machine that costs more to own and operate than the older one. Unfortunately, the owner does not know when his machine is at its lowest cost point until it is past the point. Practical difficulties aside, the theory is still sound, and is the way policy or guideline life is set.

If an operator has good records on his equipment over time, he can establish guidelines for his own use. If the records do not exist, then equipment manufacturers can recommend guidelines using their files. Such recommendations should be carefully considered before being put into action since the manufacturer's business is more closely tied to selling new machines than to extending the lives of present machines. The guideline lives are good tools for long-range planning of capital requirements and cash budgets. But they are

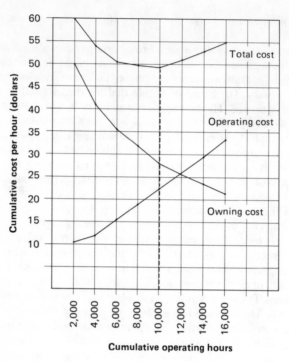

Figure 19.8. *Total cost curve—operating and owning costs.*

inadequate for making the repair-replace decision. A better tool is required.

There are at least two basic approaches to the repair-replace decision, and both call for keeping the kinds of records needed to calculate total costs and cumulative costs per operating hour. As a machine approaches its guideline life the manager must estimate the cost for the coming year and compare it with the guideline life. If the old machine's cost for the coming year is estimated to be higher than a new machine's expected lifetime average, it is time to replace. If the comparison is not favorable to the new machine, the old one is held another year. Some equipment managers limit repair to half the original purchase price during the year.

The second approach to the repair-replace decision also uses the guideline life as a tool. In the second method an equipment manager will allow any individual repair whose cost does not exceed the difference between current market value of the machine in good repair and its salvage value. The control on this type of program is that the cumulative costs of repair are not allowed to exceed some upper limit, expressed as a multiple of the purchase price—usually one to one and a half times.

Replacing a worn-out machine with an identical new machine follows the logic just described. However, replacement with a machine significantly different in price and productive capacity calls for some refinements and for consideration of the difference in price. The price difference must justify itself through savings in operating costs, added production, or some aspect of performance superior to that of the machine being replaced. There are several methods used to evaluate the new investment.

The payback method is one of the most popular and is based on the amount of time required for an investment to earn an amount of cash equal to the purchase price. The return on investment method is based on the average net income related to the value of the investment. The *present value method* is an investment decision procedure based on determining the equivalent of dollars that will be received or paid out in future time periods. This is accomplished by applying discount factors to future dollars to convert them into a present value. Both these methods recognize, among other things, the time value of money.

As already mentioned, sophisticated methods of investment evaluation, present value, and return on investment recognize that a dollar today is worth more than the promise of a dollar 10 years in the future. No great harm is done by ignoring the time value of money for relatively short-life equipment. Indeed, the approximations in estimates and record keeping may well have a greater bearing on the decision than the failure to discount costs to the present. In fact, many firms that know better, use payback period justifications in making investment decisions.

If the cost records are current, well defined, and accurate, then the investment analysis approach can be as simple or as sophisticated as desired. But if the cost records are inadequate or inaccurate, then no amount of technical finesse will lead to sound investment decisions.

Operations research techniques

Today's woods manager is faced with many problems other than determining break-even points or whether to replace a machine. He has to forecast production, cost, and profits on a setting or sales unit sometimes far in advance of harvesting and in the face of fluctuating log values and market conditions. Trucks have to be routed through transportation networks in such a manner as to achieve minimum cost of operation. The productive impact of new technology and improved equipment must be assessed. Even the queuing problems discussed in chapters 16 and 17 can be very difficult to resolve if there are many trucks hauling from many landings.

These types of complex problems and many others discussed throughout this text give rise to the application of some helpful management techniques. The techniques have names, some of which are plainly frightening to some managers, probably because they do not understand them. These techniques—linear programming, Pert-CPM, simulation, and waiting line theory—can be high-powered and extremely helpful in the decision-making process—and the manager does not have to be a mathematical genius to use them. In fact, he need only have a passing understanding of the techniques in order to benefit from them.

A yarder operator does not have to be a master mechanic to operate his machine—there are mechanics, specialists who take care of maintenance and repair. A manager does not have to be a certified public accountant to run his business—he can hire a specialist with those skills. Likewise with the techniques mentioned above, a specialist can be hired. The manager's only problem is knowing what skills are required and what technique best fits the special characteristics of his problem. The balance of the chapter is devoted to a brief introduction to operation research techniques and some of their many important applications.

The management tools mentioned earlier—linear programming, and so on— are among those which fall into the broad definition of operations research (also called *systems analysis, operations analysis,* and *management science*). The tools all have one thing in common. They employ mathematical or other explicit relationships to describe the key elements of actual physical processes. The relationships are reasonably accurate and, after analysis, are useful in drawing conclusions about the actual process. The essence of these tools is that they look only at key factors and then describe the way in which the key factors contribute to the key product.

The logging manager does much the same thing when he assesses a new setting. He looks at key factors such as slope, yarding distance, and average log size. In a very short time an experienced manager can predict production volumes and costs based on his observations. A timber faller looks at the lean of a tree, its size, and the available lays to make the decision regarding where he will fall it. In each case, the thinking process is not unlike the application of operations research techniques—look at the key factors, analyze them (mentally or otherwise), and draw conclusions.

Linear programming

Linear programming is a mathematical way of allocating limited resources to competing demands in such a way that a stated objective is satisfied within certain stated restrictions. For example, the truck fleet of a large logging firm may be the limited resource and the problem may be to allocate the use of the trucks at least cost (the objective) while transporting logs from several woods landings to several end-use points, each of which has a specific demand for logs in terms of both volume and type.

In the forest products industry, linear programming has probably been used more widely than any of the other operations research techniques (Bare 1971). Some of the best applications are in short-range allocation problems, such as the allocation of mixtures of roundwood to various converting facilities subject to many supply, demand, and operating constraints. The technique has also been applied to a number of manufacturing problems, such as log bucking, lumber manufacturing, and peeling optimum veneer mixes. Log transportation has also been the subject of linear programming solutions.

Like problem statements in other analytical methods, linear programming problem statements are abstractions and are therefore simplifications of complex conditions in the real world. They are properly supplements to management judgment rather than replacements for it. The method is deterministic, which means that all input values and all solutions are assumed to be known with certainty. All relationships between variables are simple, straightforward, and mathematically describable.

System simulation

Simulation is a technique that can be applied to a wide variety of problems ranging from inventory control to forecasting future sales or prices. It is used where there is a need to operate a complex system under varying conditions to see how it might behave in the real world. Because a *model* of the system is used, the investigation does not disrupt the real-world operation while the effects of various conditions and variables are being studied.

Simulation, unlike linear programming, does not optimize or find the best solution to a problem. It only predicts what may happen when certain describable conditions are established or changed. The relationships among the variables do not have to be simple or mathematically expressible as long as they can be tabulated. Simulation is a probabilistic rather than deterministic tool and it provides a good approach to representing uncertain and unpredictable events and conditions.

The method is used where relationships can be described by experience even when they may be too complex for mathematical formulation. It is used to study harvesting systems, inventory problems, fire behavior, and forest growth modeling.

Network analysis

Better known by the more popular names of CPM or PERT (Critical Path Method and Program Evaluation and Review Technique), network analysis is a valuable tool for planning and scheduling almost anything—from moving a highlead tower to building roads. When the tasks are simple, the solutions are equally simple—almost intuitive. However, when a manager is dealing with a project of great magnitude, such as building a new sawmill, the relationships among project elements are often not so clear. Network analysis can help a manager recognize these relationships by establishing an overall sequence of events and the timing necessary for these events to be carried out efficiently.

Network analysis uses fairly simple arithmetic and graphic display to help the manager walk through his scheduling problem. It helps the planners and managers of a project to locate the best sequence of events, or perhaps the single event or task that could delay scheduled completion of the project. In addition, it provides a means of working around the delays to shorten the total completion time for a project. Network analysis is commonly used in construction projects and is sometimes demanded when a firm bid is submitted on a large project. It is equally applicable to road construction projects and has been used to some extent in the forest products industry as a planning tool.

Operations research techniques, of which we have discussed only a few, can be used to solve both complex and not-so-complex problems. The use of linear programming in solving transportation problems has reduced transportation costs from 7 to 10 percent for user firms. Network analysis has on many occasions helped a project manager to avoid costly delays. In one case, a road construction job, the project engineer had plotted a course that would have cost him three months of extra time to complete the project. His use of network analysis allowed him to correct the schedule before any time was lost.

Naturally, any firm or manager who is of a mind to can employ the various techniques we have described. In fact, many larger forest products firms employ a full-time operations research group. However, some of the techniques can be expensive, requiring several months and several thousands of dollars for successful application. Most are computer oriented for good reason. The problems are often sufficiently complex enough to defy solution by the human mind. Because of the cost and time constraints, the techniques are at their best when the dollar stakes are relatively high—high enough to justify the time, effort, and expense required to formulate and solve problems of an extremely complex nature.

Bibliography

Adams, T. C. 1965. *High-lead logging costs as related to log size and other variables.* Research Paper PNW-23. Portland, Ore.: U.S. Department of Agriculture, Pacific Northwest Forest and Range Experiment Station.

Alder, D. 1971. Copter logs steep slopes in continuing B.C. tests. *Forest Industries* (September): 51.

Altman, J. A. 1966. Skidders are roaring onto the scene. *Pulp & Paper* (April 4): 63.

American Forest Institute. 1979. *Forest facts and figures,* revised. Washington, D.C.: American Forest Institute.

American Pulpwood Association. 1967*a. Use of articulated wheeled tractors in logging.* Technical Release 67-R-32. New York: American Pulpwood Assocation.

———. 1967*b. Technical papers of the American Pulpwood Association.* New York: American Pulpwood Association.

———. 1968. *Southern pulpwood producer survey report.* New York: American Pulpwood Association.

———. 1969. *Safeskid.* New York: American Pulpwood Association.

Anderson, H. W. 1969. Fertilization-part III: Douglas-fir region aerial fertilization makes strides in a brief span of years. *Forest Industries* (November): 30-32.

Arnold, R. K. 1969. Better trees for the South: the lay-off has begun. *Southern Lumberman* (December 15): 143-144.

Arola, L. 1972. SI-100 electronic weighting systems. *Loggers Handbook,* vol. 32. Portland, Ore.: Pacific Logging Congress.

Arthur, J. L. 1967. Beloit finds advantages with grapple skidders. Pulpwood Annual. *Pulp & Paper* (April 3): 62.

Bare, B. B. 1971. *Applications of operations research in forest management: survey.* University of Washington Center for Quantitative Science in Forestry, Fisheries, and Wildlife.

Batten, R. C. 1970. Make no mistake about it—more land restrictions eyed. *Forest Industries* (September): 34-35.

Bell, D. 1970. Interview with D. Bell, Bell Logging, Hoquiam, Washington (October).

Bell Helicopter Company. 1973. *Bell helicopter looks at logging, company report on helicopter logging and cost estimation.* Fort Worth, Texas: Bell Helicopter Company, Commercial Market Division.

Bengston, G. N. 1969. Routine fertilization of vast acreages in South may be ultimately practiced. *Forest Industries* (November): 35-37.

Benns, J. Chokerless speed skidder cuts costs. Pulpwood Annual. *Pulp & Paper* (April 3): 66.

Bent, J. H. 1970. Full tree logging system. *Pulp and Paper Magazine of Canada* (February 20): 76.

Beuter, J. H. 1971. New look in appraisal of timber value. *Forest Industries* (February): 26-28.

Bierman, H., and Drebin, A. R. 1968. *Managerial accounting: an introduction.* New York: Macmillan.

Biller, C. J. 1971. Are your log skidder tires too small and underinflated? *Forest Industries* (December): 34.

Binkley, V. W. 1965. *Economics and design of a radio-controlled skyline yarding system.* Research Paper PNW-25. Portland, Ore.: U.S. Department of Agriculture, Pacific Northwest Forest and Range Experiment Station.

——. 1966. *Engineering evaluation of skyline-crane logging systems.* Unpublished report. Portland, Ore.: U.S. Department of Agriculture, Pacific Northwest Forest and Range Experiment Station.

——. *Helicopter logging with the S64 Skycrane; report of sale.* Portland, Ore.: U.S. Department of Agriculture, Pacific Northwest Forest and Range Experiment Station.

Binkley, V. W., and Carson, W. W. 1968. *Operational test of a natural-shaped logging balloon.* Research Paper PNW-87. Portland, Ore.: U.S. Department of Agriculture, Pacific Northwest Forest and Range Experiment Station.

Binkley, V. W., and Lysons, H. H. 1968. *Planning single-span skylines.* Research Paper PNW-66. Portland, Ore.: U.S. Department of Agriculture, Pacific Northwest Forest and Range Experiment Station.

Bjerkelund, T. C. 1970. Some recent developments in the tree length systems concept. *Pulp and Paper Magazine of Canada* (February 20): 78.

Blackerby, L. H. 1969a. Genetics—Part I: man-made magic shapes forms and types of future forests. *Forest Industries* (April): 28-30.

——. 1969b. Economic contributions of forests to surrounding areas emphasized. *Forest Industries* (August): 38.

Bowen, M. G. 1978. *Canada's forest inventory—1976.* Forest Management Institute information report FMR-X-116. Ottawa, Ontario, Canada: Canadian Forestry Service, Environment Canada.

Bowen, M. G., and Garlicki, A. M. 1970. *Aerial logging: an annotated bibliography on the use of balloons and helicopters.* Report FMR-X-22. Ottawa, Canada: Canadian Forest Service, Department of Fisheries and Forestry.

Brandes, R. A. 1970. Forest residuals offer potential as source of pulp fiber in west. *Forest Industries* (March): 44-45.

Bredberg, C. J. 1970. *Woodlands report no. 24: evaluation of logging machine prototypes Timberjack 360 grapple skidder.* Montreal, Canada: Pulp and Paper Research Institute of Canada.

Brinckloe, W. D. 1969. *Managerial operations research.* New York: McGraw-Hill Book Company.

Bromley, W. S., ed. 1968. *Pulpwood productions.* Danville, Ill.: Interstate Printers & Publishers.

Brown, D. 1969. Interview with D. Brown, Mud Bay Logging Company, Sitka, Alaska (June).

Brown, N. C. 1950. *Logging—the principles and methods of harvesting timber in the United States.* New York: John Wiley and Sons.

Brown, W. B. 1970. Systems theory, organizations, and management. *Oregon Business Review* (June): 1-6.

Bryan, R. W. 1970. Environment—who is responsible? *Forest Industries* (April): 39.

——. 1971a. Attitudes on resources are not beyond changing. *Forest Industries* (April): 34-35.

——. 1971b. Environmental, forestry objectives must merge. *Forest Industries* (April):

32-33.

Bryan, R. W. 1971c. Shrinking timber base presents major obstacle. *Forest Industries* (May): 22-23.

Buck, L. 1970. Concepts of man-machine interface. *Pulp and Paper Magazine of Canada* (July): 53-56.

Buffa, E. S. 1965. *Modern production management.* New York: John Wiley & Sons.

Bussell, W. H.; Hool, J. N.; Leppert, A. M.; and Lawson, S.C.D. Pulpwood harvesting systems research at Auburn University. *Forest Engineering* 69-70.

Byrne, J. J. 1960. *Logging road handbook—the effect of road design on hauling costs.* Handbook No. 183. Washington, D.C.: U.S. Department of Agriculture, Forest Service.

Byrne, J.; Nelson, R. J.; and Googins, P. N. 1969. *Logging handbook: the effect of road design on hauling costs.* Washington, D.C.: U.S. Government Printing Office.

Caldwell, L. K. 1969. How to achieve the third forest: America must cope with challenges. *Pulp & Paper* (April): 97-99.

Campbell, C. O. 1970. *Supplement to skyline tension and deflection handbook.* Research Paper PNW-39. Portland, Ore.: U.S. Department of Agriculture, Pacific Northwest Forest and Range Experiment Station.

Campbell, D. 1966. Is the wheeled skidder a cure-all? *Pulp & Paper* (April 4): 58-59.

Campbell, J. P. 1970. Industry must respond to public's concern. *Pulp & Paper* (May): 109-113.

Canadian Forest Industries. 1969. What accidents really cost: $6 loss for each $1 profit. *Canadian Forest Industries* (February): 50.

Canadian Pulp and Paper Association. 1981. *Reference tables.* 35th ed. Montreal, Quebec, Canada: CPPA. (Data from Environment Canada, Canadian Forestry Service and Statistics Canada.)

Canadian Pulp and Paper Association, Woods Section. 1971. Environmental issues accelerate at point of concern; timber shortage is foreseen. *Forest Industries,* (May): 19-20.

Carlsson, B. Increased production with logging equipment through computerized planning. *Forest Engineering:* 66-68.

Carrier, J. M. 1970. Forest management in Canada—review and outlook. *Preprints of Papers to be presented at the 52nd Annual Meeting of the Woodlands Section, Canadian Pulp and Paper Association.* Montreal, Canada: Canadian Pulp and Paper Association Woodlands Section.

Carson, W. W., and Peters, P. A. 1971. *Gross static lifting capacity of logging balloons.* Report No. PNW-152, Portland, Ore.: U.S. Department of Agriculture, Pacific Northwest Forest and Range Experiment Station.

Caterpillar Tractor Co. 1970. *The applicator.* Peoria, Ill.

——. 1981. *Caterpillar performance handbook,* 12th ed. Peoria, Ill.

Cheek, G. C. 1970. The forest industries in the 1970's. *Technical Papers of the American Pulpwood Association.* New York: American Pulpwood Association.

Christopherson, D. 1970. Interview with D. Christopherson, DABCO Inc., Kamich, Idaho, (June).

Clephane, T. 1980. *The growth in importance of southern timber.* Text of speech to annual convention of Society of American Foresters chapter, Lufkin, Texas, April 17, 1980.

Cline, C. E. How it's done in Ontario. *Pulpwood Production.* 2ff.

Comer, G. 1970. Centralized processing to improve woods harvesting operations. *Technical Papers of the American Pulpwood Association.* New York: American Pulpwood Association.

Conway, S. 1968a. Methods improvement: a cost control tool for the logging industry. *Forest Industries* (September): 22-23.

——. 1968b. Log loading procedure serves as example for stopwatch study—methods improvement, part 2. *Forest Industries* (October): 39-41.

——. 1968c. Methods improvement, part 3, work sampling—another way to see where the time goes. *Forest Industries* (November): 34-35.

——. 1968d. Graphs help spot where time is lost. *Forest Industries* (December): 34-35.

——. 1970a. Experience plus experiments equals logging equipment. *Pacific Logger and Lumberman* (September): 1 ff.

Conway, S. 1970*b*. Logging is logging is logging—or is it? *Pacific Logger and Lumberman* (October): 4-5.

———. 1970*c*. Headaches and hard times—salvage logging. *Pacific Logger and Lumberman* (November): 1.

———. 1970*d*. PALCO executive talks about the industry. *Pacific Logger and Lumberman* (December): 2.

———. 1971*a*. Looking ahead in the timber industry. *Pacific Logger and Lumberman* (January): 1.

———. 1971*b*. Strategy gets the logs. *Pacific Logger and Lumberman* (March): 1.

———. 1971*c*. Grapple logging. *Loggers World* (October): 1 ff

———. 1971*d*. Starting in the logging business—the work you should do before the work begins. *Loggers World* (November): 53.

———. 1972. Safety—you can't impose it; but you can gain it through objectives. *Forest Industries* (August): 61-63.

———. 1973. *Timber cutting practices.* 2nd ed. San Francisco: Miller Freeman Publications.

———. 1977*a*. How to cope with smaller logs and their higher harvest costs. *Forest Industries* (September): 34-36.

———. 1977*b*. The management of productivity. Address at *Forest Industries* Business Management Clinic for Loggers, Eugene, Oregon, December 16-17.

———. 1978. *Timber cutting practices.* 3rd ed. San Francisco: Miller Freeman Publications, Inc.

Daellenbach, H. G., and Bell, E. J. 1970. *User's guide to linear programming.* Englewood Cliffs, N.J.: Prentice-Hall.

Davies, W. E. 1971. The balloon revisited; larger onion combines lift stability. *Forest Industries* (November): 48-50.

Davis, K. P. 1954. *American forest management.* New York: McGraw-Hill Book Company.

Dawson, D. H. 1969. North central FTI work encompasses many species. *Forest Industries* (June): 39-41.

Dempsey, G. W. 1970. Wood transportation at North Western Pulp and Power Ltd. *Pulp and Paper Magazine of Canada* (March): 74-75.

Dennison, W. N. 1971. Better road cost estimates—a must! *Loggers Handbook.* Portland, Ore.: Pacific Logging Congress.

DeYoung, R. 1970. Interview with Robbie DeYoung, DeYoung Logging, Montesano, Washington (May).

Dibblee, D. H. W. 1965. *The effect of some stand factors on the performance of mechanical harvesting equipment*, from a paper presented to the Woodlands Section, Canadian Pulp and Paper Association, Quebec City.

Dilworth, J. R. 1964. *Log scaling and timber cruising.* Corvallis, Ore.: O.S.U. Bookstores, Inc.

Dlesk, G. 1968. Corporate pulpwood procurement policies. *Technical Papers of the American Pulpwood Association.* New York: American Pulpwood Association.

Dyrness, C. T. 1972. *Soil surface conditions following balloon logging.* Report No. PNW-182, Portland, Ore.: U.S. Department of Agriculture, Pacific Northwest Forest and Range Experiment Station.

Edholm, R. M. 1973. *Variation in heli-logging production and cost against vehicle size for timber harvest of small diameter growth.* Commercial Market Division, Fort Worth, Texas: Bell Helicopter Company.

Federal Aviation Agency. 1965. *Basic helicopter handbook.* Washington, D.C.: U.S. Government Printing Office.

FERIC (Forest Equipment Research Institute of Canada). 1978. Technical note No. TN-20. June. Montreal, Quebec, Canada: FERIC.

Fleischer, H. O. 1968. Utilizing all species and all the tree. *Pulp & Paper* (April 1): 29-32.

Forest Club. 1971. *Forestry handbook for British Columbia.* Vancouver, B.C.: Forest Club.

Forest Industries. 1967. Shuttle logging tested on experimental basis. *Forest Industries* (October): 58.

Forest Industries. 1968*a.* Skidders overcome mud, swampy ground in delta. *Forest Industries* (April): 57.

——. 1968*b.* Grapple skidder, shear team up on southern contract operation. *Forest Industries* (May): 60.

——. 1969*a.* Timber outlook questioned by producers in Rockies; uneasiness prevails when uncle holds all the marbles. *Forest Industries* (June): 9.

——. 1969*b.* Study analyzes Douglas fir harvest program alternatives. *Forest Industries* (September): 34-35.

——. 1970*a.* More complete utilization of wood fiber sought by forest products lab. *Forest Industries* (October): 18.

——. 1970*b.* Arcata files $122-million claim for lost timberlands. *Forest Industries* (November): 13.

——. 1970*c.* Wilderness fad threatens future timber harvests. *Forest Industries* (December): 11.

——. 1971*a.* Agricultural engineers ponder forest problems. *Forest Industries* (February): 9.

——. 1971*b.* Housing predictions climb but units will be smaller. *Forest Industries* (April): 38.

——. 1971*c.* Price and supply problems are discussed. *Forest Industries* (May): 7.

——. 1979. Pulpwood: one part of the complex fiber supply scene. *Forest Industries* (June): 58.

——. 1980. Fires ruin forests all across Canada. *Forest Industries* (September): 15.

Forest Products Research Society. 1979. *Timber supply: issues and options.* Proceedings No. P-79-24. Madison, Wisconsin: Forest Products Research Society.

Foster, W. E. 1970. Interview with W. E. Foster, Royal Logging Company, Columbia Falls, Montana (June).

Garey, C. 1969. Interview with Carl Garey, Weyerhaeuser Company, Plymouth, North Carolina (June).

——. 1970. Society of American Foresters' role in developing harvesting leadership. *Technical Papers of the American Pulpwood Association.* New York: American Pulpwood Association.

Garlicki, A. M., and Calvert, W. W. 1968. Tree-length orientation and skidding forces. *Pulp and Paper Magazine of Canada* (July 21): 62.

——. 1969. A comparison of power requirements for full-tree versus tree-length skidding. *Pulp and Paper Magazine of Canada.* (July 18): 83.

——. 1970. *Skidding forces and power requirements.* Publication No. 1279. (Ottawa, Canada: The Queen's Press.

Gessel, S. P. 1969. Introduction to forest fertilization in North America. *Forest Industries* (September): 26-28.

Giordano, C. 1971. Vertical logging a challenge for Okanogan helicopters. *British Columbia Lumberman* (August): 16-19.

Glans, T., et al. 1968. *Management systems.* New York: Holt, Rinehart, and Winston.

Goodyear Aerospace Corporation. 1964. *Balloon logging systems: phase 1 – analytical studies, phase II–logistics studies.* Reports on contract 19-25, 1964. Portland, Ore.: U.S. Department of Agriculture, Pacific Northwest Forest and Range Experiment Station.

Go West. 1981. Go's official update of highway vehicle size and weight limits. *Go West* (October).

Greely, A. W. 1967. National forests can supply 'some' pulpwood. *Pulp & Paper* (April 3): 79-81.

Guttenberg, S. 1970. Economics of southern pine pulpwood pricing. *Forest Products Journal* (April): 15-18.

Hagenstein, W. D. 1970. Emotions aside, clearcutting is silviculturally sound. *Forest Industries* (December): 26-28.

Hair, D. 1970. *Meeting growing demands for wood products.* Washington, D.C.: U.S. Department of Agriculture, Forest Service.

Hair, D., and Spada, B. 1969. Hardwood timber resources of the United States. *Southern Lumberman* (December 15): 103-105.

Hallett, R. M., and McGraw, W. E. 1970. *Studies on the productivity of skidding tractors.* Publication No. 1282. Ottawa, Canada: Department of Fisheries and Forestry, Canadian Forestry Service.

Hamilton, H. R.; Bowman, Jr., R. G.; Gardner, R. W.; and Grimm, J. J. 1961. *Phase report on factors affecting pulpwood production costs and technology in the southeastern United States.* Report to the American Pulpwood Association, August, 1961. Battelle Memorial Institute, Columbus, Ohio.

Hamilton, H. R., and Grimm, J. J. 1963. *Topical report on an evaluation of alternative methods of pulpwood production.* Report to American Pulpwood Association, April 29, 1963. Battelle Memorial Institute, Columbus, Ohio.

Hammack, B. 1966. Log hauling—past and future. *Loggers Handbook*, vol. 26. Portland, Ore.: Pacific Logging Congress.

Hannaford, E. S. 1967. *Supervisors guide to human relations.* Chicago: National Safety Council.

Harsey, B. 1970. Interview with B. Harsey, Publishers Paper Co., Estacada, Oregon (July).

Hartman, R. L., and Gibson, H. G. 1971. Techniques for the wheeled skidder operator. *Pulpwood Production* (April): 58.

Hazelton, L. 1966. NE skidders outclass horses, tractors. *Pulp & Paper* (April 4): 62-63.

————. 1970. Rising labor cost will soon make longwood operations a must. *Pulpwood Annual, Pulp & Paper* (May): 125.

Hevey, R. T. Designing a logging system for mechanization. *Forest Engineering.* 76-77.

Hidy, R. W.; Hill, R. E.; and Nevins, A. 1963. *Timber and Men.* New York: Macmillan Company.

Hodges, Jr., R. D. 1969. Continuing raw material problems face us unless we all work more closely. *Pulp & Paper* (April): 99-102.

Hoelscher, L., and Beckman, J. 1974. Interview with L. Hoelscher and J. Beckman, Weyerhaeuser Company, Tacoma, Washington (January).

Holbrook, S. H. 1943. *Burning an empire.* New York: Macmillan.

————. 1949. *Tall timber.* New York: Macmillan.

Holley, Jr., D. L. 1969. Plywood's ability to compete for southern pine stumpage. *Southern Lumberman* (December 15): 177-179.

————. 1970. Factors in 1959-69 price rise in southern pine sawtimber analyzed. *Forest Industries* (April): 40-41.

Horner, A. 1970. Interview with A. Horner, Tomco, Sweet Home, Oregon (April).

Horngren, C. T. 1967. *Cost accounting—a managerial emphasis.* Englewood Cliffs, N.J.: Prentice-Hall, Inc.

Howard, H. 1968. Variables complicate even-aged monoculture. *Pulpwood Annual 1968, Pulp & Paper* (April 1): 43.

Hudson, R. 1969. Ponderosa pine improved. *Forest Industries* (April): 32.

Hunt, B. 1966. Wheeled skidders leaving their mark. *Pulp & Paper* (April 4): 53-54.

Industry Week. 1971. Safe at work. *Industry Week* (April 26): 39-43.

International Labor Office. 1968. *Guide to safety and health in forestry work.* Geneva, Switzerland: International Labour Office.

Jarck, W. 1967. Results of Appalachian rubber-tired skidder survey are reported. *Technical Papers of the American Pulpwood Association.* New York: American Pulpwood Association.

————. 1971. Fighting the squeeze of smaller trees and rising labor costs. *Pulp & Paper* (January): 98.

Johnson R. A., Kast, F. E., and Rosenwieg, J. E. 1963. *The theory and management of systems.* New York: McGraw-Hill Book Company.

Jorgensen, G. 1969. The land manager speaks for skyline logging. *Skyline Logging Symposium Proceedings,* ed. John E. O'Leary. Corvallis, Ore.: Oregon State University.

Kartheiser, G. 1970. Interview with G. Kartheiser, Canyon Logging Company, Columbia Falls, Montana (June).

King, B. 1970. Interview with B. King, King Bros. Logging, Mapleton, Oregon (December).

Kithill, D. 1970. Interview with D. Kithill, Portland, Oregon (October).

Knaf, L. 1969. Interview with L. Knaf, Great Lakes Paper Company, Fort William, Ontario Canada (July).

Knight, V. J. 1967. Manpower planning helps companies 'out of the wilderness.' *Pulp & Paper* (April): 71-73.

Koch, P. 1970. New procurement approach increases pine utilization. *Forest Industries* (March): 46.

——. 1971. Force and work to shear green southern pine logs at slow speed. *Forest Products Journal* (March): 21-26.

Lagler, J. 1970. Interview with J. Lagler, Lagler Logging, Mapleton, Oregon, (December).

Lambert, H. 1969. The third forest: where? how harvested? *Forest Industries* (April): 34.

——. 1970. PLLRC report represents a challenge to the industry. *Forest Industries* (September): 28-33.

——. 1971a Continuing problems include environment, timber supply, increased federal control. *Forest Industries* (May): 18-20.

Lateiner, A. 1965. *Modern techniques of supervision.* New York: Lateiner Publishing.

Lawrence, J. D. The effect of undergrowth on harvest costs. *Forest Engineering.* 50-51.

Ledig, F. T. 1969. FTI future is bright in Northeast: joint geneticist-industry work vital. *Forest Industries* (May): 41-43.

Letkeman, R. 1972. Feller-bunchers rapidly earn their keep in B.C. smallwood. *Canadian Forest Industries* (October): 41.

Levin, R. I., and Kirkpatrick, C. A. 1966. *Planning and control with Pert/CPM.* New York: McGraw-Hill Book Company.

Lilley, G. E. 1973. Rail transport problems serious in our industry. *Pulpwood Annual 1973, Forest Industries* (June): 6-7.

Lucia, E. 1965. *Head rig.* Portland, Ore.: Overland Press West.

Lussier, L. J. 1961. *Planning and control of logging operations.* Quebec, Canada: Laval University.

Lysons, H. H. 1966. *Compatibility of balloon fabrics with ammonia.* Report No. PNW-42. Portland, Ore.: U.S. Department of Agriculture, Pacific Northwest Forest and Range Experiment Station.

Lysons, H. H.; Binkley, V. W.; and Mann, C. N. 1966. *Logging test of a single hull balloon.* Report No. PNW-30. Portland, Ore.: U.S. Department of Agriculture, Pacific Northwest Forest & Range Experiment Station.

Lysons, H. H., and Mann, C. N. 1967. *Skyline Tension and Deflection Handbook.* Forest Service Research Report PNW-75. Portland, Ore.: U.S. Department of Agriculture, Pacific Northwest Forest and Range Experiment Station.

MacAulay, J. D. 1969. Interview with J. D. MacAulay, Timberjack Machines, Eaton Forestry Equipment Division, Woodstock, Ontario, Canada (July).

Maddex, R. L. Summary of mechanized pulpwood harvesting research. *Forest Engineering*: 39-41.

Mann, C. N. 1965. *Forces in balloon logging.* Report No. PNW-28. Portland, Ore.: U.S. Department of Agriculture, Pacific Northwest Forest and Range Experiment Station.

——. 1969. *Mechanics of running skylines.* Research Paper PNW-75, Portland, Ore.: U.S. Department of Agriculture, Pacific Northwest Forest and Range Experiment Station.

——. 1971. *Balloon-running skyline system.* Symposium on Forest Operations in Mountainous Regions, Krasnodar (USSR) August 1971.

Martin, W. H. 1971a. Pulpwood transport by river, road, and rail. *Pulp and Paper Magazine of Canada* (September): 31-40.

——. 1971b. Transportation by river drive. *Pulp and Paper Magazine of Canada* (September): 31-40.

Maslow, A. H. 1963. A theory of human motivation, in *People and Productivity*, ed. R. A. Sutermeister. New York: McGraw-Hill Book Company.

Masson, J. D. 1971. Rail transportation of pulpwood—new approach to old difficulties. *Pulp and Paper Magazine of Canada* (September): 68-76.

Matthews, D. M. 1942. *Cost control in the logging industry.* New York: McGraw-Hill Book Company.

Mayfield, F. *Skidding with crawler tractors*, paper prepared for the Caterpillar Tractor Company, Peoria, Ill.

Maynard, H. B., ed. 1963. *Industrial engineering handbook*. New York: McGraw-Hill Book Company.

McColl, B. J. 1969. *A systems approach to some industry problems*. A paper presented at the Second Woodhandling Symposium jointly sponsored by the Technical and Woodlands Section, Canadian Pulp and Paper Association, Ottawa, Canada, September 15-17, 1969. Ottawa, Canada: Pulp and Paper Institute.

McGraw, W. E. 1966. Skidder developments are on increase. *Pulp & Paper* (April 4): 53.

McGraw, W. E. and Hallett, R. M. 1970. *Studies on the productivity of skidding tractors*. Department of Fisheries and Forestry, Canadian Forestry Service Publication No. 1282. Ottawa: The Queen's Press.

McIntosh, J. A. 1968. Production analysis of balloon logging. *Forest Products Journal* (October).

McIntosh, J. A., and Wright, D. M. 1970. Haul roads in the sky. *Canadian Forest Industries* (January).

Meier, R. C.; Newell, W. T.; and Pazer, H. L. 1969. *Simulating in business and economics*. Englewood Cliffs, N.J.: Prentice-Hall.

Mills, R. *Balloon logging, through the crystal ball.*

Muench, Jr., J. 1969. Another look at private woodlands. *Southern Lumberman* (December 15): 123-25.

Mulligan, P. 1973. An interview with P. Mulligan, FMC Corp., San Jose, California (October).

National Safety Council. 1961. *Supervisors Safety Manual*. Chicago: National Safety Council.

Nearing, J. 1970. Interview with J. Nearing, Nearing Logging, Coeur d'Alene, Idaho (June).

Nelson, Jr., A. W. 1969. The third forest: how to meet expanded timber demands for 2000. *Pulp & Paper* (April): 93-97.

Nelson, E. D. 1965. *Logging costs, considerations and estimation*. Unpublished Report for School of Forestry, Oregon State University, Corvallis, Oregon.

O'Leary, J. E., ed. 1969. *Skyline logging symposium proceedings*. Corvallis, Ore.: Oregon State University.

——, ed. *Skyline logging*. Corvallis, Ore.: Oregon State University.

O'Leary, J. E., and Mosher, F. 1971. *Balloon logging in the Douglas fir region of North America*. Symposium on Forest Operations in Mountainous Regions, Krasnodar (USSR), August 1971.

Optner, S. L. 1965. *Systems analysis for business and industrial problem solving*. Englewood Cliffs, N.J.: Prentice-Hall.

Pacific Logging Congress. 1966. *Loggers Handbook*. Portland, Ore.: Pacific Logging Congress.

——. 1967. *Loggers Handbook*. Portland, Ore.: Pacific Logging Congress.

——. 1968. *Loggers Handbook*. Portland, Ore.: Pacific Logging Congress.

——. 1969. *Loggers Handbook*. Portland, Ore.: Pacific Logging Congress.

——. 1970. *Loggers Handbook*. Portland, Ore.: Pacific Logging Congress.

——. 1971. *Loggers Handbook*. Portland, Ore.: Pacific Logging Congress.

Parker, H. W. 1966. Training and maintenance for skidders. *Pulp & Paper* (April 4): 59-62.

Paterson, W. G. 1971. Transport on forest roads–1980. *Pulp and Paper Magazine of Canada* (September): 42-63.

Pease, D. A. 1969. Forest resource study reviewed; evaluation of services authorized. *Forest Industries* (June): 34-36.

——. 1971a. Production up from '69, price situation desperate, but 1971 outlook is good. *Forest Industries* (January): 36-39.

——. 1971b. Helicopter undergoes private test in logging. *Forest Industries* (March): 34-37.

——. 1971c. 1970 lumber production. *Forest Industries* (May): 10-13.

Pease, D. A., and Davis, W. E. 1971. Helicopter yarding is costly but trial operations expand. *Forest Industries* (September): 48-50.

Perotto, R. 1967. Two skidding grapples cover many needs. *Pulp & Paper* (April 3): 62.

Peters, P. A. 1973*a*. *Estimating production of a skyline yarding system*. Seattle, Wash.: U.S. Department of Agriculture, Pacific Northwest Forest and Range Experiment Station.

———. 1973*b*. Balloon logging: a look at current operating systems. *Journal of Forestry* (September).

Peters, P. A.; Lysons, H. H.; and Shindo, S. Aerodynamic coefficients of four balloon shapes at high attack angles. *Proceedings Seventh AFCRL Scientific Balloon Symposium*. Seattle, Wash.: U.S. Department of Agriculture, Pacific Northwest Forest and Range Experiment Station.

Peterson, H. W. 1969. A logging manager looks at skyline logging. *Skyline Logging Symposium Proceedings*, ed. J. E. O'Leary. Corvallis, Ore.: Oregon State University.

———. 1971. Main threat to clearcutting is public misunderstanding. *Pulp & Paper* (June): 16-17.

Phelps, R., and Hair, D. 1976. *The demand and price situation for forest products 1975-76*. Washington, D.C.: U.S. Department of Agriculture, Forest Service.

———. 1977. *The demand and price situation for forest products 1976-77*. Washington, D.C.: U.S. Department of Agriculture, Forest Service.

Pollitzer, S. 1971. Land withdrawals lead environmental problems. *Forest Industries* (May): 21

Pope, C. L. 1954. How to control costs for crawler tractors. *The Timberman* (August): 66.

Popradi, L. 1971. Copter yarding for B.C. smallwood. *Canadian Forest Industries* (June).

Potters, P. A. 1967. *Analysis of tree length harvesting operations using rubber-tired skidders*. Technical Paper 67-5: (4.3132) from the spring meeting of the APA Appalachian Technical Division, May 25-26, 1967, Chillicothe, Ohio.

Prater, J. D. 1970. *Logging and environmental quality—How*. A speech delivered to the Oregon Logging Conference, February, 1970, in Eugene, Oregon.

Pritchett, W. L., and Hanna, H. 1969. Fertilization II: results to date, potential gains in Southeast generate cautious optimism. *Forest Industries* (October): 26-28.

Public Land Law Review Commission. 1970. *One third of the nation's land*. Washington, D.C.

Pulp and Paper Magazine of Canada. 1969. Study on the movement of pulpwood past dams. *Pulp and Paper Magazine of Canada* (July): 43-47.

Pulpwood Production. 1969*a*. Automatic choker added to one-man logger. *Pulpwood Production* (April): 56.

———. 1969*b*. The MF Treever: it's the best thing we've found to get some wood. *Pulpwood Production* (June): 8.

———. 1970*a*. J. M. Newman's production secret: scatter felling and pre-decking. *Pulpwood Production* (June): 24.

———. 1970*b*. Consider the small woodlot as a continuing source of supply for wood-using industries. *Pulpwood Production* (March): 4.

———. 1970*c*. A new approach to logging in northeast Georgia. *Pulpwood Production* (December): 6.

———. 1970*d*. South's 1969 pulpwood production up sharply. *Pulpwood Production* (December): 26.

———. 1971*a*. Shortwood or tree-length—which costs less? *Pulpwood Production* (May): 22.

———. 1971*b*. A look at Georgia-Pacific's pine operation in Florida. *Pulpwood Production* (September): 40.

Rennie, P. 1969. Skagit skyline equipment. *Skyline Logging Symposium Proceedings*, ed. J. E. O'Leary. Corvallis, Ore.: Oregon State University.

Rich, S. U. 1970. *The marketing of forest products*. New York: McGraw-Hill Book Company.

Richardson, B. Y. 1971. Trafficability tests of an articulated log skidder. *Forest Products Journal* (January): 31.

Roche, L. 1969. Genetic variations, introduced species assessed in central, eastern Canada work. *Forest Industries* (June): 43-44.

Ruttan, R. 1963. Operation skylift. *British Columbia Lumberman* 47 (11).

Schabas, W. 1979. The KFF and home-made delimber in promising QNS experiment. *Pulp and Paper Canada* (January).

Schillings, P. L. 1969. *Selecting crawler skidders by comparing relative operating costs.* Research Paper INT-59. Ogden, Utah: U.S. Department of Agriculture. Intermountain Forest and Range Experiment Station.

Schmitt, D. 1969. Genetics—part III: softwood and hardwood programs conducted in the South by United States Forest Service. *Forest Industries* (June): 38-39.

Shaw, C. L. 1970. Producers look at their top market—the United States. *Forest Industries* (June): 37.

Shaw, R. G. Machinery in British forests: appendix II: timber extraction by helicopter. Institute of Agricultural Engineers. *England* 17(4): 112-113.

Silversides, R. C. 1967. Abitibi analyzes chokerless skidding. *1967 Pulpwood Annual, Pulp & Paper* (April 3): 57-60.

Sinclair, S. 1969. Log export record set last year. *Forest Industries* (February): 38.

Soderberg, P. 1969. Interview with P. Soderberg. Clear Creek Logging Company, Sitka, Alaska (June).

State of Washington. 1972. *Safety standards for logging operations.* Olympia, Washington: State of Washington, Department of Labor and Industries, Division of Safety (September 1), p. 59.

Stenzel, G., and Pearce, K. J. 1972. *Logging and pulpwood production.* New York: Ronald Press.

Stewart, L. L. 1969. Forest products industry trends. *Southern Lumberman* (December 15): 92-94.

Stillings, F. I., and Gerson, W. M. 1966. *Report on an administrative study of skyline logging.* San Francisco: U.S. Department of Agriculture, Forest Service, Region Five.

Strand, R. F., and Miller, R. E. 1969. Douglas-fir growth can be increased report from Pacific Northwest shows. *Forest Industries* (October): 29-31.

Sundberg, U. 1962. *A pilot study to use balloon in cable skidding.* Forest Research Institute of Sweden, Technical Report.

Swan, H. S. D. 1969. Eastern Canada probes less costly wood—trees on good sites, near mills, fertilized. *Forest Industries* (November): 32-35.

Taras, M. A., and Schroeder, J. G. 1969. Needed: new quality standards and grading methods for the South's changing forest industry. *Southern Lumberman* (December 15): 151-153.

Tennus, M. E.; Ruth, R. H.; and Bertsen, C. M. 1955. *An analysis of production and costs in high-lead yarding.* Portland, Ore.: U.S. Department of Agriculture, Forest Service, Pacific Northwest Forest and Range Experiment Station.

TEREX Corp. 1981. *Production and cost estimating of material movement with earthmoving equipment.* Hudson, Ohio.

The Timberman. 1949. Lidgerwood versus Mundy—origin of steel spar skidder. *The Timberman* (October): 54.

———. 1949. Railroad logging in the West. *The Timberman* (October): 63.

Thomas, G. 1970. Interview with G. Thomas. Alfred Logging, Estacada, Oregon (September).

Traczewitz, O. G. 1969. How 'the third forest' in the South will be managed. *Pulp & Paper* (April): 111-112.

Trebett, J. 1966. Effect of large log trucks on hauling performance. *Loggers Handbook*, vol. 26. Portland, Ore.: Pacific Logging Congress.

———. 1970. Logging trends. *British Columbia Lumberman* (January): 28-30.

Trestrail, R. W. 1969. Whither softwood stumpage values? *Pulp & Paper* (August): 129-132.

Trimble, Jr., G. R. 1969. More intensive forest management ahead in Northeast and Appalachians. *Pulp & Paper* (April): 105-106.

Tufts, D. *Planning a logging operation with analyses of some pulpwood logging systems.* A report prepared for Vocational Agriculture Teachers on Pulpwood Production.

———. 1968. Planning a pulpwood logging operation. *Technical Papers of the American Pulpwood Association* (April): 12.

Tull, F. *Rubber tired skidders.* A presentation by F. Tull of the Edward Hines Lumber Company at the 32nd annual Oregon Logging Conference.

Turnbull, J. R. 1971. Industry's goals for land use must be stated, and fought for. *Pulp & Paper* (June): 15.

Ulrich, A. 1978. *The demand and price situation for forest products 1977-78.* Washington, D.C.: U.S. Department of Agriculture, Forest Service.

U.S. Department of Agriculture, Forest Service. 1965. *Timber trends in the United States.* Washington, D.C.: U.S. Government Printing Office.

———. 1969. *Lumber and plywood supply, a situation report.* Washington, D.C.: U.S. Department of Agriculture, Forest Service.

———. 1970. *Single-span skyline timber volume survey.* Portland, Ore.: U.S. Department of Agriculture, Pacific Northwest Forest and Range Experiment Station.

———. 1977. *Forest statistics of the U.S.* A review draft issued in 1978 by the U.S. Department of Agriculture, Forest Service. Washington, D.C.: U.S. Government Printing Office.

———. 1980. *An assessment of the forest and range land situation in the United States.* FS-345. Washington, D.C.: U.S. Government Printing Office.

U.S. Department of Commerce, Bureau of the Census. 1977. *Projections of the population of the United States: 1977 to 2050.* Washington, D.C.: U.S. Government Printing Office.

U.S. Department of Interior, Bureau of Land Management. *Forest engineering handbook, a guide for logging, planning, and forest road engineering.*

U.S. Department of Labor. Various years. *Handbook of labor statistics.* Washington, D.C.: U.S. Government Printing Office.

U.S. Department of Labor, Bureau of Labor Statistics. 1979. *Employment and earnings, United States, 1909-1978.* Bulletin 1312-11. Washington, D.C.: U.S. Government Printing Office.

———. 1980. *Occupational injuries and illnesses in 1978: summary, U.S. Department of Labor statistics.* Report 586. March.

Viele, D. 1969. Interview with D. Viele, International Paper Company, Camden, Arkansas (June).

Wackerman, A. E.; Hagenstein, W. D.; and Michell, A. S. 1966. *Harvesting timber crops.* New York: McGraw-Hill Book Company.

Westell, Jr., C. E. 1969. Positive planning for high yield forestry needed in lake states. *Pulp & Paper* (April): 108-110.

Western Timber Industries. 1971. FS westside fir overappraisals total millions. *Western Timber Industries* (July): 1.

Weyerhaeuser Company. 1971. *Clearcutting vital tool in forest management.* Tacoma, Washington. Weyerhaeuser Company.

Wheat, J. W. 1969. Genetics—part II: forest tree improvement work advances in Douglas fir region. *Forest Industries* (May): 38-40.

Wheeler, P. R. 1969. Implementation of South's 'third forest' plan. *Southern Lumberman* (December 15): 157-158.

White, D. P. 1969. Research and development in forest fertilization in Northeast and Midwest. *Forest Industries* (September): 29-31.

White, W. B. 1970. Rail transportation of pulpwood and chips. *Pulp & Paper Magazine of Canada* (June): 111-118.

Wiest, J. D., and Levy, F. K. 1969. *A management guide to Pert/CPM.* Englewood Cliffs, N.J.: Prentice-Hall.

Williams, C. G. 1970. *Labor economics.* New York: John Wiley & Sons, Inc.

Williams, J. P. 1971. Barge transportation of pulpwood. *Pulp & Paper Magazine of Canada* (July): 94-99.

Wilson, A. W. 1971. Wood shortage after 1980? *Forest Industries* (March): 26.

Winer, H. I. 1967. Skidding productivity study's first report. *Pulpwood Annual, Pulp & Paper* (April 3): 68.

Wolf, J. 1965. Comparing skidding costs for crawler tractors. *Technical Papers of the American Pulpwood Association.* Paper 65-22, 4.3131. New York: American Pulpwood Association.

Worrell, A. C. 1959. *Economics of American forestry*. New York: John Wiley and Sons.

Wren, T. E. 1966. Rubber-tired skidders: terrain factors. *Pulp & Paper* (April): 30-32.

Zobul, B. 1969. Industrial tree improvement in the South. *Forest Industries* (April): 30-32.

Index

About the
Author

Steve Conway, well-known author of *Timber Cutting Practices,* began his prolific career in the woods industry in 1956, when, as a seventeen-year-old boy, he got a job as a choker setter for Weyerhaeuser Company in Coos Bay, Oregon. Conway then went on to gain invaluable experience as a rigging slinger, busheler, and bullbuck in the woods operations of the Pacific Northwest.

Conway's vast practical knowledge of the forest products industry has grown out of a continuing spiral of job responsibility. During his career he has viewed the logging business from the perspective of a woods worker, supervisor, manager and administrator. After working shifts and as a foreman for Weyerhaeuser operations while attending college, Conway became corporate raw materials operations manager in Federal Way, Washington. In 1975 he moved to North Carolina to be woods manager for Weyerhaeuser's holdings in that region. Little more than a year later, he returned to corporate headquarters as vice president for raw materials operations. He subsequently accepted responsibilities for labor relations and productivity projects in the company. In 1980 he became region vice president in southwest Oregon, directing all activities there, from forestry through logging, raw materials management and manufacturing. Now vice president of timberlands for Scott Paper Company in Philadelphia, Pennsylvania, Conway is responsible for over three million acres of timberland and all solid wood operations in the United States and Canada.

Conway's rare ability to communicate his concepts of efficient and sound timber harvesting methods is demonstrated not only by *Timber Cutting Practices* and this volume, but by his more than 80 articles on production techniques and industry concerns published in leading forest products magazines. He served some years ago as editor of *Pacific Logger and Lumberman.*

Conway holds an A.S. certificate in woods industries technology from Southwestern Oregon Community College and a B.B.A. degree with a major in production management from the University of Oregon.

With many external forces impinging on the industry, Conway feels there is strong incentive to get back to the basics of logging. Large and small companies alike can successfully confront challenging times by applying these basics forcefully and consistently. Conway has revised *Logging Practices,* a book of practical methods and principles, for loggers, that special breed of men who work with a vital renewable resource—timber.

2838